Lecture Notes in Civil Engineering

Volume 169

Lecture Notes in Civil Engineering (LNCE) publishes the latest developments in Civil Engineering - quickly, informally and in top quality. Though original research reported in proceedings and post-proceedings represents the core of LNCE, edited volumes of exceptionally high quality and interest may also be considered for publication. Volumes published in LNCE embrace all aspects and subfields of, as well as new challenges in, Civil Engineering. Topics in the series include:

- Construction and Structural Mechanics
- Building Materials
- Concrete, Steel and Timber Structures
- Geotechnical Engineering
- Earthquake Engineering
- Coastal Engineering
- Ocean and Offshore Engineering; Ships and Floating Structures
- Hydraulics, Hydrology and Water Resources Engineering
- Environmental Engineering and Sustainability
- Structural Health and Monitoring
- Surveying and Geographical Information Systems
- Indoor Environments
- Transportation and Traffic
- Risk Analysis
- Safety and Security

To submit a proposal or request further information, please contact the appropriate Springer Editor:

- Pierpaolo Riva at pierpaolo.riva@springer.com (Europe and Americas);
- Swati Meherishi at swati.meherishi@springer.com (Asia - except China, and Australia, New Zealand);
- Wayne Hu at wayne.hu@springer.com (China).

All books in the series now indexed by Scopus and EI Compendex database!

More information about this series at http://www.springer.com/series/15087

Nikolai Vatin
Editor

Proceedings of STCCE 2021

Selected Papers

Springer

Editor
Nikolai Vatin
Peter the Great St. Petersburg
Polytechnic University
Saint-Petersburg, Russia

ISSN 2366-2557 ISSN 2366-2565 (electronic)
Lecture Notes in Civil Engineering
ISBN 978-3-030-80102-1 ISBN 978-3-030-80103-8 (eBook)
https://doi.org/10.1007/978-3-030-80103-8

This Springer imprint is published by the registered company Springer Nature Switzerland AG
The registered company address is: Gewerbestrasse 11, 6330 Cham, Switzerland

Preface

Proceedings contains reports of the Second International Scientific Conference on Socio-Technical Construction and Civil Engineering, STCCE-2021 (April 21–29, 2021, Kazan State University of Architecture and Engineering, Kazan, Russia).

Proceedings includes reports of scientists, representatives of the construction industry, and postgraduate students from different technical universities not only from the Russian Federation (28 Russian institutes and universities). The conference aroused great interest among scientists, postgraduate students, and specialists from various fields of science.

Representatives from more than 20 countries of the world attended the conference (Australia, Armenia, the Great Britain, Vietnam, Germany, Greece, Egypt, Israel, India, Iraq, Italy, Kazakhstan, Canada, Kirgizia, Lithuania, the USA, Taiwan, Uzbekistan, the Ukraine, Czech Republic, Sweden, South Korea, Japan, etc).

Such a wide geography of the conference participants is valuable for the exchange of experience and knowledge on the issues of construction, building materials, as well as their application in various fields of construction industry and confirms the relevance of topics of the conference.

Within the framework of the conference, following main branches were discussed:

- building constructions, buildings, and structures
- bridges, roads, and tunnels
- building materials and products
- construction technology and organization
- energy efficiency and thermal protection of buildings
- heat supply, ventilation, air conditioning, gas supply, and lighting
- technological complexes and automated systems in construction and mechanical engineering
- transport system development.

The reports of the conference touched upon the issues of research in the field of calculating the roof truss nodes, made of both from steel and from composite materials; experimental studies in the field of the stress state of a multilayer massif under slab foundations and field tests of pile foundations for cyclic loading; developments in the field of the stress state of hinged thin-walled structures; research in the field of work of combined floors as stiffness diaphragms; analytical and numerical calculations of quasi-stable heat transfer mode in shielding constructions of building and installations; flow characteristics, a low-pressure ejector installation; research in the field of the geometric modeling of coil heat exchanger based on spring-twisted channel; design characteristics of a horizontal light tube construction considering requirements of lighting and thermal protection of enclosing structures; development of effective materials for the construction of structural elements of the coatings of bridge structures; asphaltene as the main element of oil dispersion systems; influence on the structure formation of asphaltenes, etc. Thus, the choice of the declared directions of the conference is quite logical and justified since the topics discussed touch upon the main problems in the construction industry.

All articles included in the collection have been reviewed by highly qualified scientists in the relevant scientific fields. The 40 best scientific manuscripts corresponding to the profile of the conference and reflecting the results of theoretical and experimental studies of the authors are recommended for publication.

Annual conference is to be organized on the main topics of the construction industry.

Vatin Nikolay Ivanovich

Contents

Investigation of Portland Cement in 3D Concrete Printing 1
Rustem Mukhametrakhimov and Liliya Lukmanova

Polymer Mixtures Based on Polyvinyl Chloride for the Production of Construction Materials 14
Karina Khuziakhmetova, Lyaylya Abdrakhmanova, and Rashit Nizamov

Vibrational Spectra of p-Carboxylate and p-Sulfonate Azocalix[4]arene 22
Victor Furer, Alexandr Vandyukov, Elena Popova, Svetlana Solovieva, and Igor Antipin

Implementation of the Decision-Making Algorithm in the Bridge Management System 31
Karim Nurmuhametov, Tagir Zinnurov, and Denis Sadykov

Alkali Activation of Russian Calcined Medium-Grade Clay: Influence of NaOH Concentration 38
Nailia Rakhimova, Vladimir Morozov, and Aleksey Eskin

Cost Modeling of a Land Plot for Private Housing Construction 47
Olga Borovskikh, Elvira Shagiakhmetova, Adilya Nizamova, and Tatiana Kazymova

Locally Concrete Filled Reinforced Joints of RHS and SHS Trusses . . . 58
Linur Gimranov and Ivan Kuznetsov

Parametric Vibrations of Viscoelastic Rectangular Plates with Concentrated Masses 72
Mirziyod Mirsaidov, Nikolay Vatin, Rustamkhan Abdikarimov, Dadakhan Khodzhaev, and Bakhodir Normuminov

Analysis of Light Comfort and Thermal Protection of a Building Taking into Account Changes in the Geometry of the Window Slope 80
Aleksey Ivantsov and Vadil Sirazitdinov

Improvements of Techniques for Determining Hazardous Area Near Facilities Under Construction 90
Almaz Shaehov, Lyudmila Koklugina, Ruslan Ibragimov, and Yuliya Evstigneeva

Analysis of the Functional Planning Development of Cities in the USSR on the Example of Kazan 101
Maria Grishina

Prediction of Creep for a Reinforced Concrete Beam Strengthened with an External Reinforcement System Using the Stepped Isothermal Method 113
Almaz Shakirov and Alfred Sulejmanov

Economic Aspects of Infrastructure Projects Implementation in Towns and Medium-Sized Cities 122
Yulia Medyanik, Elvira Shagiakhmetova, Liliya Gimadieva, and Dilyara Vakhitova

Carbonate Phase in the Formation of Binding Substance in Dolomite Cement 135
Nikolay Shelikhov and Ruslan Sagdiev

Improving the Dynamic Properties of Pedestrian Overpasses by Including a Reinforced Concrete Slab in Structural Behavior 142
Valery Eremeev, Gennady Shmelev, Pavel Eremeev, and Daniil Eremeev

Method of Calculating the Strength of Clay Soils Under Triaxial Regime Loading 150
Ilizar Mirsayapov

Additional Settlement of the Raft-Pile Foundation, Taking into Account the Deformations of the Pile Under Cyclic Loading 160
Ilizar Mirsayapov

Calculation of the Endurance of Reinforced Concrete Bending Elements by the Method of Limit Stresses 167
Ilizar Mirsayapov

Field Tests of Combined Pile Raft Foundation Under Cyclic Loading 175
Ilizar Mirsayapov, Marat Shakirov, and Danil Sabirzyanov

Influence of Inelastic Deformations of Reinforcement on the Stress-Strain State of Reinforced Concrete Bending Elements Under Cyclic Loading 183
Ilizar Mirsayapov

Geometric Modeling of Coil Heat Exchanger Based on Spring-Twisted Channel 194
Yakov Zolotonosov, Iraida Krutova, and Ekaterina Vachagina

Deformation Features of Raft-Pile Foundation Models Under Cyclic Loading 203
Ilizar Mirsayapov and Marat Shakirov

Numerical Study of the Flow in a Symmetrical Ventilation Junction Tee with a Baffle Vane 213
Arslan Ziganshin, Svetlana Eremina, Guzel Safiullina, and Konstantin Logachev

Polymer Multilayer Films for Packaging of Building Materials 223
Zhanna Gerkina, Valentina Serova, and Viktor Stroganov

Air Diffuser for Ventilation and Air Conditioning Systems with Quantitative Control 230
Rinat Safiullin and Vladimir Posokhin

Model of Degradation of Composite Materials of Building Structure's Load-Bearing Elements 239
Rashit Kayumov, Alfred Sulejmanov, and Dmitry Strakhov

Design Characteristics of a Horizontal Light Tube Construction Considering Requirements of Lighting and Thermal Protection of Enclosing Structures 250
Artem Petrov and Maksim Kolchin

Foamed Wood-Polymer Composites Based on Polyvinyl Chloride 261
Anvar Islamov and Venera Fakhrutdinova

Modification of Styrene-Acrylic Coatings with Carbon-Containing Shungite Filler 269
Victor Stroganov, Maxim Amel'chenko, Evgenii Vdovin, Radmilla Tabaeva, and Eduard Kraus

Effect of the Viscogel Additive on the Rheological Parameters of Bitumen 279
Maxim Lashin, Marina Vysotskaya, and Evgeniy Vdovin

Biostability of Road Cement-Mineral Materials Reinforced with Dispersed Fibre 289
Evgenii Vdovin, Victor Stroganov, Maxim Amel'chenko, Eduard Kraus, and Ilshat Fazleev

Alternative Mineral Powders for Asphalt Concrete 297
Marina Vysotskaya, Evgenii Vdovin, Dmitry Kuznetsov, and Artem Shiryaev

Modification by Zeolite-Containing Additive the Road-Building Materials Based on Carbonate Crushed Stone-Sand Mixtures and Cements 308
Evgenii Vdovin, Lenar Mavliev, and Nikita Konovalov

Coupling Agents Based on Single-Walled CNTs for Polyvinylchloride Wood-Polymer Composites 318
Ayaz Khantimirov, Lyaylya Abdrakhmanova, and Vadim Khozinv

Nanomodified Polymer-Bitumen Binders 325
Damir Ayupov, Dmitry Makarov, and Rauf Kazakulov

The Effect of Additives of Mechanically Activated Mineral Fillers on the Properties of Composite Gypsum Binders 334
Marat Khaliullin and Alsu Gilmanshina

About Wear Courses and Concrete Pavements Abrasion of Highways 343
Viktor Ushakov, Mikhail Goryachev, Grigory Diakov, and Sergey Yarkin

Modification of Harsh Cement Pavement Concretes with Bitumen Emulsion 351
Aidar Garipov, Dmitry Makarov, Vadim Khozin, and Sergey Stepanov

Numerical Study of the Influence of the Inlet Geometric Parameters on the Jet Characteristics 364
Juliya Kareeva, Raushan Zakieva, and Kseniya Bliznjakova

Gypsum-Fiber Radioprotective Facing Materials 372
Albert Galautdinov, Rustem Mukhametrakhimov, and Valery Kupriyanov

Author Index 383

Investigation of Portland Cement in 3D Concrete Printing

Rustem Mukhametrakhimov(✉) and Liliya Lukmanova

Kazan State University of Architecture and Engineering, Kazan 420043, Russia

Abstract. The right choice of the binder type in design of mixes for 3D concrete printing (3DCP) will contribute to sustainable development of this technology, reduce risks of defect formation and inconsistencies in finished products. The results of influence of Portland cements with different mineralogical composition on the rate of plastic strength of set cement, the main rheological characteristics of mixes and the physical and mechanical properties of 3D printed hardened composites are presented. The rate of plastic strength of set cement was determined in accordance with requirements of ASTM C403 using a pocket penetrometer C194. The yield stress of mix was determined using a simple viscometer, which is a hollow polypropylene cylinder 200 mm high and 105 mm in inner diameter. Physical and mechanical properties (average density, flexural and compression strength, softening coefficient – the ratio of wet strength to dry compression strength, water absorption) of 3D printed hardened composites were determined in accordance with Russian standards. It was found both cements without mineral additives and cements with additives are most expedient to use in developing of concrete and mortar mixes for 3DCP from the position of availability of cements on the market. From the point of view of mineralogical composition of cements, it was found cements with a high content of clinker minerals C_3S and C_3A are the most expedient to use because of providing a quick set of strength of the freshly formed concrete mix at the initial time of hardening in 3DCP. From the point of view of the rate of plastic strength, mixes based on CEM II/A-P 32.5 N and CEM II/A-S 32.5 R are the most effective, which are characterized by earlier initial set compared to the rest studied compositions. From the point of view of obtaining the most optimal rheological indicators and high physical and mechanical properties, mortar and concrete mixes based on Portland cement CEM I 42.5 N, sand with fineness modulus Mk = 3, cement-to-sand (C/S) ratio of 1:3 and mobility class Pk 2 (according to Russian standard GOST 5802-86) are the most advisable to use in 3DCP. Nevertheless, it is possible to use other studied Portland cements in 3DCP, which, to a lesser extent, contribute to obtaining optimal rheological properties, in conjunction with modifying additives.

Keywords: Concretes · Cements · Mortars · 3D concrete printing · 3DCP · Additive manufacturing · Rheology

1 Introduction

3D concrete printing (3DCP) is a rapidly developing technology, as evidenced by the large number of projects that are being implemented around the world in this area from

N. Vatin (Ed.): STCCE 2021, LNCE 169, pp. 1–13, 2021.
https://doi.org/10.1007/978-3-030-80103-8_1

1997 to 2021: USA (University of Southern California, TotalKustom), UK (Loughborough University), China (WinSun), Netherlands (Delft University of Technology, Eindhoven University of Technology), France (XtreeE), etc. [1, 2]. At present, elements of buildings and structures [3–5], bridge structures [6, 7], small-scale architectural structures [8, 9] and interior items are being erected using this technology.

However, with all the advantages of 3DCP, it should be noted that there remain problems that limit the production of high-quality construction products [10–13], which is associated with the use of not adapted raw mixes as «ink» due to various raw materials, the main of which are mineral binders and aggregates. The chemical and mineralogical composition of these components, aggregate size, cement-to-sand (C/S) ratio significantly affect rheological and technological properties of mix, geometry of products and structures, buildability of mix, rate of structure formation and the physical and mechanical characteristics of the composites [14]. Therefore, correct choice of the type of binder in design of mixes for 3DCP will contribute to sustainable development of this technology, reduce risks of defect formation and inconsistencies in finished products.

Cement is most widely used binder among mineral binders in 3DCP [12, 15–21]. Availability of cement on the construction material market [22], ensuring the rapid strength gain of freshly formed concrete mix in the initial hardening period [23], ensuring the stability of the properties of the mixes both in the 3D printing process and in the process operation of the obtained composites, the ability to regulate properties by active mineral and chemical additives are among the defining criteria when choosing Portland cement in 3DCP.

Analysis of the global volume of the cement industry market according to the US Geological Survey [24] indicates the relative stability of the cement volume production in the world (Fig. 1). Over the past ten years, cement production in the world is up by 24.2% – from 3.3 to 4.1 billion tons.

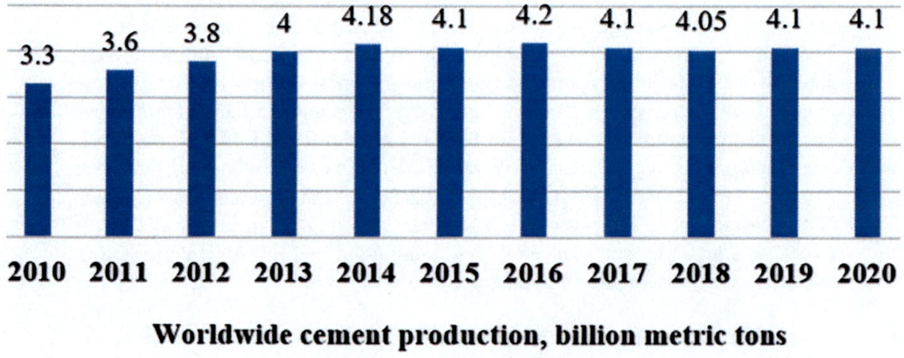

Fig. 1. Dynamics of worldwide cement production in 2010–2020 [24].

Further to 2019 and 2020 the worldwide cement production amounted to 4.1 billion tons [24]. Nevertheless, such countries as China, India and Vietnam are still leaders, which account for 54%, 8% and 2%, respectively, of the entire world market. Dynamics of worldwide cement production in 2019–2020 by country is shown in Fig. 2 [24].

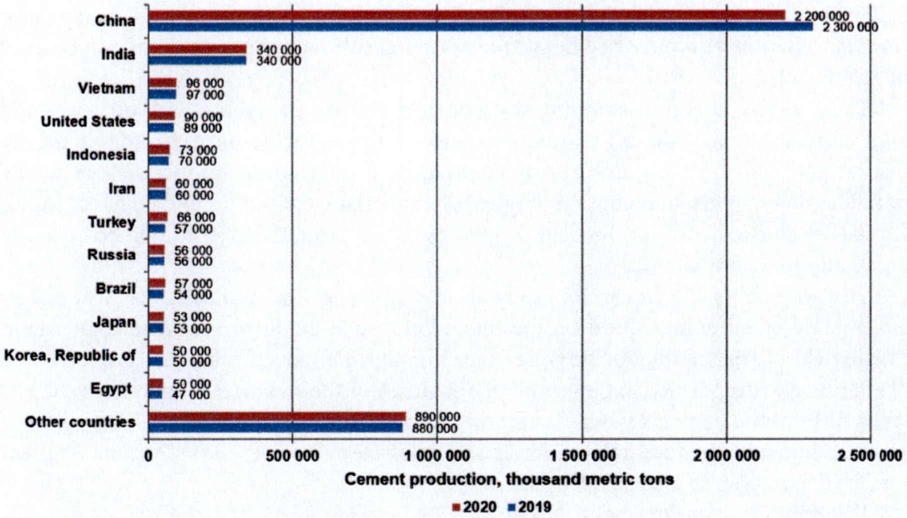

Fig. 2. Dynamics of worldwide cement production in 2010–2020 by country [24].

An analysis of the official information of Russian Federal State Statistics Service on the cement production market in Russian Federation [25] indicates a decrease in the share of cement production compared to the previous year. Thus, at the end of 2020, it was produced 55 984.7 thousand tons of cement in Russian Federation (97.1% compared to the same period of the previous year). At the same time, the bulk of cement production in the country in 2020 peaked in Portland cements without mineral additives. The volume of output of this product amounted to 34 511 thousand tons (61.6% of the total Russian cement production), which is twice as much as Portland cement with additives, which ranks second in terms of production – 18 911 thousand tons (33.8% of the total volume cement production in the country). Analyzing the market of production of cements with mineral additives, it is worth noting according to data for 2010–2018 there is a trend that indicates a decrease in the share of their output (2016 – 44%, 2018 – 41%, 2018 – 40%) [26]. From the position of availability of cements on the market it indicates that study and develop compositions mainly based on Portland cements without and with mineral additives would be appropriate in 3DCP.

The authors of studies [12, 16, 17, 27, 28] point to that Portland cements with ordinary early strength, indicated by N, and Portland cements with high early strength, indicated by R according to EN 197-1 can be effective in 3DCP. Providing a quick set of strength of freshly formed concrete mix at the initial time of hardening can be achieved by using of cements with a high content of clinker minerals C_3S and C_3A, which harden especially quickly and are used in the manufacture of rapid hardening Portland cements. Among the varieties of rapid hardening cements, which include fast-hardening ($C_3S + C_3A =$ 60 … 65%), extra-fast-hardening ($C_3S = 60 \ldots 65\%$, $C_3A \leq 8\%$), ultra-fast-hardening ($C_3S = 65 \ldots 68\%$, $C_3A \leq 8\%$ + additions of C_4AF and $CaCl_2$ when grinding clinker) [29]. These types of rapid hardening Portland cements have alite clinker, the content of C_3S in them is more than 60%. Thus, the study and application of the above types of Portland cements is relevant in 3DCP. It should be noted that, in addition to these cements, alternative types of cements, for example, sulfoaluminate cements can be used in 3DCP [30].

Thus, works aimed at studying the effect of Portland cements with ordinary and high early strength with and without mineral additives, as the most affordable on the market and providing a quick set of strength of freshly formed concrete mix at the initial hardening period, on the rheological characteristics of mortar and concrete mixes for 3DCP, physical and mechanical properties of 3D printed hardened composites, are especially relevant.

The purpose of the research is to study the effect of Portland cements with different mineralogical composition on the rate of plastic strength of set cement, the main rheological characteristics of mixtures and the physical and mechanical properties of 3D hardened composites, to determine the features of their choice from among studied types of Portland cements when designing of compositions for 3DCP.

The object of the research is mortar and concrete mixes based on Portland cement for 3DCP and composites based on them.

The subject of the research is the rheological characteristics of mortar and concrete mixes for 3DCP, physical and mechanical properties of 3D hardened composites depending on the type of Portland cement.

2 Materials and Methods

The research was carried out in the laboratory of additive manufacturing in construction (Kazan State University of Architecture and Engineering, Russia).

The following materials were used in the research process:

a) Portland cements with different strength classes and mineralogical composition:

   - CEM I 42.5 N (manufacturer «Aziya Tsement») complying with GOST 31108-2016 (Russian standard);
   - CEM II/A-S 32.5 R (manufacturer «SLK Cement») complying with GOST 31108-2016 (Russian standard);
   - CEM II/A-P 32.5 N (manufacturer «MORDOVTSEMENT») complying with GOST 31108-2016 (Russian standard);
   - CEM II/B-P 32.5 N (manufacturer «Sengileevsky tsementny zavod») complying with GOST 31108-2016 (Russian standard);

b) fine aggregate:

   - quartz sand with a fineness module Mk = 1.2, Mk = 2.3, Mk = 2.4, Mk = 3 complying with GOST 8736-2014 (Russian standard);

c) tap drinking water complying with GOST 23732-2011 (Russian standard).

Material and chemical compositions of Portland cements are given in Tables 1, 2.

Table 1. Material composition of Portland cements.

Types of common cement	Composition (percentage by mass)						Minor additional constituents
	Clinker				Blastfurnace slag	Pozzolana (silica clay)	
	C_3S	C_2S	C_3A	C_4AF			
1	2	3	4	5	6	7	8
CEM I 42.5N	68.1	9.4	7.2	11.0	–	–	4.1
CEM II/A-S 32.5R	62.6	14.2	5.5	12.6	16.0	–	2.9
CEM II/A-P 32.5N	63.3	13.6	6.6	13.5	–	18.5	–
CEM II/B-P 32.5N	57.0	14.0	4.5	13.0	–	28.0	–

Table 2. Chemical compositions of Portland cements.

Types of common cement	Composition (percentage by mass)							
	SiO_2	Al_2O_3	Fe_2O_3	CaO	MgO	Insoluble residue	SO_3	Cl
1	2	3	4	5	6	7	8	9
CEM I 42.5N	23.58	4.89	3.48	61.1	1.1	–	2.92	0.014
CEM II/A-S 32.5R	23.6	5.5	3.6	58.7	3.3	0.8	3.0	0.01
CEM II/A-P 32.5N	21.38	5.31	4.44	65.8	1.23	–	3.12	0.015
CEM II/B-P 32.5N	21.1	4.4	4.3	64.2	0.9	–	2.9	0.008

Normal consistency and setting time of the cement paste were determined according to GOST 310.3-76 (Russian standard).

The rate of plastic strength of set cement was determined in accordance with requirements of ASTM C403 using a pocket penetrometer C194 in the time interval 0–8 h.

Table 3. Characteristics of the mixes.

Mix no.	Cement type	Fineness modulus of sand	C/S	Mobility class/depth of immersion of the cone, cm
1	2	3	4	5
Mix 1	CEM I 42.5N	Mk 2.3	1:3	Pk 3 / 8.9
Mix 2		Mk 2.4		Pk 3 / 8.9
Mix 3		Mk 3		Pk 2 / 7.0
Mix 4	CEM II/A-S 32.5R	Mk 1.2	1:2	Pk 2 / 6.5
Mix 5				Pk 3 / 8.7
Mix 6				Pk 4 / 12.0
Mix 7	CEM II/A-P 32.5N	Mk 3	1:3	Pk 2 / 7.0
Mix 8	CEM II/B-P 32.5N			

Eight compositions based on various Portland cements and fineness modulus of sand, with different C/S and mobility classes Pk 2-Pk 4 were prepared to determine rheological characteristics of mortar and concrete mixes and the physical and mechanical properties of composites (Table 3). The mobility of the compositions was regulated by changing the water content.

Mixing of the components of the raw materials was carried out in a forced action concrete mixer for 10 min until homogeneous mass was obtained.

The mobility of the mortar and concrete mixes was determined to GOST 5802-86 (Russian standard) by the depth of immersion in it of the reference cone.

The yield stress of mix was determined using a simple viscometer, which is a hollow polypropylene cylinder 200 mm high and 105 mm in inner diameter.

The value of yield stress was determined in accordance with the method [31] according to Eq. (1):

$$\tau_0 = \frac{hd^2\rho}{kD^2}, \qquad (1)$$

where τ_0 – yield stress of mix, Pa;

h и d– high and diameter of viscometer, respectively, m;

ρ – mixture density, kg/m^3;

D – spreading diameter of the mixture, m;

k – coefficient taking into account the redistribution of stresses in viscoplastic materials, equal to 2.

The sample was formed from a sand-cement mix by extrusion in workshop 3D printer «AMT S-6044» «SPETSAVIA LLC» (Yaroslavl, Russia), organized by a portal system, in accordance with a specified digital three-dimensional model (G-code).

The printed samples hardened for 28 days in natural conditions: temperature (20 ± 2) °C, relative humidity (50 ± 20) %.

3DCP-samples for testing were prepared by cutting them into prisms with dimensions of 40 × 40 × 160 mm.

Fig. 3. Flexural test of 3D printed hardened composite using testing machines SI-2-100-UHL4.2.

The average density of hardened cement-sand mortars was determined according to GOST 5802-86 (Russian standard), concretes – in accordance with GOST 12730.1-78 (Russian standard).

Water absorption of mortars was determined according to GOST 5802-86 (Russian standard), concretes – in accordance with GOST 12730.3-78 (Russian standard).

The flexural and compression strength of the hardened mortars and concretes was determined on specimens-beams with dimensions of 40 × 40 × 160 mm and halves of prism specimens obtained after flexural tests, according to GOST 5802-86 (Russian standard) (Fig. 3) using testing machines SI-2-100-UHL4.2, IP-1000-0, MII-100. The choice of this method for the investigated fine-grained concretes is due to the absence of coarse aggregate (more than 5 mm) in them, which makes it possible to produce samples of smaller sizes, regulated by GOST 10180-2012 (Russian standard) for concretes, the complexity of forming samples on a 3D printer with the dimensions specified in GOST 10180-2012 (Russian standard).

3 Results and Discussions

At the first stage of the research, the effect of Portland cement type on the normal consistency, the setting time of the cement paste was studied (Table 4).

As can be seen from Table 5, cement type has a different effect on normal consistency and setting time of the cement paste. The highest water demand is striking for cement with the highest content of sedimentary rock (silica clay) – CEM II/B-P 32.5 N, which is caused by a very developed specific surface of the mineral additive, requiring a significant amount of water to wet it. The lowest water demand is striking for CEM II/A-S 32.5

Table 4. Normal consistency, setting time of Portland cements.

Cement type	Normal consistency, %	Setting time	
		Initial set	Final set
CEM I 42.5 N	32.00	3 h 51 min	4 h 26 min
CEM II/A-S 32.5 R	27.60	3 h 09 min	4 h 24 min
CEM II/A-P 32.5 N	33.88	2 h 34 min	4 h 40 min
CEM II/B-P 32.5 N	40.50	3 h 38 min	5 h 38 min

R. Initial set of cement paste based on Portland cements with mineral additives occurs 13–77 min earlier than the composition based on Portland cement with no additives. Final set of the cement paste based on CEM II/A-S 32.5 R occurs 2 min earlier than mix based on Portland cement with no additives, the rest of the mixes slow down the final set of the cement paste by 14–72 min compared to the mix based on Portland cement with no additives.

At the second stage, the influence of the Portland cement type on the rate of plastic strength of set cement over time was studied (Fig. 4).

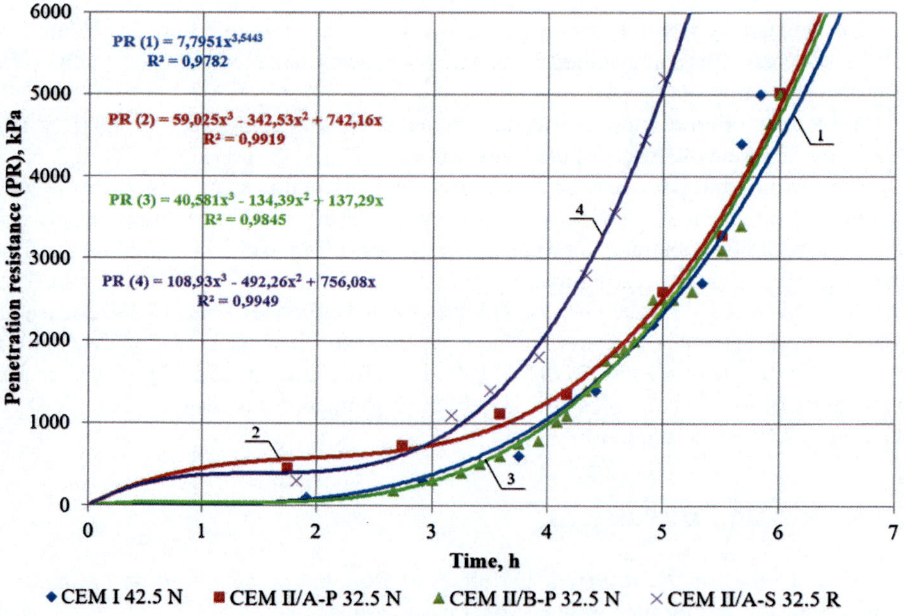

Fig. 4. Penetration resistance values of different cement pastes versus time: 1 – CEM I 42.5 N; 2 – CEM II/A-P 32.5 N; 3 – CEM II/B-P 32.5 N; 4 – CEM II/A-S 32.5 R.

As can be seen from Fig. 4 in the interval of 0–4 h, the greatest increase in the values of penetration resistance is characteristic of compositions based on CEM II/A-P 32.5N and CEM II/A-S 32.5 R, which is consistent with the results obtained (Table 5) – initial set of these mixes occurs earlier compared to other mixes. In the long-term setting time (4–6.5 h), there is a significant acceleration of the kinetics of the plastic strength of set cement based on CEM II/A-S 32.5 R in comparison with compositions based on CEM I 42.5 N, CEM II/A-P 32.5 N, CEM II/BP 32.5 N, which have an identical rate of plastic strength development in the studied time interval.

The results of experimental studies reflecting the influence of Portland cement type, fineness modulus of sand, C/S and the mobility class of mixes on the rheological, physical and mechanical characteristics of 3D printed hardened composites are given in Table 5.

From Table 5 it can be seen that Portland cement type, fineness modulus of sand, C/S and mobility class of the mix have a different effect on formation of the rheological, physical and mechanical characteristics of 3D printed hardened composites. Among the considered Portland cement types, the highest indicators of average density, compressive strength (CS), softening coefficient – the ratio of wet strength to dry compression strength (K_p) and low water absorption (W) are possessed by 3D printed hardened composites based on CEM I 42.5N in combination with sand with fineness modulus Mk = 3, mobility class – Pk 2. The use of sand with a fineness modulus Mk = 2.3 and Mk = 2.4 in conjunction with Portland cement CEM I 42.5N with the same mobility class leads to close values of the ultimate yield stress, softening coefficient and water absorption of 3D printed hardened composites.

When using Portland cement CEM II/A-S 32.5 R in combination with sand with fineness modulus Mk = 1.2, 3D printed hardened composites of mobility class Pk 3 have the most optimal rheological and physical and mechanical characteristics. With a decrease of mix mobility from Pk 3 to Pk 2 there is a significant increase in yield stress, a decrease in the average density of the samples, flexural strength (FS) and compression strength, softening coefficient. This is due to the peculiarity of the 3DCP, which leads to increased air entrainment of inactive mortar mix during its rotation and movement by screw in bunker of 3D printer, and as a result structural defects and a decrease in the physical and mechanical properties of composites are happened. We also observed this phenomenon on samples of another composition with a similar mobility [32].

The use of Portland cement CEM II/B-P 32.5 N leads to a decrease in flexural strength by 23.8%, in compression strength – by 16.1%, an increase in softening coefficient – by 7.5%, water absorption – by 16.8% compared to 3D printed hardened composites based on CEM II/A-P 32.5 N with the same fineness modulus of sand, C/S and mix mobility. However, yield stress of the mix and the average density of the samples based on these compositions differ insignificantly. The revealed decrease in physical and mechanical properties when using Portland cement CEM II/B-P 32.5N in comparison using Portland cement CEM II/A-P 32.5N is caused by an increased content of silica clay in the first one (up to 28%).

Table 5. Rheological, physical and mechanical characteristics of 3D printed hardened composites depending on the Portland cement type, fineness modulus of sand, C/S and the mobility class of mixes.

Cement type	Fineness modulus of sand	C/S	Mobility class/depth of immersion of the cone, cm	Yield stress, Pa	Average density, kg/m^3	FS, MPa	CS, MPa	K_p	W, %
1	2	3	4	5	6	7	8	9	10
CEM I 42.5 N	Mk 2.3	1:3	$\frac{Pk\,3}{8.9}$	77	1980	5.0	32.1	0.75	8.4
	Mk 2.4		$\frac{Pk\,3}{8.9}$	78	2050	7.0	34.4	0.76	8.4
	Mk 3		$\frac{Pk\,3}{7.0}$	94	2060	6.0	41.2	0.82	8.3
CEM II/A-S 32.5 R	Mk 1.2	1:2	$\frac{Pk\,2}{6.5}$	208	1890	4.1	25.7	0.65	10.6
			$\frac{Pk\,3}{8.7}$	104	2000	4.5	27.9	0.70	11.2
			$\frac{Pk\,4}{12.0}$	29	1950	4.2	26.5	0.67	12.8
CEM II/A-P 32.5 N	Mk 3	1:3	$\frac{Pk\,2}{7.0}$	113	2010	4.2	31.6	0.53	9.5
CEM II/B-P 32.5 N				116	2020	3.2	26.5	0.57	11.1

4 Conclusions

1. Analysis of the worldwide cement production market over the past ten years (2010–2020) indicates its relative stability. At the end of 2020, Russia appears eighth in worldwide cement production. Cements without mineral additives and cements with additives account for a significant share of cement production in Russia in 2020, accounting for 61.6% and 33.8% of the total Russian cement production, respectively. Due to the availability of these cements, it is most expedient to use them in development of concrete and mortar mixes for 3DCP.
2. From the point of view of the mineralogical composition of cements, the use of cements with a high content of clinker minerals C_3S and C_3A are the most expedient to use because of providing a quick set of strength of the freshly formed concrete mix at the initial time of hardening in 3DCP. This requirement is met by rapid hardening cements that have alite clinker (content of C_3S is more than 60%). At the same time, the analysis of works in the field of 3DCP indicates that, in addition to rapid hardening cements, cements with ordinary early strength can also be effective in 3DCP.
3. From the point of view of the rate of plastic strength development over time, mixes based on CEM II/A-P 32.5 N and CEM II/A-S 32.5 R, which are characterized by

an earlier initial set compared to the rest of the studied mixes are the most effective. In 3DCP, this factor is of particular importance, since it allows printing products of a higher height without technological interruptions. The studied Portland cements types have a different effect on normal density and setting time of cement paste. The highest water demand is typical for Portland cement CEM II/B-P 32.5 N – 40.5%, the lowest – for CEM II/A-S 32.5 R (27.6%).

4. Among the considered Portland cement types, from the point of view of obtaining the most optimal rheological (yield stress) and high physical and mechanical properties (average density, flexural and compression strenght, softening coefficient, water absorption), mortar and concrete mixes based on Portland cement CEM I 42.5N (C_3S – 68.1%, C_2S – 9.4%, C_3A – 7.2%, C_4AF – 11%), sand with fineness modulus Mk = 3, C/S = 1:3 and mobility class Pk 2 are the most expedient to use in 3DCP.
5. The use of Portland cements with mineral additives (CEM II/A-S 32.5 R, CEM II/A-P 32.5 N, CEM II/B-P 32.5 N) in 3D CP necessitates their further study in order to improve the rheological characteristics of mixes, physical and mechanical properties of 3D printed hardened composites based on them, which is possible through the use of various modifying additives.
6. At the same time, it is possible to use other investigated Portland cements in 3DCP, including those with mineral additives (CEM II/A-S 32.5 R, CEM II/A-P 32.5 N, CEM II/B-P 32.5 N), which are less conducive to obtaining optimal rheological properties. For their effective use, including non-additive cements, directed regulation of the rheological properties of concrete mixes is necessary, which can be carried out by regulating the water-cement ratio, the size and amount of aggregate, as well as modification with active mineral and chemical additives.

Acknowledgments. This research was funded by President of Russia Scholarship for young scientists and graduate students (SP-1051.2021.1), «Civil Society Foundation» and supported by «3D-Stroy LLC».

References

1. Buswell, R.A., Leal de Silva, W.R., Jones, S.Z., Dirrenberger, J.: 3D printing using concrete extrusion: a roadmap for research. Cem. Concr. Res. **112**, 37–49 (2018). https://doi.org/10.1016/j.cemconres.2018.05.006
2. Mukhametrakhimov, R., Vakhitov, I.: Additivnaya tekhnologiya vozvedeniya zdaniy i sooruzheniy s primeneniyem stroitelnogo 3D-printera. Izvestiya KGASU **4**, 350–359 (2017)
3. Anton, A., Reiter, L., Wangler, T., Frangez, V., Flatt, R.J., Dillenburger, B.: A 3D concrete printing prefabrication platform for bespoke columns. Autom. Constr. **122**, 103467 (2021). https://doi.org/10.1016/j.autcon.2020.103467
4. He, Y., Zhang, Y., Zhang, C., Zhou, H.: Energy-saving potential of 3D printed concrete building with integrated living wall. Energy Build. **222**, 110110 (2020). https://doi.org/10.1016/j.enbuild.2020.110110
5. Hager, I., Golonka, A., Putanowicz, R.: 3D printing of buildings and building components as the future of sustainable construction? Procedia Eng. **151**, 292–299 (2016). https://doi.org/10.1016/j.proeng.2016.07.357

6. Salet, T.A.M., Ahmed, Z.Y., Bos, F.P., Laagland, H.L.M.: 3D printed concrete bridge. In: Proceedings of the International Conference on Progress in Additive Manufacturing, pp. 2–9 (2018)
7. Salet, T.A.M., Ahmed, Z.Y., Bos, F.P., Laagland, H.L.M.: Design of a 3D printed concrete bridge by testing. Virtual Phys. Prototyp. **13**, 222–236 (2018). https://doi.org/10.1080/17452759.2018.1476064
8. Mukhametrakhimov, R.K., Lukmanova, L.V., Gorbunova, P.S.: 3D-printed S-shaped bench for urban landscaping (2020)
9. Mukhametrakhimov, R.K., Lukmanova, L.V.: O-shaped fiber-reinforced concrete bench made on a 3D printer (2020)
10. Mukhametrakhimov, R., Lukmanova, L.: Influence of the technological properties of cement-sand mortar on the quality of 3D printed products. In: IOP Conference Series: Materials Science and Engineering, vol. 890, p. 012082 (2020). https://doi.org/10.1088/1757-899x/890/1/012082
11. Mukhametrakhimov, R., Lukmanova, L.: Influence of cement-sand mortar mobility on the quality of 3D printed hardened composite. Constr. Unique Build. Struct. **94**, 9404 (2021). https://doi.org/10.4123/CUBS.94.4
12. Le, T.T., Austin, S.A., Lim, S., et al.: Hardened properties of high-performance printing concrete. Cem. Concr. Res. **42**, 558–566 (2012). https://doi.org/10.1016/j.cemconres.2011.12.003
13. Marchment, T., Sanjayan, J., Xia, M.: Method of enhancing interlayer bond strength in construction scale 3D printing with mortar by effective bond area amplification. Mater. Des. 107684 (2019). https://doi.org/10.1016/j.matdes.2019.107684
14. Vdovin, E.A., Stroganov, V.F.: Properties of cement-bound mixes depending on technological factors. Mag. Civ. Eng. **93**, 147–155 (2020). https://doi.org/10.18720/MCE.93.12
15. Alghamdi, H., Nair, S.A.O., Neithalath, N.: Insights into material design, extrusion rheology, and properties of 3D-printable alkali-activated fly ash-based binders. Mater. Des. **167**, 107634 (2019). https://doi.org/10.1016/j.matdes.2019.107634
16. Vaitkevičius, V., Šerelis, E., Kerševičius, V.: Effect of ultra-sonic activation on early hydration process in 3D concrete printing technology. Constr. Build. Mater. **169**, 354–363 (2018). https://doi.org/10.1016/j.conbuildmat.2018.03.007
17. Di Maio, L., Coppola, B., Courard, L., Michel, F., Incarnato, L., Scarfato, P.: Data on thermal conductivity, water vapour permeability and water absorption of a cementitious mortar containing end-of-waste plastic aggregates. Data Br. **18**, 1057–1063 (2018). https://doi.org/10.1016/j.dib.2018.03.128
18. Zhang, Y., Zhang, Y., Liu, G., Yang, Y., Wu, M., Pang, B.: Fresh properties of a novel 3D printing concrete ink. Constr. Build. Mater. **174**, 263–271 (2018). https://doi.org/10.1016/j.conbuildmat.2018.04.115
19. Kruger, J., Zeranka, S., van Zijl, G.: 3D concrete printing: A lower bound analytical model for buildability performance quantification. Autom. Constr. **106**, 102904 (2019). https://doi.org/10.1016/j.autcon.2019.102904
20. Ma, G., Li, Z., Wang, L.: Printable properties of cementitious material containing copper tailings for extrusion-based 3D printing. Constr. Build. Mater. **162**, 613–627 (2018). https://doi.org/10.1016/j.conbuildmat.2017.12.051
21. Shakor, P., Sanjayan, J., Nazari, A., Nejadi, S.: Modified 3D printed powder to cement-based material and mechanical properties of cement scaffold used in 3D printing. Constr. Build. Mater. **138**, 398–409 (2017). https://doi.org/10.1016/j.conbuildmat.2017.02.037
22. Khozin, V.G., Khokhryakov, O.V., Nizamov, R.K.: Low water demand carbonate cements are promising binders for concrete. Concr. Reinf. Concr. **601**, 15–28 (2020)

23. Jianchao, Z., Zhang, T., Faried, M., Wengang, C.: 3D printing cement-based ink, and it's application within the construction industry. In: ASCMCES-17, p. 02003. MATEC Web of Conferences (2017)
24. U.S. Geological Survey: Cement Statistics and Information (2021)
25. Russian Federal State Statistics Service: Information on the socio-economic situation of Russia in 2020
26. Zaytseva, Y.V.: Development of a scientific and methodological base for substantiation and comprehensive planning of strategies for the development of mining processing industries, taking into account the innovative component (2020)
27. Sikora, P., et al.: The effects of nanosilica on the fresh and hardened properties of 3D printable mortars. Constr. Build. Mater. **281**, 122574 (2021). https://doi.org/10.1016/j.conbuildmat.2021.122574
28. Chen, Y., et al.: Improving printability of limestone-calcined clay-based cementitious materials by using viscosity-modifying admixture. Cem. Concr. Res. **132**, 106040 (2020). https://doi.org/10.1016/j.cemconres.2020.106040
29. Rakhimova, N.R.: Modern hydraulic binders. KSUAE, Kazan (2014)
30. Mohan, M.K., Rahul, A.V., De Schutter, G., Van Tittelboom, K.: Early age hydration, rheology and pumping characteristics of CSA cement-based 3D printable concrete. Constr. Build. Mater. **275**, 122136 (2021). https://doi.org/10.1016/j.conbuildmat.2020.122136
31. Bazhenov, Y.M., Demyanova, V.S., Kalashnikov, V.I.: Modifitsirovannyye vysokokachestvennyye betony. Izdatelstvo assotsiatsii stroitelnykh vuzov, Moscow (2006)
32. Mukhametrakhimov, R., Lukmanova, L.: Structure and properties of mortar printed on a 3D printer. Mag. Civ. Eng. **102** (2021). https://doi.org/10.34910/MCE.102.6

Polymer Mixtures Based on Polyvinyl Chloride for the Production of Construction Materials

Karina Khuziakhmetova(✉), Lyaylya Abdrakhmanova, and Rashit Nizamov

Kazan State University of Architecture and Engineering, 420043 Kazan, Russia

Abstract. The work is devoted to the development of rigid polyvinyl chloride (PVC) compositions for obtaining materials for construction with new properties. Acrylonitrile-butadiene styrene copolymer (ABS) was added to the mixed composition in different concentrations (from 10 to 40 parts per 100 parts of PVC resin). The performance and structural parameters of the extruded flat profiles were considered. Increasing the concentration of ABS leads to changes in the technological and operational properties of the composites. This work shows the influence of the features of the supramolecular structure of composites on the properties when the concentration of copolymer is increased: weight reduction of samples, increase of high elastic deformation, hardness and bending strength, as well as increase of heat resistance and decrease of the flowing temperature.

Keywords: PVC · ABS · Modification · Polymer blends · Recycling

1 Introduction

The operation of construction polymeric products requires compliance with the requirements due to the functional purpose of the materials [1–4]. Modification of polymeric building materials allows to regulate many surface and structural properties [5–8]. Among the most responsible building polymers is polyvinyl chloride (PVC) [9]. The combination with a huge number of technological additives due to its polarity, as well as its cost, makes PVC a versatile building material [10–15]. A frequent problem in the use of PVC is the complexity of processing [16, 17], which can be solved by using an acrylic-based modifier – acrylonitrile-butadiene styrene (ABS) [18, 19].

ABS is one of the most widely used polymers [20]. Butadiene is the core, which has a hardening effect, acrylonitrile-styrene copolymer is the shell, which is grafted onto the butadiene surface, which increases the compatibility of the blends and enhances interfacial adhesion [21, 22]. ABS has excellent overall characteristics but low flame resistance, which limits its application [23, 24]. However, together with PVC, many problems begin to be solved. In addition, PVC/ABS blends are of interest due to the compatibility of the styrene-acrylonitrile phase in the block copolymer of ABS and PVC depending on the phase structure of the blend. Therefore, PVC is often used to improve process properties [25–28].

Most studies have focused on ABS/PVC blends in which the matrix polymer is ABS [29–31]. However, polymer-polymer blends based on PVC with a high concentration of

N. Vatin (Ed.): STCCE 2021, LNCE 169, pp. 14–21, 2021.
https://doi.org/10.1007/978-3-030-80103-8_2

ABS have special properties [32]. Therefore, this paper will consider the performance and structural properties of PVC/ABS polymer-polymer blends with a high concentration of ABS in the PVC matrix.

2 Materials and Methods

Suspension PVC of S-6359-M grade, complex stabilizer – bicarbonate lead stearate and stabilizer-lubricant – calcium stearate were used in the research. Powdered impact strength and recyclability modifier ABS-20P was provided by Uzlov plant Plastik JSC (Russia). ABS has a softening point of 97 °C and a melt flow index at 220 °C and 10 kgf not less than 5–12 g/10 min. The average particle size of ABS was 199.5 μm with a characteristic bimodal particle size distribution. The composition of the experimental mixtures per hundredth of PVC resin (phr) is presented in Table 1.

Table 1. Composition of experimental mixtures.

Component	Concentration, phr			
PVC	100	100	100	100
Bicarbonate lead stearate	5	5	5	5
Calcium stearate	3	3	3	3
ABS	0	20	30	40

Mixtures of different compositions were prepared in a laboratory mixer LDU-3 MPR with a propeller nozzle for 4 min at 700 rpm. A Lab Tech Scientific LTE 16-40 laboratory twin-screw extruder with a flat-slot die was used to produce PVC compositions in the form of flat profiles. Extrusion heating was regulated in ten zones of the barrel at a screw speed of 16–30 rpm (depending on the formulation), the optimum engine load was 30–50% of the maximum. The compound was in the extruder cylinder for 8–10 min. Samples at the outlet of the die were flat profiles of 2 × 30 mm or more (depending on the swelling coefficient of the melt) and 20–25 cm long.

Tensile strength was studied using a tensile machine TM-250 at a tensile speed of 100 mm/min on five prepared samples of size 150 × 15 mm.

The bending strength was determined using a tensile machine TM-250 at a loading rate of 50 mm/min on five prepared samples of 80 × 10 × 3 mm in size.

The density of the samples was determined by hydrostatic weighing on torsion scales TS-500 weighing up to 0.5 g, as well as by measuring and weighing.

Brinell hardness test was carried out on a WPM LEIPZIG 300/436 hardness tester. The test specimens were flat-parallel plates with a thickness of 5 mm and a width of 15 mm. Hardness determination was based on the indentation of a metal ball (5 mm diameter) under a given load (100 kg/cm^2) into the surface of the sample and measuring the depth of the imprint.

Thermo mechanical analysis was carried out on the device, working on the principle of constant loading in conditions of compression at a load of 1 N and a constant heating rate of 3 °C/min up to 220 °C on samples with a diameter of 8 ± 0.5 mm and a thickness of 3 ± 0.1 mm.

Scanning electron microscopy was performed on a Carl Zeiss Merlin high-resolution scanning electron microscope at a 5 kV primary electron accelerating voltage and a 300 pA probe current to minimize exposure to the test object. Samples were placed in liquid nitrogen, followed by chipping. The chipped samples were fixed on a holder and placed in a chamber of Quorum Q 150TES vacuum unit. The conductive layer was applied by cathodic spraying with Au/Pd alloy in an 80/20 ratio. The thickness of the applied layer was 15 nm.

Using an energy dispersion spectrometer, the elemental composition of the samples at different sections of the composite with nitrogen fixation as a reference element in the composition of ABS and chlorine showing the presence of PVC was investigated.

3 Results

Tensile and bending strength values are presented in Table 2. As can be seen from the data presented, the strength properties are at the level of unmodified samples. A slight tendency to a decrease in the results is quite acceptable for the indicators of the corresponding profiles and moldings made of rigid PVC.

Table 2. Tensile and bending strength values.

ABS concentration, phr	Tensile strength, MPa	Bending strength, MPa
0	51.5	65.2
20	49.3	50.2
30	49.2	51.5
40	49.3	53.2

The density of the compositions with increasing concentration of ABS begins to decrease from 1.27 to 1.22 g/cm^3 (Fig. 1). The low density of ABS copolymer allows lightening the weight of PVC products.

The hardness of the compositions with increasing concentration of ABS increases linearly from 76.5 to 80.3 kg/cm^2 (Fig. 2), while remaining less than the hardness values of unmodified PVC.

Figure 3 shows the thermo mechanical curves of PVC samples. They show that the glass transition temperature (T_g) begins to increase with increasing ABS concentration, and the flow temperature (T_f) decreases. As a consequence, the region of high elasticity begins to decrease and transitions to the fluid state at lower temperatures takes place. The region of high elasticity in the blended samples is characterized by a greater degree of difference from the horizontal change in strain.

Analysis of the supramolecular structure of the samples allows to some extent to explain the revealed changes in the properties of polymer-polymer mixtures. The microphotographs are presented in Fig. 4.

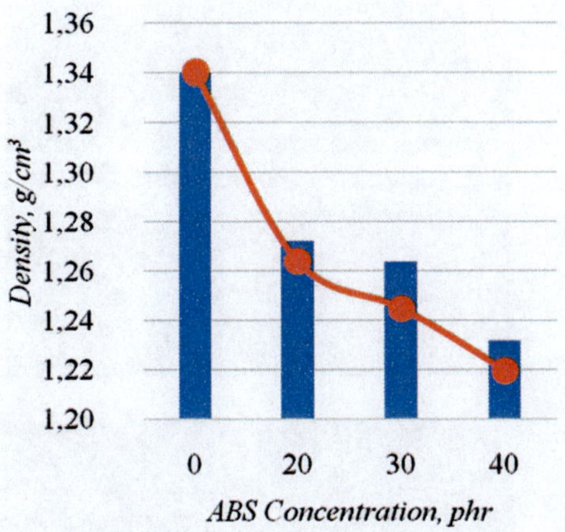

Fig. 1. Dependence of density on ABS concentration determined by: ■– measuring and weighing; ●– hydrostatic weighing.

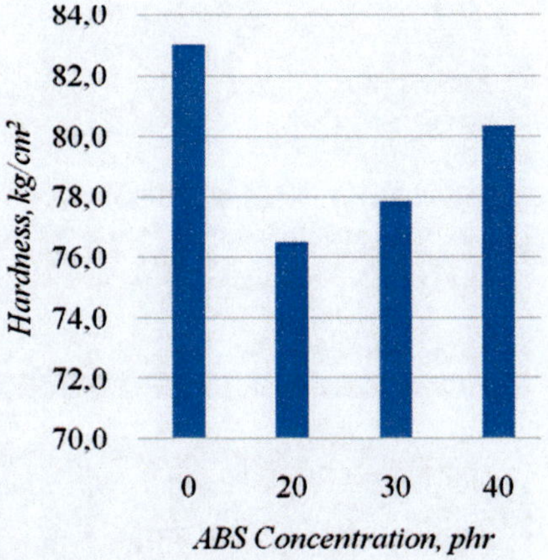

Fig. 2. Dependence of hardness on ABS concentration.

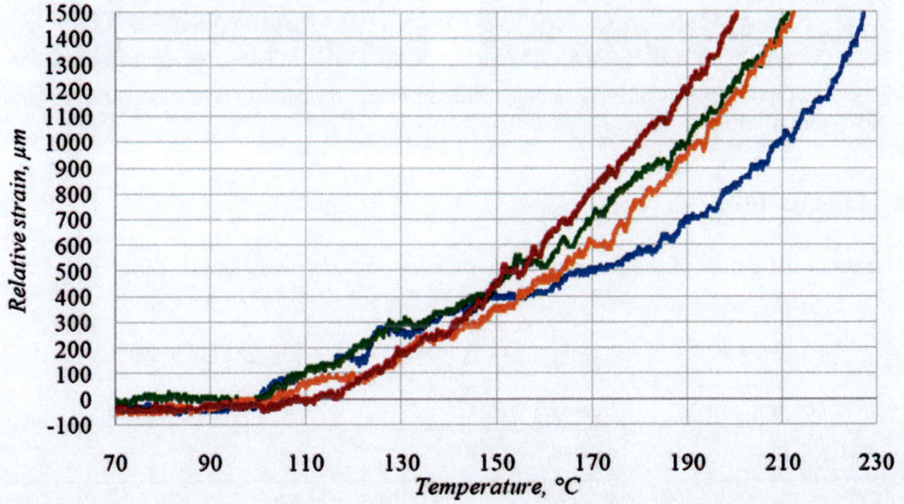

Fig. 3. Thermo mechanical curves of PVC samples with different concentrations of ABS (phr): ■– 0; ■– 20; ■– 30; ■– 40.

As the ABS concentration increases, there is a tendency for the transition from structures containing dispersed ABS-plastic inclusions to continuous phase structures.

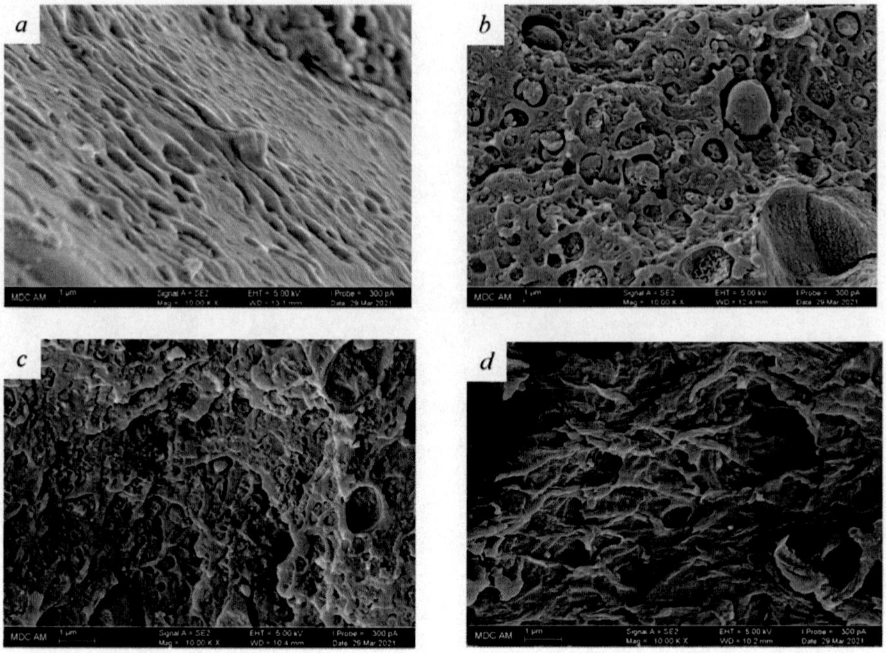

Fig. 4. Microphotographs of PVC samples with different concentrations of ABS (phr): a – 0; b – 20; c – 30; d – 40.

4 Discussion

The processing data of the thermo mechanical curves are presented in Table 3.

Table 3. Processing data of thermomechanical curves of PVC samples.

ABS concentration, phr	Temperature, °C		Area of high elasticity, °C
	$T_g^{initial} - T_g^{final}$	T_f	
0	97–126	206	80
20	96–133	186	53
30	100–133	185	52
40	110–133	180	47

Styrene-coated ABS gives the polymer-polymer mixture good fluidity, resulting in a more pronounced degree of macromolecule thawing at the boundary layer of the mixture. This fact is due to the contribution of segmental mobility of heterogeneous macromolecules of PVC and ABS polymers. In general, it can be observed that an increase in ABS concentration leads to a decrease in the area of high elasticity.

Although the glass transition temperature of ABS is slightly higher than that of PVC, some increase in this temperature is observed for mixed samples. This phenomenon is important in terms of increasing the heat resistance of the samples. This fact may be due to the possibility of formation of interpenetrating structures with increasing ABS concentration.

Obviously, an increase in ABS concentration leads to structural changes in the PVC-based polymer-polymer mixture. The solubility of ABS shell in the form of acrylonitrile-styrene copolymer in the PVC matrix contributes to active interaction with the butadiene core. The results are confirmed by energy-dispersive analysis performed during structural measurements. The data of energy dispersion spectra processing are presented in Table 4.

Table 4. Content of nitrogen atoms in the elements of the structure.

ABS concentration, phr	% weight/% atomic	
	Cl	N
0: in the dispersed phase in the dispersed medium	 22.14/8.99 16.90 /6.56	 – –
20: in the dispersed phase in the dispersed medium	 20.19/8.35 20.24/8.05	 1.47/1.53 –
30: in the dispersed phase in the dispersed medium	 21.70 /8.83 18.26/5.85	 1.14/1.17 –
40: in the dispersed phase in the dispersed medium	 24.65/10.58 25.49/10.30	 8.16/8.58 2.29/2.50

The element nitrogen was detected in the samples at 40 phr not only on the surface of the dispersed phase, but also in the dispersion medium. The data indicate that at a concentration of 40 phr ABS in the PVC matrix, the structure is essentially free of phase in the form of dispersed inclusions of various sizes. A joint structure of two continuous interpenetrating structures is observed. Obviously, further increase of ABS should lead to phase inversion, when the matrix becomes a continuous phase of ABS plastic.

5 Conclusion

Thus, the changes in the properties of the blended composites are generally determined by the type of the formed supramolecular structure. Increasing the ABS concentration up to 40 phr leads to an increase in hardness and bending strength up to 80.3 kg/cm^2 and 5.4 MPa, respectively. At the same time the tensile strength of the modified samples changes insignificantly. A decrease in density to 1.22 g/cm^3 and an increase in heat resistance of the composites were also found. The increase of ABS concentration leads to narrowing of the area of high elastic deformation. The type of the composite structure formed depends on the ABS content as well as the processing temperature regimes due to the decrease in the flow temperature of the composites to a large extent.

References

1. Ellen, MacArthur Foundation (2017). https://doi.org/10.1103/PhysRevB.74.035409
2. Liu, Y., Zhou, C., Li, F., Liu, H., Yang, J.: Stocks and flows of polyvinyl chloride (PVC) in China: 1980–2050. Resour. Conserv. Recycl. **154**, 104584 (2020). https://doi.org/10.1016/j.resconrec.2019.104584
3. Nizamov, R.K.: Stroit. Mater. **68** (2006)
4. Galeev, R., Nizamov, R., Abdrakhmanova, L., Khozin, V.: IOP Conference Series: Materials Science and Engineering (2020). https://doi.org/10.1088/1757-899X/890/1/012111
5. Joshi, P., Marathe, D.: Experimental investigation of mechanical properties of impact modified polyvinyl chloride-fly ash composites. J. Miner. Mater. Charact. Eng. **07**(01), 34–47 (2019). https://doi.org/10.4236/jmmce.2019.71003
6. Islamov, A., Fakhrutdmova, V.: IOP Conference Series: Materials Science and Engineering (2020). https://doi.org/10.1088/1757-899X/890/1/012083
7. Ashrapov, A.K., Abdrakhmanova, L.A., Nizamov, R.K., Khozin, V.G.: Nanotekhnologii v Stroit. Nauchnyy Internet-Zhurnal **3**, 13 (2011)
8. Galeev, R., Abdrakhmanova, L., Nizamov, R.: Solid State Phenom. (2018). https://doi.org/10.4028/www.scientific.net/SSP.276.223
9. Ciacci, L., Passarini, F., Vassura, I.: The European PVC cycle: in-use stock and flows. Resour. Conserv. Recycl. **123**, 108–116 (2017). https://doi.org/10.1016/j.resconrec.2016.08.008
10. Karaś, R.: The technology of designing plastic windows and their transport from Poland since 1990. Polimery **66**(1), 21–29 (2021). https://doi.org/10.14314/polimery.2021.1.3
11. Gholizadeh, H., Fazlollahtabar, H., Khalilzadeh, M.: J. Intell. Fuzzy Syst. **40** (2021). https://doi.org/10.3233/JIFS-190718
12. Putrawan, I.D.G.A., Azharuddin, A., Adityawarman, D., Ar Rahim, D., Tek Kim, J.: Indones **18** (2020). https://doi.org/10.5614/jtki.2019.18.2.3
13. Jemii, H., Bahri, A., Boubakri, A., Hammiche, D., Elleuch, K., Guermazi, N.: On the mechanical behaviour of industrial PVC pipes under pressure loading: experimental and numerical studies. J. Polym. Res. **27**(8), 1–13 (2020). https://doi.org/10.1007/s10965-020-02222-1
14. Fu, Z., Yang, Z., Rong, Y., Deng, L., Wu, J.: J. Vinyl Addit. Technol. (2021). https://doi.org/10.1002/vnl.21812
15. Allen, N.S., Edge, M.: J. Vinyl Addit. Technol. **27** (2021). https://doi.org/10.1002/vnl.21807
16. Abbas-Abadi, M.S.: The effect of process and structural parameters on the stability, thermo-mechanical and thermal degradation of polymers with hydrocarbon skeleton containing PE, PP, PS, PVC, NR, PBR and SBR. J. Therm. Anal. Calorim. **143**(4), 2867–2882 (2020). https://doi.org/10.1007/s10973-020-09344-0
17. Galeev, R., Nizamov, R., Abdrakhmanova, L.: E3S Web Conference (2020). https://doi.org/10.1051/e3sconf/202016414018
18. Gilbert, M., et al.: Brydson's Plast. Mater. (Eighth Edition, Edited by M. Gilbert, (Butterworth-Heinemann, 2017), pp. 631–652 (2017)

19. Lavrov, N.A., Belukhichev, E.V.: Plast. Massy (2020). https://doi.org/10.35164/0554-2901-2020-3-4-55-59
20. Sabah, F., En-Naji, A., Wahid, A., Ghorba, M.El., Chakir, H.: Key Eng. Mater. (2019). https://doi.org/10.4028/www.scientific.net/KEM.820.40
21. Lu, G., et al.: ACS Omega **5** (2020). https://doi.org/10.1021/acsomega.0c02803
22. Simionescu, T.M., Spiridon, I., Varganici, C.D., Darie-Nita, R.N., Minea, A.A.: Environ. Eng. Manag. J. **19** (2020). https://doi.org/10.30638/eemj.2020.073
23. Hu, D., Zhou, Q., Zhou, K.: J. Appl. Polym. Sci. **136** (2019). https://doi.org/10.1002/app.48220
24. Bano, S., Ramzan, N., Iqbal, T., Mahmood, H., Saeed, F.: Polish J. Chem. Technol. **22** (2020). https://doi.org/10.2478/pjct-2020-0029
25. Kurek, A.P., Dotto, M.E.R., de Araújo, P.H.H., Sellin, N.: J. Appl. Polym. Sci. (2017). https://doi.org/10.1002/app.44571
26. Li, Y., et al.: Polymer (Guildf), 190 (2020). https://doi.org/10.1016/j.polymer.2020.122198
27. Jaidev, K., Suresh, S.S., Gohatre, O.K., Biswal, M., Mohanty, S., Nayak, S.K.: Waste Manag. Res. **38** (2020). https://doi.org/10.1177/0734242X19890918
28. Ehsan, K.: Fine Chem. Eng. (2020). https://doi.org/10.37256/fce.122020476
29. Matseevich, A., Matseevich, T., Askadskii, A.: MATEC Web Conference (2018). https://doi.org/10.1051/matecconf/201819604069
30. Chen, F., Liang, H., Yin, S., Huang, S., Tang, Q.: Fabrication of novel resinous diamond composites with acrylonitrile butadiene styrene/polyvinyl chloride/dioctyl phthalate/diamond by hot pressing molding. J. Mater. Res. **34**(10), 1734–1743 (2019). https://doi.org/10.1557/jmr.2019.79
31. Xiong, C.W., Ho, C.Y., Qiao, D.: Medziagotyra **26** (2020). https://doi.org/10.5755/j01.ms.26.1.20310
32. Zhou, L.L., Lin, Y.S., Yang, J.Y., Wu, Q.Y.: Zhongguo Suliao/China Plast. **15**, 27 (2001)

Vibrational Spectra of p-Carboxylate and p-Sulfonate Azocalix[4]arene

Victor Furer[1](✉), Alexandr Vandyukov[2], Elena Popova[2], Svetlana Solovieva[2], and Igor Antipin[3]

[1] Kazan State University of Architecture and Engineering, 420043 Kazan, Russia
furer@kgasu.ru
[2] A.E. Arbuzov Institute of Organic and Physical Chemistry, RAS, Arbuzov st., 420088 Kazan, Russia
[3] Kazan Federal University, Kremlevskaya st., 420008 Kazan, Russia

Abstract. The vibrational spectra of calixarenes with azobenzene units having sulfonate (*p-SAC*) and carboxylate (*p-CAC*) groups in the para position were recorded and analyzed. The optimization of the structure and analysis of normal vibration for calixarenes were performed using the DFT method. The calculated geometric parameters, the frequencies of the harmonic oscillations, the band intensities in the IR spectra, and the Raman scattering activity of the calixarenes are consistent with the experimental data. The *p-CAC* and *p-SAC* calix[4]arene molecules are in conical conformation due to the ring system of H-bonds along the low rim. The H-bond is weaker in the *p-SAC* molecule. The orientation of the aromatic moieties depends on the type of terminal functional groups. The energy differences between the *E*- and *Z*-forms of azobenzene groups in *p-CAC* and *p-SAC* are 60.8 and 62.6 kcal/mol, respectively. The experimental vibrational spectra of the calixarenes were interpreted using the potential energy distribution. Bands characteristic of trans and cis conformations of azobenzene fragments have been assigned. The HOMO and LUMO frontal molecular orbitals of the calixarene molecules are localized. The studied calixarenes can be used as antioxidants for thermal and light stabilization of polymer building materials.

Keywords: Calixarenes · IR spectroscopy · Raman spectroscopy · Hydrogen bonding · DFT · Normal modes

1 Introduction

Calixarenes are widely used in biology, medicine, chemistry, and materials science [1–4]. For biological experiments and drugs, solubility in water is particularly important.

The addition of the carboxylate and sulfonate groups along the upper rim of the calixarenes makes it possible to obtain water-soluble substances [1–4]. Calixarenes with carboxylate and sulfonate groups interact with biological objects and are used as sensors for bacteria, metal ions, and water purification [5].

The inclusion of azobenzene bonds in calixarene molecules enables them to be used as optical sensors for ions of various metals, which play important roles in medicine

N. Vatin (Ed.): STCCE 2021, LNCE 169, pp. 22–30, 2021.
https://doi.org/10.1007/978-3-030-80103-8_3

and ecology [6]. When forming complexes of calixarene molecules with metal ions, a color change of solutions often occurs, which can be used for analytical purposes [7]. Under the influence of electromagnetic waves in the UV region, azobenzene groups are isomerized, and the shape of molecules and their ability to form complexes change [8]. Earlier, we reported the study of IR spectra of calixarenes [9, 10]. Analysis of IR spectra of calixarenes in combination with DFT calculations is important for studying the supramolecular properties of calixarenes.

In this work, we tried to identify the structural features of calixarenes containing azobenzene units, carboxylate, and sulfonate groups by comparing their vibrational spectra. A comparison of the experimental vibrational spectra of calixarenes with different functional groups opens the possibility of separating the bands of aromatic fragments, azobenzene units, and terminal groups. The results obtained will provide a better understanding of the spectral characteristics of these virtually important compounds and their dynamic properties.

Materials that combine high strength with low average density are called polymer construction materials. They are characterized by increased efficiency. The area of their application is limited by the disadvantages common for polymer materials: low heat resistance, flammability, and a decrease in strength under prolonged loading due to creep. The studied calixarenes can be used as antioxidants for thermal and light stabilization of polymer building materials.

2 Materials and Methods

The production procedure and the main parameters of *p-CAC* and *p-SAC* calixarenes have been described previously [6] (Fig. 1).

Crystalline samples were pressed into KBr tablets, and IR spectra were recorded using Bruker Vector 22 spectrometer in a region of 4000 to 400 cm^{-1} with a resolution of 4 cm^{-1}. 64 spectral scans have been accumulated. Raman spectra were stimulated by a 1064 nm Nd: YAG laser line with a power of 50 mW in the region of 3500–100 cm^{-1} and were recorded by a Bruker FT-Raman RAM II module of the FTIR Vertex 70 spectrometer.

The calculation of vibrations of the calixarenes was performed using the gradient-correlated DFT with a B3LYP functional [11, 12] and 6-31G(d, p) basis set and the Gaussian09 program [13]. As a first approximation of the geometries of *p-CAC* and *p-SAC*, we used experimental data obtained by X-ray diffraction [14]. The attribution of the oscillations was fulfilled using the potential energy distribution [15]. To assess the ability of calixarenes to form complexes, the chemical potential and the electrophilicity index was calculated [16, 17].

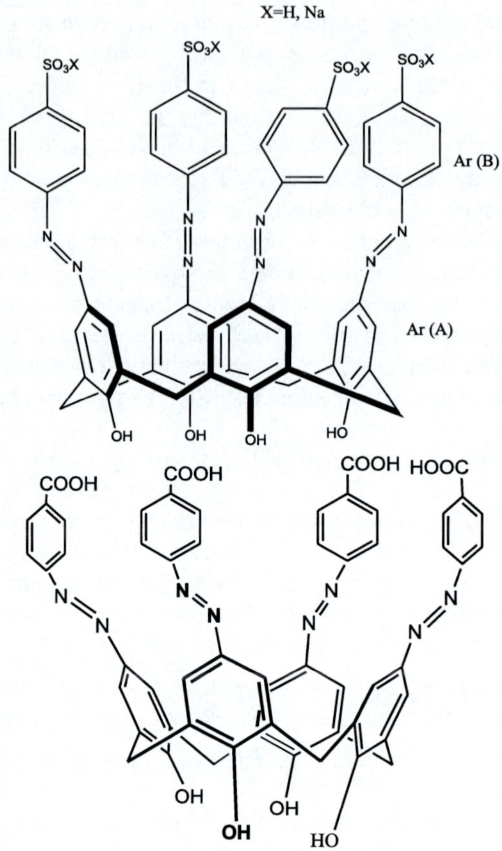

1 2

Fig. 1. Structure of p-SAC (1) and p-CAC (2).

3 Results

The results of geometric optimization of calixarene molecules for different isomers of azobenzene units are shown in Fig. 2. In *p-CAC* and *p-SAC* molecules, the most stable form is the *E*-form of azobenzene groups. The energy differences between the *E*- and *Z*-forms of *p-CAC* and *p-SAC* are 60.8 and 62.6 kcal/mol.

The calculated structures of *p-CAC* and *p-SAC* molecules are consistent with experimental X-ray data [18]. Although the distances between neighboring oxygen atoms differ slightly from each other in calculation and experiment, the four H-bonds forming ring system in *p-CAC* and *p-SAC* can be considered equivalent. The mean values of the distances calculated between neighboring oxygen atoms in *p-CAC* and *p-SAC,* are respectively 2.65 and 2.66 Å and correspond to the experiment [18]. The mean values of the theoretical angles O–H···O in *p-CAC* and *p-SAC* are equal to 162.8 and 162.3°.

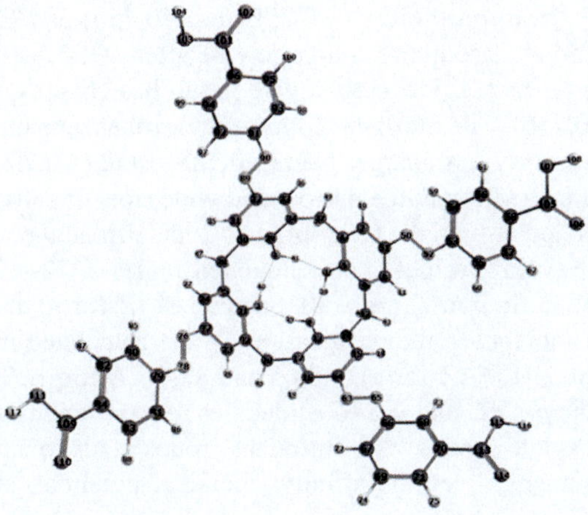

1

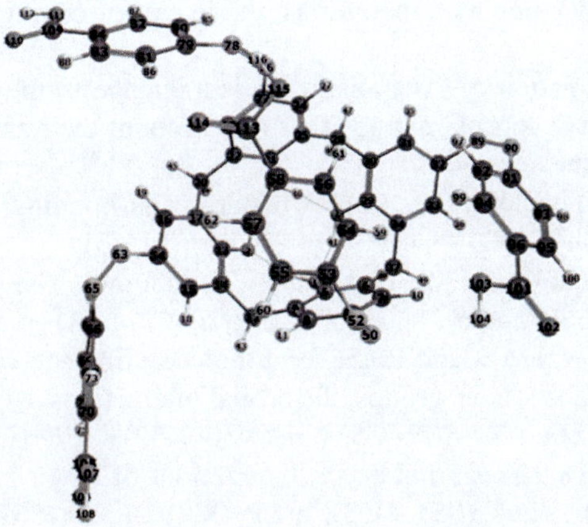

2

Fig. 2. Optimal geometry and numbering of atoms of the forms *E*- (1) and *Z*- (2) for *p-CAC*.

The most stable and polar is the conformation of the *cone* with four intramolecular cooperative H-bonds (Fig. 2). Theoretical dihedral angles C(71)–N(70)–N(72)–C(73) and N(70)–N(72)–C(73)–C(75) of the *p-SAC* molecule were equal to 179.3 and 167.5° (form *E*), and 11.4 and 138.3° (form *Z*) are consistent with experiment [18]. The theoretical dihedral angles C(51)–N(50)–N(52)–C(53) and N(50)–N(52)–C(53)–C(55) of the *p-CAC* molecule were equal to 180.0° and 179.9° (form *E*) and 12.3 and 133.6° (form *Z*). Theoretical lengths of N(70)–N(72) in the *E*- and *Z*-forms of *p-SAC* are 1.262 and 1.248 Å, which corresponds to experiment 1.260 and 1.253 Å [18].

Electrostatic interactions dominated the formation of complexes of calixarene with metal ions. Therefore, we calculated the charges on the atoms of the functional groups. Accordingly, the *p-CAC* and *p-SAC* molecules contain polar OH bonds with natural atomic charges [17] (in e) on the O1 (−0.752 and −0.750) and H38 (0.530) atoms, respectively. The sulfonate group has charges on the atoms O50 (−0.938), S51 (2.406), O52 (−0.916), O53 (−0.921). The carboxylate group has charges on the atoms O102 (−0.600), O103 (−0.716). The nitrogen atoms of the azobenzene unit of *p-CAC* and *p-SAC* calixarenes have fairly low charges N50 (−0.188), N52 (−0.207). There are active sites of reaction and interaction in the narrow and wide rims of calixarene molecules.

The angles of torsion around the CH_2−Ar bonds determine the configuration of the calixarenes [19]. The average values of the calculated angles φ(C7−C6−C37−C34) and χ(C6−C37−C34−C33) in *p-SAC* are −90.0 and 93.1° (*Z*-form) and −91.9 and 89.1° (*E*-form). In *p-CAC* molecules, the mean values of the calculated parameters φ and χ are equal to −91.3 and 91.0° (*Z*-form), −89.5 and 88.4° (*E*-form). The structure of the calixarene bowl in the *p-CAC* and *p-SAC* molecules remains unchanged.

Replacing carboxylate groups with sulfonate groups leads to an increase in dipole moment, ionization energy, electron affinity, chemical potential, and electrophilicity index.

The taking into account of the molecular orbitals of the *p-CAC* and *p-SAC* molecules shows that the conjugation includes fragments of azobenzene. During the formation of complexes of calixarenes with metal ions, the length of conjugated fragments and the color change.

During the formation of H-bonds, there is a change in the electron density in the molecules. The NBO approximation takes into account the transfer of electron density from the donor to the acceptor.

The *p-CAC* molecule has a complex pattern of H-bonding interactions: n(LP_1O1) → σ^*_1(O4–H41), n(LP_2O1) → σ^*_1(O4–H41), n(LP_2O1) → σ^*_1(C5–C9) with stabilization energies equal to 15.42, 18.12 and 4.73 kcal/mol. The interactions σ_2(C5–C6) → σ^*_2(C7–C51), σ_2(C5–C6) → σ^*_2(C8–C9), σ_2(C7–C51) → σ^*_2(C8–C9) with energies equal to 22.27, 17.03 and 20.56 kcal/mol describe the conjugation in aromatic fragments. In the azobenzene groups, the orbital interactions σ_2(C7–C51) → σ^*_2(N50–N52), σ_1(N50–N52) → σ^*_2(C7–C51), σ_2(N50–N52) → n*(LP_1C53) with the energies 21.37, 10.44, 18.92 kcal/mol occur. Interactions of lone electron pairs of sulfonate groups n(LP_2O50) → σ^*_1(S51–O53), n(LP_2O52) → σ^*_1(S51–C78), n(LP_3O52) → σ^*_1(O51–S51) occur with energies 16.36, 19.53, 21.66 kcal/mol. Interactions of lone electron pairs of carboxylate groups n(LP_2O102) → σ^*_1(C96–C101), n(LP_2O102) → σ^*_1(C101–C103), n(LP_2O103) → σ^*_2(C101–O102) occur with energies 18.67, 33.72, 46.12 kcal/mol.

4 Discussion

Figures 3, 4 show the vibrational spectra of *p-CAC* and *p-SAC*. Our calculations reproduce the intensity of the most visible bands in the vibrational spectra of *p-CAC* and *p-SAC*.

The characteristics of the studied vibrational spectra of *p-CAC* and *p-SAC* are their similarity (Fig. 3, 4). The intensity of the bands in the vibrational spectra of *p-CAC* and *p-SAC* vary more strongly than the frequencies. A comparison of the *p-CAC* and *p-SAC* spectra distinguishes the bands of the carboxylate and sulfonate groups (Fig. 3, 4).

In the calixarenes studied, a complex system of intramolecular and intermolecular H-bonds is realized, which is manifested in the IR spectra. In the IR spectra of *p-CAC* and *p-SAC,* there are no bands of free OH groups in the region of 3600 cm^{-1} (Fig. 3), which means that all hydroxyl groups participate in the formation of H-bonds. The hydroxyl bands have maxima at 3430 cm^{-1} for *p-SAC* and at 3400 cm^{-1} for *p-CAC*. It can be assumed that these bands are due to hydroxyl groups on the lower rim of calixarene molecules involved in the ring system of intramolecular H-bonds.

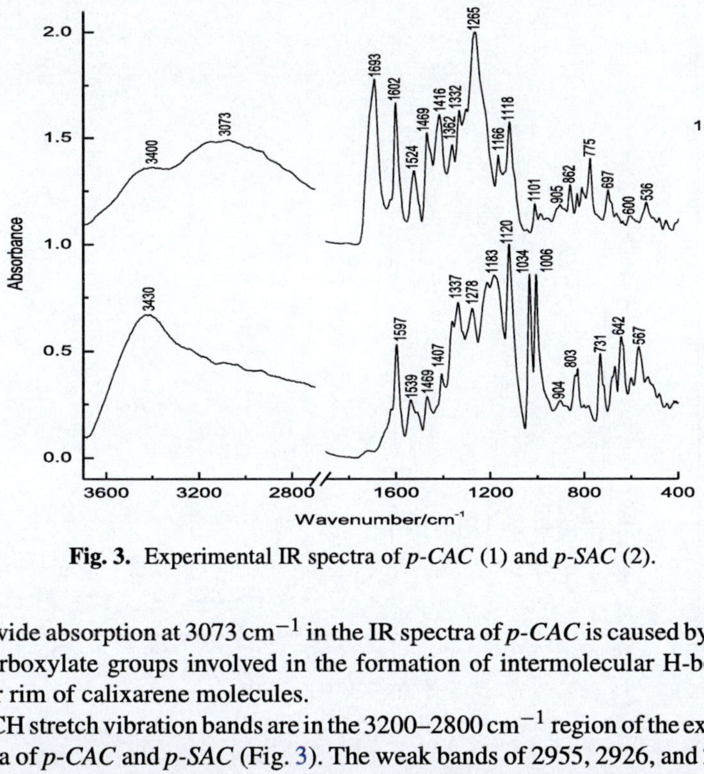

Fig. 3. Experimental IR spectra of *p-CAC* (1) and *p-SAC* (2).

The wide absorption at 3073 cm^{-1} in the IR spectra of *p-CAC* is caused by vibrations of the carboxylate groups involved in the formation of intermolecular H-bonds along the upper rim of calixarene molecules.

The CH stretch vibration bands are in the 3200–2800 cm^{-1} region of the experimental IR spectra of *p-CAC* and *p-SAC* (Fig. 3). The weak bands of 2955, 2926, and 2856 cm^{-1} observed in the IR spectrum of *p-SAC* are connected with the CH stretching of the methylene groups. In the Raman spectra of *p-CAC* and *p-SAC* calixarenes, a weak line of 3077 cm^{-1} stretching vibrations of CH bonds of aromatic moieties and a line of 2938 cm^{-1} stretching vibrations of CH_2 groups were discovered.

The absorption at1692 cm^{-1} is due to stretching vibrations of the carboxylate in the IR spectrum of *p-CAC.* The shift of this band in the low-frequency region is due to the H-bonds.

Bands of 1626, 1598 cm^{-1} in Raman *p-SAC* spectra and lines at 1627, 1607 cm^{-1} in Raman *p-CAC* spectra have been attributed to pulsations of benzene groups (Fig. 4). The asymmetric deformation vibrations of the methylene groups give a peak at about 1470 cm^{-1} in the vibrational spectra of *p-CAC* and *p-SAC.*

Stretch vibrations of aromatic CC bonds, HCH, and COH deformation vibrations cause the 1362 cm^{-1} band in the IR spectrum of *p-SAC.* Stretch vibrations of the CC and CO bonds show the medium intensity absorption in the range 1280–1160 cm^{-1} in the IR spectra of *p-SAC* and *p-CAC*.

Absorption peaks in the range 1120–1000 cm^{-1} in the IR spectrum of *p-SAC* are caused by SO and SC bond stretching vibrations and SOH bending vibrations. CC bonds stretching vibrations and CCH bending vibrations give the bands 905, 845, and 830 cm^{-1} in the IR spectrum of *p-SAC* and the bands 905, 862, and 832 cm^{-1} in the IR spectrum of *p-CAC*. The bands of 801, 782 cm^{-1} in the IR spectrum of *p-SAC* and the 775 cm^{-1} band in the IR spectrum of *p-CAC* refer to non-planar vibrations of aromatic groups.

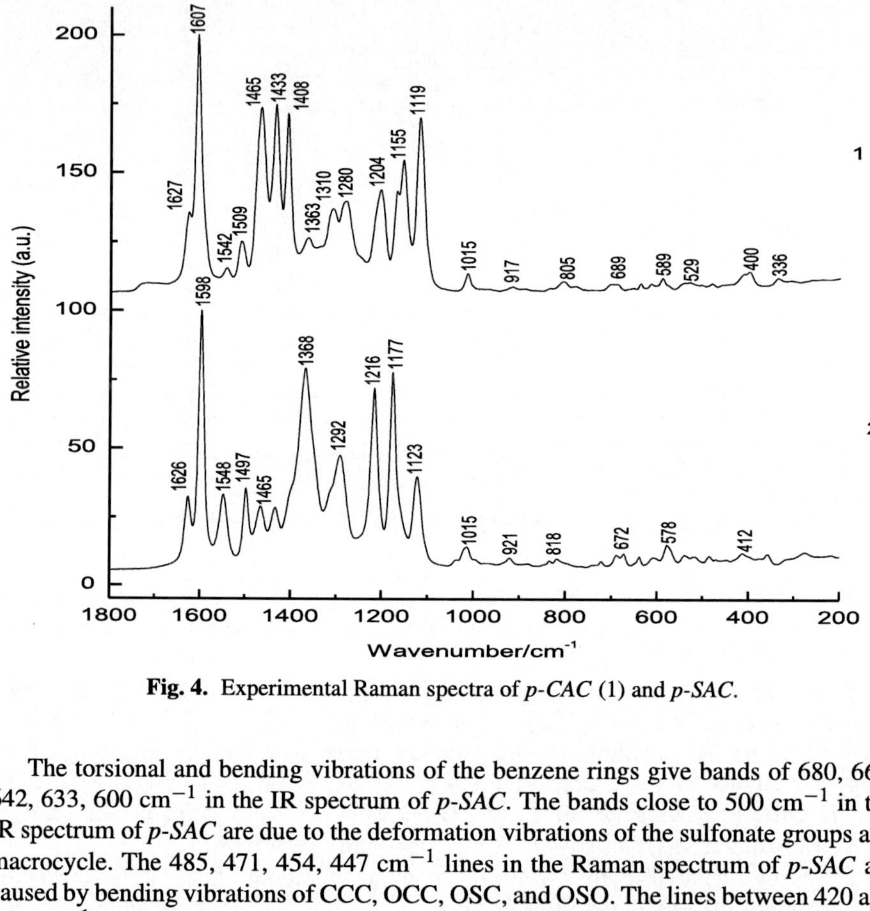

Fig. 4. Experimental Raman spectra of *p-CAC* (1) and *p-SAC*.

The torsional and bending vibrations of the benzene rings give bands of 680, 668, 642, 633, 600 cm^{-1} in the IR spectrum of *p-SAC*. The bands close to 500 cm^{-1} in the IR spectrum of *p-SAC* are due to the deformation vibrations of the sulfonate groups and macrocycle. The 485, 471, 454, 447 cm^{-1} lines in the Raman spectrum of *p-SAC* are caused by bending vibrations of CCC, OCC, OSC, and OSO. The lines between 420 and 160 cm^{-1} of the Raman spectrum of *p-SAC* have been assigned to torsional vibrations around the CC and SO bonds.

Comparison of the vibrational spectra of *p-CAC* and *p-SAC* allows selecting bands of sulfonate (1213, 1183, 1120, 1034, 1006 cm^{-1}) and carboxylate (1692, 1265, 775 cm^{-1}) groups.

The vibrational spectra of *E*- and *Z*-forms of *p-CAC* and *p-SAC* molecules differ from each other in the region 1510–1390 cm^{-1}.

The vibrational spectra of *p-CAC* and *p-SAC* differ considerably (Figs. 3, 4). Spectral analysis of calixarenes makes it possible to distinguish the 1607 cm^{-1} band in the vibrational spectra of *p-CAC* due to stretching vibrations the benzene rings. Sulfonate groups have 1120, 1034, 1006 cm^{-1} bands associated with $S - O$ stretching vibrations. The carboxyl groups exhibit strong bands characteristic of 1692 and 1265 cm^{-1} in the IR spectrum of *p-CAC*. The stretching vibrations of the $N = N$ azobenzene bonds cause a band near 1410 cm^{-1} in the vibrational spectra of *p-CAC*.

5 Conclusion

The vibrational spectra of azocalixarenes with sulfonate and carboxylate groups were recorded. Theoretical analysis of the spectra was carried out. The geometric and electronic parameters of calixarenes were calculated for different isomers of azobenzene units.

A comparative study of the spectra of azocalixarenes containing carboxylate and sulfonate groups makes it possible to identify the bands characterizing the vibrations of the different functional groups. The structure of the calixarene bowl in the *p-CAC* and *p-SAC* molecules remains unchanged.

In *p-CAC* and *p-SAC* molecules, the most stable is the *E*-form of azobenzene groups. The energy differences between the *E*- and *Z*-forms of *p-CAC* and *p-SAC* are 60.8 and 62.6 kcal/mol. Taking into account the molecular orbitals of the *p-CAC* and *p-SAC* molecules shows that the conjugation includes fragments of azobenzene. During the formation of complexes of calixarenes with metal ions, the length of conjugation fragments and the color change.

The studied calixarenes can be used as antioxidants for thermal and light stabilization and extension of the service life of polymer building materials and corresponding building structures.

References

1. Gutsche, C.D.: Calixarene. An Introduction. The Royal Society of Chemistry, Cambridge (2008)
2. Ovsyannikov, A.S., Solovieva, S.E., Antipin, I.S., Ferlay, S.: Coordination polymers based on calixarene derivatives: structures and properties. Coord. Chem. Rev. **352**, 151–186 (2017). https://doi.org/10.1016/j.ccr.2017.09.004
3. Vicens, J., Harrowfield, J., Baklouti, L. (eds.): Calixarenes in the Nanoworld. Springer, Heidelberg (2007). https://doi.org/10.1007/978-1-4020-5022-4
4. Solovieva, S.E., Burilov, V.A., Antipin, I.S.: Thiacalix[4]arene's lower rim derivatives: synthesis and supramolecular properties. Macroheterocycles **10**(1), 134–136 (2017). https://doi.org/10.6060/mhc170512a
5. Perret, F., Coleman, A.W.: Biochemistry of anionic calix[n]arenes. Chem. Commun. **47**(26), 7303–7319 (2011). https://doi.org/10.1039/c1cc11541c
6. Kashapov, R.R., et al.: Self-aggregation and solubilizing properties of the supramolecular system based on azobenzenesulfonate calix[4]arene and CTAB. Macroheterocycles **10**(4–5), 454–459 (2017). https://doi.org/10.6060/mhc171146k

7. Bonvallet, P.A., Mullen, M.R., Evans, P.J., Stoltz, K.L., Story, E.N.: Improved functionality and control in the isomerisation of a calix[4]arene-capped azobenzene. Tetr. Lett. **52**(10), 1117–1120 (2011). https://doi.org/10.1016/j.tetlet.2010.12.107
8. Banadara, H.M.D., Burdette, S.C.: Photoisomerization in different classes of azobenzene. Chem. Soc. Rev. **41**(5), 1809–1825 (2012). https://doi.org/10.1039/c1cs15179g
9. Furer, V.L., Potapova, L.I., Kovalenko, V.I.: DFT study of hydrogen bonding and IR spectra of calix[6]arene. J. Mol. Struct. **1128**, 439–447 (2017). https://doi.org/10.1016/j.molstruc.2016.09.010
10. Furer, V.L., Vandyukov, A.E., Zaripov, S.R., Solovieva, S.E., Antipin, I.S., Kovalenko, V.I.: FT-IR and FT-Raman study of hydrogen bonding in p-alkylcalix[8]arenes. Vibr. Spec. **95**(1), 38–43 (2018). https://doi.org/10.1016/j.vibspec.2018.01.006
11. Becke, A.D.: Density-functional thermochemistry. III. The role of exact exchange. J. Chem. Phys. **98**(7), 5648–5652 (1993). https://doi.org/10.1063/1.464913
12. Lee, C., Yang, W., Parr, R.G.: Development of the Colle-Salvetti correlation-energy formula into a functional of the electron density. Phys. Rev. B **37**(2), 785–789 (1988). https://doi.org/10.1103/PhysRevB.37.785
13. Frisch, M.J.: Gaussian 09 Revision C.01. Gaussian Inc., Wallingford (2010)
14. Ehlinger, N., Perrin, M.: Structure of p-tetrakis-(4-nitrophenylazo)calix[4]-arene-4-picoline (1:4) complex. J. Incl. Phenom. **22**(1), 33–40 (1955). https://doi.org/10.1007/BF00706496
15. Sipachev, V.A.: Calculation of shrinkage corrections in harmonic approximation. J. Mol. Struct. (THEOCHEM) **121**(1), 143–151 (1985). https://doi.org/10.1016/0166-1280(85)80054-3
16. Parr, R.G., Pearson, R.G.: Absolute hardness: companion parameter to absolute electronegativity. J. Am. Chem. Soc. **105**(26), 7512–7516 (1983). https://doi.org/10.1021/ja00364a005
17. Glendening, E.D., Landis, C.R., Weinhold, F.: Natural bond orbital methods. Comput. Mol. Sci. **2**(1), 1–42 (2012). https://doi.org/10.1002/wcms.51
18. Atwood, J.L., Orr, G.W., Hamada, F., Vincent, R.L., Bott, S.G., Robinson, K.D.: Calixarenes as second-sphere ligands for transition metal ions. Synthesis and crystal structure of $[(H_2O)_5Ni(NC_5H_5)]_2(Na)$[calix[4]arene sulfonate]•3.5 H_2O and $[(H_2O)_4Cu(NC_5H_5)_2]$-$(H_3O_3)_3$ [calix[4]arene sulfonate] • 10 H_2O. J. Incl. Phenom. **14**(1), 37–46 (1992)
19. Ugozzoli, F., Andretti, G.D.: Symbolic representation of the molecular conformation of calixarenes. J. Incl. Phenom. **13**(4), 337–348 (1992). https://doi.org/10.1007/BF01133233

Implementation of the Decision-Making Algorithm in the Bridge Management System

Karim Nurmuhametov, Tagir Zinnurov(✉), and Denis Sadykov

Kazan State University of Architecture and Engineering, 420043 Kazan, Russia

Abstract. Traditionally, for the full functioning of bridge management systems, they contain: a block for collecting information about structures, a block for making decisions, and a block for implementing decisions. However, in most of these systems, the decision-making block does not take into account the significance of the structure, and the priority of restoration work is determined only based on data about their technical conditions. In conditions of a large number of bridges and limited funding, the task of taking into account the significance of structures when forming a strategy for their restoration becomes relevant. The authors propose a decision-making algorithm that takes considers this feature and has been implemented using the example of Apastovsky and Pestrechinsky municipal districts of the Republic of Tatarstan, the Russian Federation. Firstly, the technical condition of the structures under consideration was taken into account. Then their significance was assessed, considering the traffic intensity of these objects and the functional classification of the highways on which they are located. Next, an approximate estimate of the cost of restoring these objects was made, and the necessary costs were distributed over several years. As a result of the work done, an optimal strategy for bringing 48 bridges to standard was formed.

Keywords: Strategy · Bridge · Management systems · Algorithm · Location

1 Introduction

Bridges are an important component of any developed transport infrastructure. The uninterrupted operation, comfort and safety of traffic depend on the technical condition of bridge structures. Based on these reasons, the task of ensuring the safety of bridge structures is always relevant.

A bridge management system is a means of managing bridges throughout their life cycle (design, construction, operation, and maintenance). Bridge management systems help road authorities perform tasks such as bridge inventory (database), systematic planning of maintenance, repair and restoration activities, optimizing the allocation of financial resources, and improving the safety of bridge users.

Many scientific papers [1–9] are devoted to bridge control systems. Some of them summarize the world experience of using such systems [1–3], some talk about the experience of using them in specific regions [4–10]. Scientific research in the field of development of bridge management systems is developing in several directions. The authors

N. Vatin (Ed.): STCCE 2021, LNCE 169, pp. 31–37, 2021.
https://doi.org/10.1007/978-3-030-80103-8_4

Hanley, Lee, Wang, Shim, Wattan, and Jeong monitor and collect information to form databases or create models [11–16]. Other researchers, on the contrary, use the information obtained as a forecast or assessment of the technical condition of bridges. [17–24]. For example, Valenzuela proposes to introduce an indicator of the technical condition of the bridge [17], and Gao and Zambon scientists predict the condition and operation of concrete bridges [20, 23]. The most interesting are the works of the authors Fang, Sun, Inkoom, which offer a probabilistic approach to the functioning of a system of elements or bridge structures [25, 26]. There are other works in this area of research [27–31].

Having considered approaches to the development and use of bridge management systems, we have come to the conclusion that in most systems, when determining the priority of repair work at the decision-making stage, the significance of the structure is not taken into account. However, the task of accounting for the significance of structures becomes especially relevant in the conditions of their large number and limited financial resources. It is especially important to take into account the uncertainty in the development of the operating conditions and the importance of construction in the road network when forming strategies for the restoration of bridges located on roads of regional significance in the Russian Federation. Therefore, the purpose of this study is to build an adapted algorithm for the functioning of regional bridge management systems.

2 Materials and Methods

Earlier, in [29], the authors proposed a decision-making algorithm, which was refined and presented in Fig. 1 with changes.

The algorithm consists of the following main stages:

- Estimation of Technical Condition;
- Assessment of the Importance;
- Assessment of the Cost.

In this paper, in order to demonstrate the principles of the algorithm, bridges located on regional highways in Apastovsky and Pestrechinsky districts of the Republic of Tatarstan are considered.

For the algorithm to work effectively the following initial information about bridge structures was needed:

- name of the structure;
- technical condition of the structure (emergency, pre-accident, unsatisfactory, satisfactory, good, design);
- year of the last diagnosis or inspection of the structure;
- category of road where the structure is located;
- traffic intensity on the road where the structure is located;
- potential number of users of the bridge structure;
- the possibility and the length of the bypass;
- type of work required to bring the structure into a standard condition (construction to replace the existing one), reconstruction, major repairs, repairs, maintenance);
- bridge surface area.

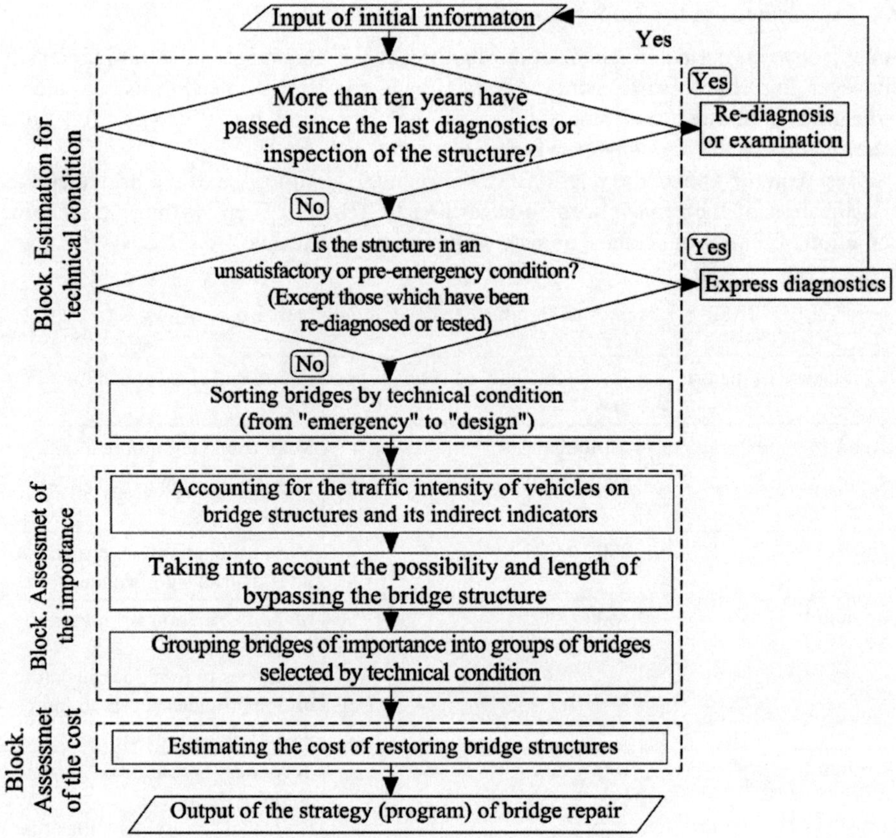

Fig. 1. Decision-making algorithm.

3 Results and Discussion

3.1 Estimation of Technical Condition

In the Apastovsky district of the Republic of Tatarstan, there are 29 bridges on regional roads. Of these, 7 are in unsatisfactory condition, 18 are in satisfactory condition and 5 are in good technical condition. There are 19 bridges on regional roads in Pestrechinsky district of the Republic of Tatarstan. Of these, 1 is in emergency condition, 4 in unsatisfactory condition, 10 in satisfactory condition and 4 in good technical condition.

This estimation of the technical condition of bridges is based on the data of their diagnostics, which was carried out in accordance with current Russian regulatory documents. For further operation of the algorithm bridges are sorted from the emergency state to the design state.

3.2 Assessment of the Importance

An objective numerical indicator of the significance of the bridge is the traffic intensity. However, this characteristic is not always known, so there is a need to use its indirect indicators. An indirect indicator of the traffic intensity on the bridge can be a functional classification of the highway on which it is located.

Based on the above, the relationship between the significance of the bridge and the classification of the road is formed according to Table 1. Then, within the technical condition of the set of bridges, objects with higher significance are selected.

Table 1. Degrees of significance depending on other parameters.

Significance of the bridge	Traffic intensity of vehicles per day	Functional class of a automobile road
High I	>14000	Main trunk automobile roads
High II	6000–14000	Secondary trunk automobile roads
High III	2000–6000	Main trunk automobile roads of inter-district significance
Medium I	1500–2000	Main distribution automobile roads Entrances to rural localities (the bridge provides access to more than 10 localities)
Medium II	1000–1500	Secondary distribution automobile roads Entrances to rural localities (the bridge provides access to 6–9 localities)
Medium III	200–1000	Entrances to rural localities (the bridge provides access to 4–5 localities)
Low I	100–200	Entrances to rural localities (the bridge provides access to 2–3 localities)
Low II	50–100	Entrances to rural localities (the bridge provides access to 1 locality)
Low III	<50	Distribution automobile roads of inter-district significance Entrances to rural localities (the bridge is located behind all or most of localities and access to them is not provided)

3.3 Assessment of the Cost

Estimating the cost of restoring bridges is a time-consuming task and, in order to optimize it, the cost of work is determined in Russian rubles, based on the cost of restoring similar objects on the territory of the Republic of Tatarstan in 2016–2019.

Based on the results of the algorithm, the priority of work on the restoration of bridges in the Apastovsky and Pestrechinsky districts of the Republic of Tatarstan was determined.

The total amount of funding required for these works was about 12.35 billion rubles. This amount of necessary funding is distributed over several years (Table 2).

Table 2. Strategy for restoring bridges in Apastovsky and Pestrechinsky districts.

Year of the strategy implementation	Number of objects	Necessary expenses, thousand rubles
1	11	308592.28
2	14	305265.26
3	6	309859.36
4	9	311440.16

This example shows that it is possible to take simultaneously into account the technical condition, significance and cost of bridges at the stage of determining the priority of their restoration in accordance with the decision-making algorithm proposed by the authors.

The proposed approach with the necessary adjustments can be integrated not only into existing road bridge management systems, but also into newly created ones.

4 Conclusions

In both domestic and foreign bridge management systems, when determining the priority of repair work at the decision-making stage, the significance of the structure is not taken into account.

The authors propose a decision-making algorithm that allows forming an effective repair strategy that determines the priority of work in the conditions of a large number of bridges and a limited amount of funding.

The proposed algorithm was tested in the Apastovsky and Pestrechinsky districts of the Republic of Tatarstan. Based on the results of modeling, a strategy was formed for bringing the bridge structures in operation to a standard state.

In order to optimize the algorithm and increase its efficiency, there is a need for further study and optimization of methods for evaluating the technical condition, significance and cost of bridge structures.

References

1. Nigamatova, O.I., Ovchinnikov, I.G.: Control systems for the condition of bridges. Internet-Journal Naukovedenie **7**(3) (2015). https://doi.org/10.15862/09TVN315
2. Nigamatova, O.I., Ovchinnikov, I.G.: International experience of using expert systems for assessing the state of bridge structures. Internet-journal Naukovedeni **8**(1) (2016). https://doi.org/10.15862/66TVN116
3. Jeong, Y., Kim, W.S., Lee, I., Lee, J.: Bridge inspection practices and bridge management programs in China, Japan, Korea, and U.S. J. Struct. Integr. Maintenance **3**(2) (2018). https://doi.org/10.1080/24705314.2018.1461548
4. Fruguglietti, E., Pasqualato, G., Spallarossa, E.: Bridge management system implementation in Italy: pontis and other BMS application in Italy. Bridge Maintenance, Safety, Management, Resilience and Sustainability, 1st edn. CRC Press, Italy (2012). https://doi.org/10.1080/15732479.2014.889178
5. Zhai, L.: Research on Information Maintenance and Management System of Bridge Group over Wei-River Proceedings of the International Forum on Energy, Environment and Sustainable Development. Atlantis Press, China (2016). https://doi.org/10.2991/ifeesd-16.2016.16
6. Gholamia, M., Mohd Sama, A., Yatima, J.: Assessment of bridge management system in Iran. Procedia Eng. **54** (2013). https://doi.org/10.1016/j.proeng.2013.03.052
7. Rybalov, Y.V.: Automated information and analytical system for artificial structures on highways. SAPR i GIS Avtomobilnyh Dorog **2**(5) (2015). https://doi.org/10.17273/CADGIS.2015.2.19
8. Akgul, F.: Bridge management in Turkey: a BMS design with customised functionalities. Struct. Infrastruct. Eng. **12**(5) (2016). https://doi.org/10.1080/15732479.2015.1035284
9. Marasabessy, E.: Implementation of bridge management system on interurban bridge in Maluku province. J. Civil Eng. Forum **1**(2) (2017). https://doi.org/10.22146/jcef.23999
10. Harywijaya, W., Afifuddin, M., Isya, M.: Penilaian kondisi jembatan menggunakan bridge management system (BMS) dan bridge condition rating (BCR). Jornal Arsip Rekayasa Sipil Dan Perencanaan **3**(1) (2020). https://doi.org/10.24815/jarsp.v3i1.16462
11. Hanley, C., Matos, J.C., Kelliher, D., Pakrashi, V.: Integrating multivariate techniques in bridge management systems. J. Struct. Integr. Maintenance **2**(3) (2017)
12. Lee, J.H., Kim, S.C.: Mobile technology in the field data collection and management. J. Struct. Integr. Maintenance **1**(3) (2016). https://doi.org/10.1080/24705314.2016.1211233
13. Wan, C., et al.: Development of a bridge management system based on the building information modeling technology. Sustainability **11**(17) (2019). https://doi.org/10.3390/su11174583
14. Shim, C.Su., Kang, H., Dang, N.S., Lee, D.: Development of BIM-based bridge maintenance system for cable-stayed bridges. Smart Struct. Syst. **20**(6) (2017). https://doi.org/10.12989/sss.2017.20.6.697
15. Wattan, S.A., Al-Bakri, M.: Development of bridges maintenance management system based on geographic information system techniques. J. Eng. **25**(7) (2019). https://doi.org/10.31026/j.eng.2019.07.02
16. Jeong, S., Hou, R., Lynch, J.P., Sohn, H., Law, K.H.: An information modeling framework for bridge monitoring. Adv. Eng. Softw. **114** (2017). https://doi.org/10.1016/j.advengsoft.2017.05.009
17. Valenzuela, S., De Solminihac, H., Echaveguren, T.: Proposal of an integrated index for prioritization of bridge maintenance. J. Bridge Eng. **15**(3) (2010)
18. Wen, Q., Qiang, M., Xia, B., An, N.: Discovering regulatory concerns on bridge management: an author-topic model-based approach. Transp. Policy **75** (2019). https://doi.org/10.1016/j.tranpol.2017.04.012

19. Markiz, N., Jrade, A.: Integrating fuzzy-logic decision support with a Bridge Information Management System (BRIMS) at the conceptual stage of bridge design. J. Inf. Technol. Constr. **23** (2018). https://doi.org/10.1016/j.ijsbe.2014.08.002
20. Zambon, I., Vidovic, A., Strauss, A., Matos, J.: Condition prediction of existing concrete bridges as a combination of visual inspection and analytical models of deterioration. Appl. Sci. **9**(1) (2019). https://doi.org/10.3390/app9010148
21. Rashidi, M., Samali, B., Sharafi, P.: A new model for bridge management: Part A: condition assessment and priority ranking of bridges. Aust. J. Civil Eng. **14**(1) (2016)
22. Rashidi, M., Samali, B., Sharafi, P.: A new model for bridge management: Part B: decision support system for remediation planning. Aust. J. Civ. Eng. **14**(1), 46–53 (2016)
23. Gao, Z., Liang, R.Y., Xuan, T.: VIKOR method for ranking concrete bridge repair projects with target-based criteria. Results Eng. **3** (2019). https://doi.org/10.1016/j.rineng.2019.100018
24. Apriani, W., Megasari, S.W., Putri Loka, W.A.: Penilaian kondisi jembatan rangka baja di riau dengan metode bridge management system. Siklus: Jurnal Teknik Sipil **4**(2) (2018). https://doi.org/10.31849/siklus.v4i2.1706
25. Sun, J.W., Park, K.H., Lee, M.J.: A multi-level asset management decision method considering the risk and value of bridges. J. Asian Architect. Build. Eng. **16**(1) (2017). https://doi.org/10.3130/jaabe.16.163
26. Fang, Y., Sun, L.: Developing a semi-markov process model for bridge deterioration prediction in Shanghai. Sustainability **11**(19) (2019). https://doi.org/10.3390/su11195524
27. Inkoom, S., Sobanjo, J.: Availability function as bridge element's importance weight in computing overall bridge health index. Struct. Infrastruct. Eng. **14**(12) (2018)
28. Yashnov, A.N., Zubko, A.V.: On the issue of improving the reliability of bridge structures on highways. Dorogi i mosty **40**(2) (2018)
29. Nurmuhametov, K.A., Zinnurov, T.A., Sadykov, D.M.: Decision making algorithm in the regional bridges management system. News KSUAE **3**(49) (2019)
30. Khozin, V., Khokhryakov, O., Nizamov, R.: IOP Conference Series: Materials Science and Engineering (2020)
31. Mavliev, L., Vdovin, E.: E3S Web Conference (2019)

Alkali Activation of Russian Calcined Medium-Grade Clay: Influence of NaOH Concentration

Nailia Rakhimova[1](✉), Vladimir Morozov[2], and Aleksey Eskin[2]

[1] Kazan State Univeristy of Architecture and Engineering, 420043 Kazan, Russia
[2] Kazan Federal University, 420008 Kazan, Russia

Abstract. The object of this study was to investigate how 5–14 M NaOH concentrations affect the mechanical properties, amorphous-crystalline phase transformations, and composition of the reaction product of alkali-activated calcined kaolinite/montmorillonite cement with reactive phases of 48%, and with the SiO_2/Al_2O_3 ratio 1.62. The hardened pastes based on calcined clays showed the highest compressive strength following their activation with 8 M NaOH. The zeolite content in the alkali-activated calcined clay cements peaked at 11 M NaOH, and the corresponding Na_2O/SiO_2 and Na_2O/Al_2O_3 molar ratios were 0.6 and 0.93, respectively. The main reaction products in the studied alkali-activated calcined clay cements were N-A-S-H, zeolite A, faujasite-Na, analcime, and sodium aluminium silicate hydrate. The obtained properties and reaction products facilitate the use of the proposed alkali-activated calcined clay materials for controlled low-strength materials, matrices for immobilization of radioactive wastes.

Keywords: Clay · Calcination · Cement · Alkali · Zeolite

1 Introduction

The continuous progress in production and use of inorganic binders and materials based on them to meet increasing environmental and technical requirements requires changes in the raw materials base [1–4]. Calcined clays with good performance and being more available source compared to traditional supplementary cementitious materials like blast-furnace slag and fly ash have attracted great attention and begun to be applied in the development and implementation of both sustainable low-clinker Portland and non-clinker cements including alkali-activated binder systems [5–8]. In terms of composition, clays are extremely diverse materials and differ in their chemical compositions, percentage, types of clay and non-clay minerals and impurities present. High quality kaolin clays with a 1:1 kaolinite content of more than 65% are considered the most valuable resource for obtaining pure and highly reactive amorphous metakaolin for various purposes. However, among the major global industrial mineral resources, 90% of the clays employed in production are 2:1 layer-lattice minerals, whereas only 10% are kaolin or bentonite clays [9]. To expand the range of available clay resources, a number of studies have recently been conducted. Such studies have demonstrated the suitability

N. Vatin (Ed.): STCCE 2021, LNCE 169, pp. 38–46, 2021.
https://doi.org/10.1007/978-3-030-80103-8_5

of various thermally treated clays, including medium - (40–65%) and low-grade (<40%) kaolin clays, as well as multimineral clays consisting of kaolin and/or non-kaolin clay minerals and non-clay minerals as precursors for alkali-activated materials [10–14] and as sources for the synthesis of zeolites during the alkali hydrothermal process [15, 16]. The reaction rate, structure, and result properties of alkali-activated calcined clay-based cements are due to: chemical-mineralogical composition of clays, purity, crystallinity of clays, temperature and time of calcination, particle size, type and concentration of the alkaline activator, and curing conditions. These aspects for metakaolin-based systems were investigated in detail in a big number of studies [17–22].

The study is focused on the influence of 5–14 M NaOH solution alkalinity on the compressive strength, amorphous-crystalline phase transformations, and composition of the reaction products of alkali-activated cements based on calcined medium grade clay in accordance with Na_20/SiO_2 and Na_20/Al_20_3.

2 Materials and Methods

The clay used to produce alkali activated cements was obtained from the deposit in Russian Federation (Republic of Tatarstan). The chemical composition of the clay is presented in Table 1.

Table 1. Chemical composition of clays.

Component (mass % as oxide)								
SiO_2	Al_2O_3	Fe_2O_3	CaO	Na_2O	K_2O	MgO	Other oxides	LOI
53.8	33.2	0.7	0.5	0.3	0.7	0.3	1.3	9.2

Specific surface area of clay after calcination and grinding was 6200 cm^2/g (Blaine).

The Sodium hydroxide pellets with alkaline activator were obtained from Meterra (Russian Federation). The NaOH solution with a molarity of 5, 8, 11, 14 was prepared by dissolving of NaOH pellets in deionized water and cooling down to room temperature during 24 h before use.

The chemical composition of the clay was determined by using atomic emission spectrometry (OPTIMA 4300 DV). The particle size distribution of the clays was measured using a laser particle size analyzer (Horiba La-950V2). Clay was dispersed by ultrasound in ethanol and dispersion medium. 68% of the clay particles were <45 μm and 23% were <10 μm.

The clays were calcined at 600, 700, and 800 °C for 2 h using heating rates of 1.7, 2.5 and 3.3 °C/min, respectively, and milled in a planetary mill MPL-1.

The calcined clay samples were kneaded for about 10 min with alkaline activators solution at liquid/ratio of 0.45 to form pastes. The pastes were manually cast into 25 × 25 × 25 mm cube moulds and were vibrated to remove entrained air. The compressive strength of the cubes was tested after steam curing, following 4-h treatment programme to reach the desired temperature, 20 h of dwell time at 85 °C and 3-h cooling. Compression tests were performed by applying a load between the two surfaces that were vertical during casting. Each strength determination quoted is based on an average of six measurements from the same cast.

X-ray diffraction (XRD) and thermal analyses (TG/DSC) were performed on finely ground clays and samples of hardened pastes that had been steam cured. The XRD data were collected using a D2 Phaser X-ray diffractometer in a Bragg-Brentano θ–2θ configuration with a Cu Kα radiation, operating at 40 kV and 30 mA. The data analysis was performed using the DIFFRAC plus Evaluation Package EVA Search/Match and the PDF-2 ICDD database. The mineralogical composition analysis of clays was estimated by analyzing the X-ray diffractograms in the software product Diffrac.eva V3.2. The STA 443 F3 Jupiter simultaneous thermal analysis apparatus was used for TG/DSC. The samples were heated from 30 °C to 1000 °C at a heating rate of 10 °C/min. The data was analysed using Netzsch Proteus Thermal Analysis.

The reactive SiO_2/Al_2O_3 ratios and potentially reactive phase content in the calcined clays were obtained according to the procedure outlined in [23].

3 Results and Discussions

3.1 Compressive Strength of Hardened Pastes

Preliminary studies demonstrated that the samples based on the clay that was calcined at 800 °C had higher compressive strengths than one calcined at 600 °C and 700 °C. Thus, the data below are presented for cements made from clay heat treated at 800 °C. The effects of the NaOH solution concentration and calcined clay type on the compressive strengths of hardened alkali-activated clay pastes are listed in Table 2. According to the results, the maximum compressive strength was observed for hardened pastes obtained by 8M NaOH.

Table 2. Compressive strength of hardened alkali-activated clay cement pastes (MPa).

Clay	Concentration of NaOH solution (M)	Compressive strength (MPa)
Clay 1	5	0.7
	8	7.3
	11	6.9
	14	2.2

3.2 Characteristics of Calcined Clays

X-ray and thermal analyses were carried out to reveal the transformations in the clay composition after heat treatment. XRD patterns for the raw and heat-treated clay are shown in Fig. 1.

The minerals identified in raw clays were kaolinite-48% $Al_2(Si_2O_5)(OH)_4$ (PDF 01-078-1996), montmorillonite-4% $Na,Ca_{0.3}Al_2(Si,Al)_4O_{10}(OH)_2{\cdot}xH_2O$ (PDF 00-060-0316), quartz-33% SiO_2 (PDF 01-079-1910), and muscovite-15% $K_{0.77}Al_{1.93}(Al_{0.5}Si_{3.5})O_{10}(OH)_2$ (PDF 01-070-1869). The thermal treatment of the clay led to the disappearance of the peaks related to clay minerals, which indicates their disordering. Figure 2 presents the thermal analysis results. Endothermic peaks were detected at 50 °C – corresponding to the loss of adsorbed water, in the 440–540 °C range – corresponding to the dehydroxylation of kaolinite, and in the 700–850 °C range – corresponding to the dehydroxylation of montmorillonite. Weight mass loss for raw clay was 11.1%, calcined clay was 1.69%. The amount of reactive phase and reactive SiO_2/Al_2O_3 were 48% and 1.62.

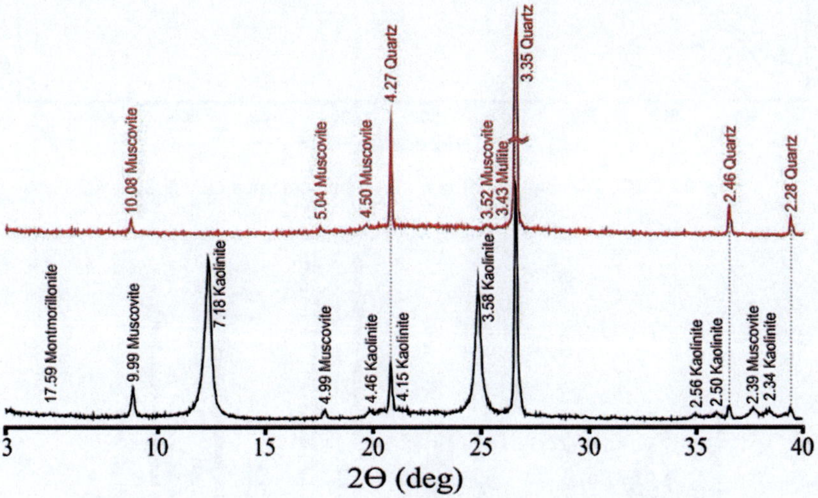

Fig. 1. X-ray patterns of raw clay (a) and calcined clay at 800 °C (b).

3.3 Characteristics of Hardened Pastes

The reacted 8-14M NaOH alkali-activated calcined clay systems were analyzed by XRD and thermal analyses. The XRD patterns of the samples of hardened pastes are shown in Fig. 3.

Shift of diffuse halo in the 25–30°2θ relative to that found in the XRD diffraction pattern of the original calcined clays indicates the formation of N-A-S-H binder gel [24–26]. Clay activation with NaOH also leads to the creation of the newly formed crystalline reaction products belonging to group of zeolites – mainly zeolite A $Na_2Al_2Si_{1.85}O_{7.7}{\cdot}5.1H_2O$ (PDF 00-038-0241) and small amounts (less than 1–3%) of faujasite-K (PDF 00-039-1380), analcime $Na_{10}Mg_3Al_{16}Si_{32}0_{96}{\cdot}25H_20$(PDF 00-042-1378), sodium aluminum silicate hydrate$Na_6(AlSiO_4)_6{\cdot}4H_2O$(PDF 00-042-0216), calcite $CaCO_3$ (PDF 00-005-0586).

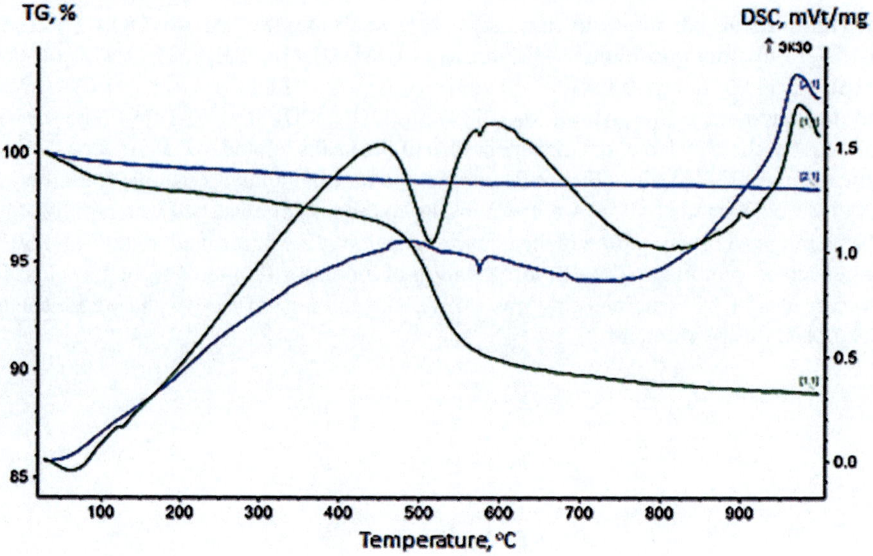

Fig. 2. DSC, TG analyses of raw clay 1 and calcined clay at 800 °C.

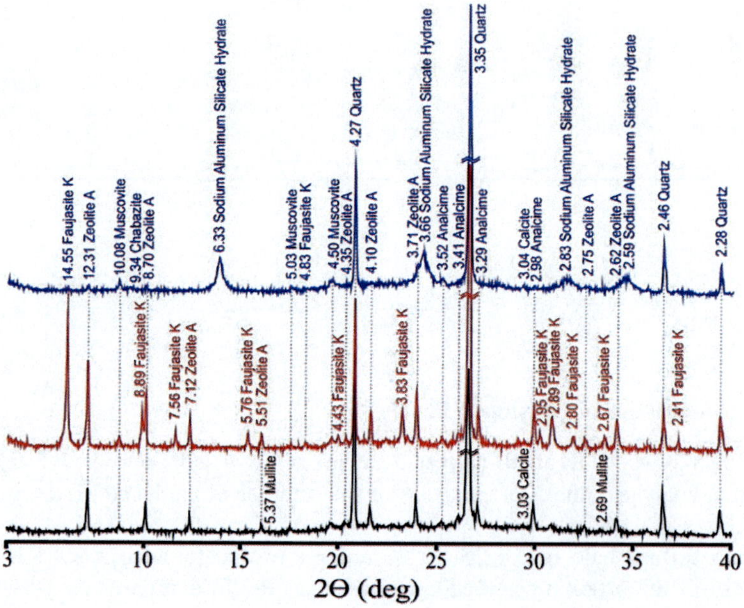

Fig. 3. X-ray diffractograms of AACC1 hardened pastes.

Such minerals as quartz SiO_2(PDF 01-079-1910) and muscovite $K_{0.77}Al_{1.93}(Al_{0.5}Si_{3.5})O_{10}(OH)_2$(PDF 01-070-1869 were also identified.

Thermal analysis results are shown in Fig. 4. The weight loss in the range of 30–200 °C are due to free evaporable water in gel pores and therefore can be considered as a characteristic of the formed binder gel volume. For 8M, 11M, 14M NaOH activated pastes the weight loss was 8.07%, 10.25%, 9.77%, respectively. It can be concluded from these results that the activation of calcined clay by 11M NaOH provides maximum formation of binder gel with cementitious properties.

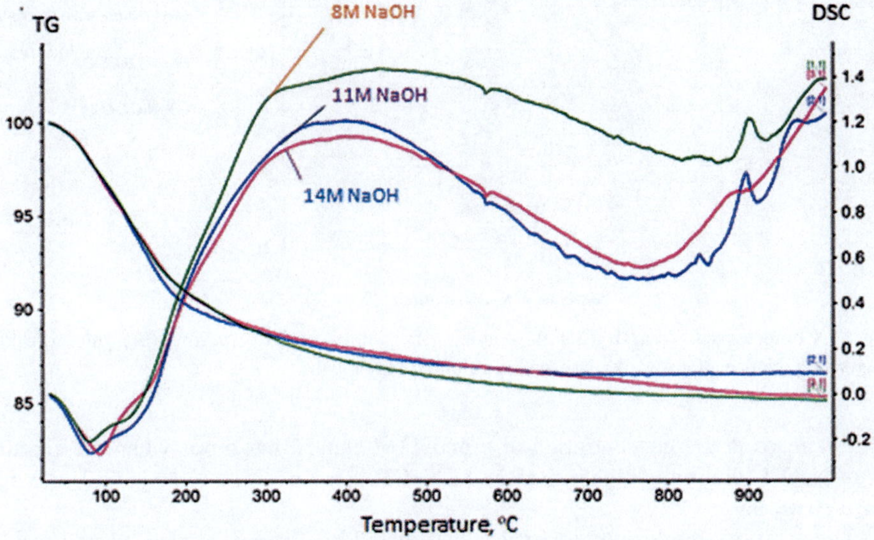

Fig. 4. DSC, TG analyses of hardened pastes.

The content of zeolite A and total crystalline phase as indicators of the reaction in systems calcined clay – 8, 11, 14 M NaOH were calculated based on XRD data. The results were summarized with compressive strength, $Na_20/Si0_2$, Na_20/Al_20_3 in Fig. 5.

As it was reviewed by Liew [26] SiO_2/Al_2O_3 ratio providing satisfactory mechanical performance of geopolymers should be 3–5. It is the reason for low compressive strength of the hardened pastes under study. The compressive strength decrease with the increase in the alkalinity of NaOH solution is associated with the change in the ratios of sodium oxide to reactive silica and alumina oxides which are critical factors and, as a consequence, in amorphous-crystalline transformations and zeolitisation degree. Molar ratios $Na_20/Si0_2$ in system calcined clay – 8–14 MNaOH were in a range 0.44–0.77 and Na_20/Al_20_3 – 0.67–1.17. The increase in $Na_20/Si0_2$ over 0.44 and Na_20/Al_20_3 over 0.67 in clay-based pastes led to the decrease in strength caused by the decrease of N-A-S-H gel, which was confirmed by XRD and thermal analyses and the increase of degree of zeolitisation of alkali activated system. This finding is in accordance with results of Nasab et al. [27] who stated that the low $Na_20/Si0_2$ providing high silicate species caused greater formation of amorphous geopolymer products while high $Na_20/Si0_2$ ratios showed high crystallization of zeolite materials, which are porous and have poor mechanical performance. As can be seen from Fig. 5 the amount of zeolites increases

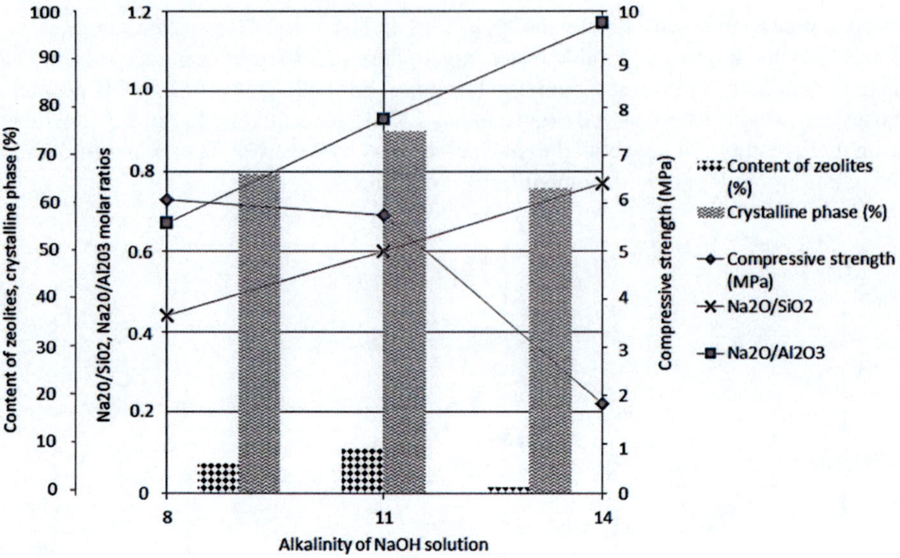

Fig. 5. Compressive strength (MPa), content of zeolites and crystalline (%), molar ratios $Na_20/Si0_2$, $Na0/Al_20_3$ for alkali-activated calcined clay pastes.

with increase in the concentration of 8 and 11M NaOH, but drops when the alkalinity of NaOH solution is higher than 11 M NaOH corresponding to $Na_20/Si0_2$ 0.6 and Na_20/Al_20_3 0.93.

These ratios are close to optimal values stated by Duxson et al. [28] and results analyzed by Liew [26]. According to Duxson et al. [28] at Na_20/SiO_2 below 1.0 the majority of the aluminium eleased from dissolution was expected to be incorporated in geopolymer network. As to Na_20/Al_20_3 optimal is 1 [29] to serve electrical neutrality of IV-fold coordinated $Al^{3+}$and Na_2O [29, 30]. Thus, for calcined clay based samples with low reactive phase portion 48% alkali molarity 11 M NaOH was found to be threshold beyond which the amount of zeolites crystallized drastically decreased.

4 Conclusion

Amorphous-crystalline phase transformations, reaction product compositions, and compressive strengths of two thermally treated clays with reactive phase contents of 48%, and with the Si/Al molar ratios of 1.62, were investigated for the 5–14 M NaOH solution concentrations. Alkali-activated calcined clay cement demonstrated compressive strength maxima upon activation with 8 M the NaOH solution. The zeolite content in the alkali-activated calcined clay cements peaked at 11 M NaOH, and the corresponding molar ratios of $Na_2O/SiO_2$and Na_2O/Al_2O_3 were 0.6 and 0.93, respectively. The main reaction products in the studied alkali-activated calcined clay cements were N-A-S-H, zeolite A, faujasite-Na, analcime, and sodium aluminium silicate hydrate.

The obtained properties and reaction products favor the use of such materials in ion-exchange processes, controlled low-strength materials (8.3 MPa or less according to

ACI 229R), and matrices for the solidification of toxic and radioactive wastes. Especially considering the fact that the compressive strength required for cementitious waste forms varies significantly across different countries, i.e., 0.35 MPa in France, 3.4 MPa in the US, 5 MPa in the Russian Federation, 7 MPa in the UK, and 10 MPa in Switzerland [31].

References

1. Scrivener, K., Martirena, F., Bishnoi, S., Maity, S.: Calcined clay limestone cements. Cem. Concr. Res. **114**, 49–56 (2018). https://doi.org/10.1016/j.cemconres.2017.08.017
2. Juenger, M.C.G., Snellings, R., Bernal, S.A.: Supplementary cementitious materials: new sources, characterization, and performance insights. Cem. Concr. Res. **122**, 257–273 (2019). https://doi.org/10.1016/j.cemconres.2019.05.008
3. Khalifa, A.Z., et al.: Cem. Concr. Res. **132**, 106050 (2020). https://doi.org/10.1016/j.cemconres.2020.106050
4. Rakhimova, N.R.: Geosyst. Eng. **23**(5), 287–298 (2020)
5. Khaliullin, M., Faizrakhmanov, I.: The effect of ground limestone on the properties of a composite gypsum binder using thermally activated clay as a pozzolan component. Izvestiya KGASU **3**(49), 190–196 (2019)
6. Shelikhov, N., Rakhimov, R., Sagdiev, R., Zareznov, D.: The current state of production and use of foam glass for thermal insulation. Izvestiya KGASU **4**(46), 319–325 (2018)
7. Rakhimov, R., Rakhimova, N., Gaifullin, A., Bikmukhametov, A., Morozov, V.: Influence of the composition and temperature of roasting on the additives of kaolin and polymineral clays in Portland cement on the properties of cement stone. Izvestiya KGASU **3**(49), 172–180 (2019)
8. Khaliullin, M., Dimieva, A.: Composite gypsum binder under introducing thermally activated clay as a pozzolanic component and adding ground limestone. In: IOP Conference Series: Materials Science and Engineering, vol. 890, no. 1, p. 012093 (2020). https://doi.org/10.1088/1757-899X/890/1/012093
9. Galán, R.E., Ferrell, R.E.: Genesis of clay minerals. In: Bergaya, F., Lagaly, G. (eds.) Handbook of Clay Science, pp. 83–126. Amsterdam, Elsevier (2013)
10. Dietel, J., Warr, L.N., Bertmer, M., Steudel, A., Grathoff, G.H., Emmerich, K.: The importance of specific surface area in the geopolymerization of heated illitic clayAppl. Clay Sci. **139**, 99–107 (2017). https://doi.org/10.1016/j.clay.2017.01.001
11. D'Elia, A., Pinto, D., Eramo, G., Giannossa, L.C., Ventruti, G., Laviano, R.: Effects of processing on the mineralogy and solubility of carbonate-rich clays for alkaline activation purpose: mechanical, thermal activation in red/ox atmosphere and their combination Appl. Clay Sci. **152**, 9–21 (2018). https://doi.org/10.1016/j.clay.2017.11.036
12. Zhang, Z., Wang, H., Yao, X., Zhu, Y.: Effects of halloysite in kaolin on the formation and properties of geopolymers. Cem. Concr. Comput. **34**(5), 709–715 (2012). https://doi.org/10.1016/j.cemconcomp.2012.02.003
13. Rakhimova, N., Rakhimov, R.: Advances in development of calcined clays as supplementary cementitious materials. In: IOP Conference Series: Materials Science and Engineering, vol. 890, no. 1, p. 012085 (2020). https://doi.org/10.1088/1757-899X/890/1/012085
14. Rakhimova, N., Rakhimov, R., Morozov, V., Eskin, A.: Calcined low-grade clays as sources for zeolite containing material. Period. Polytechn. Civ. Eng. **65**(1), 204–214 (2021). https://doi.org/10.3311/PPci.14084
15. Ma, Y., Yan, C., Alshameri, A., Qiu, X., Zhou, C., Li, D.: Synthesis and characterization of 13X zeolite from low-grade natural kaolin. Adv. Powd. Techn. **25**(2), 495–499 (2014). https://doi.org/10.1016/j.apt.2013.08.002

16. Mezni, M., Hamzaoui, A., Hamdi, N., Srasra, E.: Synthesis of zeolites from the low-grade Tunisian natural illite by two different methods. Appl. Clay Sci. **52**(3), 209–218 (2011)
17. Autef, A., Joussein, E., Gasgnier, G., Rossignol, S.: J. Non-Cryst. Sol. **366**, 13–21 (2013)
18. Deng, G., He, Y., Lu, L., Hu, S.: The effect of activators on the dissolution characteristics and occurrence state of aluminum of alkali-activated metakaolin. Constr. Build. Mater. **235**, 117451 (2019). https://doi.org/10.1016/j.conbuildmat.2019.117451
19. Koleżyński, A., Król, M., Żychowicz, M.: J. Molec. Str. **1163**, 465–471 (2018)
20. Lahoti, M., Wong, K.K., Yang, E.H., Tan, K.H.: Effect of alkali cation type on strength endurance of fly ash geopolymers subject to high temperature exposure. Ceram. Int. **44**(5), 5726–5734 (2018). https://doi.org/10.1016/j.matdes.2018.05.023
21. Qian, X., Wang, J., Wang, L., Fang, Y.: Enhancing the performance of metakaolin blended cement mortar through in-situ production of nano to sub-micro calcium carbonate particles. Constr. Build. Mater. **196**(1), 681–691 (2019). https://doi.org/10.1016/j.conbuildmat.2018.11.134
22. Wan, Q., Rao, F., Song, S., Cholico-González, D.F., Ortiz, N.L.: Combination formation in the reinforcement of metakaolin geopolymers with quartz sand. Cem. Concr. Comput. **80**, 115–122 (2017). https://doi.org/10.1016/j.cemconcomp.2017.03.005
23. Ruiz-Santaquiteria, C., Fernández-Jiménez, A., Skibsted, J., Palomo, A.: Clay reactivity: Production of alkali activated cements. Appl. Clay Sci. **73**, 11–16 (2013). https://doi.org/10.1016/j.clay.2012.10.012
24. Fernández-Jiménez, A., Monzó, M., Vicent, M., Barba, A., Palomo, A.: Structure of calcium silicate hydrates formed in alkaline-activated slag: influence of the type of alkaline activator. Micropor. Mesopor. Mater. **108**(1–3), 41–49 (2008). https://doi.org/10.1111/j.1151-2916.2003.tb03481.x
25. Duxson, P., Fernandez-Jimenez, A., Provis, J.L., Lukey, G.C., Palomo, A., van Deventer, J.S.J.: Geopolymer technology: the current state of the art. J. Mater. Sci. **42**, 2917–2933 (2007). https://doi.org/10.1007/s10853-006-0637-z
26. Liew, Y.M., Heah, C.Y., Al Bakri, M.M., Hussin, K.: Progr. Mater. Sci. **83**, 595–629 (2016)
27. Nasab, G.M., Golestanifard, F., MacKenzie, K.J.D.: The effect of the SiO2/Na2O ratio in the structural modification of metakaolin-based geopolymers studied by XRD, FTIR and MAS-NMR. J. Ceram. Sci. Technol. **05**, 185–192 (2014). https://doi.org/10.4416/JCST2014-00007
28. Duxson, P., Provis, J.L., Lukey, G.C., Mallicoat, S.W., Kriven, W.M., van Deventer, J.S.J.: Coll. Surf. A: Physicochem **269**, 47–58 (2005)
29. Davidovits, J.: Inorganic polymeric new materials. J. Therm. Anal. **37**, 1633–1656 (1991)
30. Yunsheng, Z., Wei, S., Zongjin, L.: Appl. Clay Sci. **47**(3–4), 271–275 (2010)
31. Abdel Rahman, R.O., Rakhimov, R.Z., Rakhimova, N.R., Ojovan, M.I.: Cementitious Materials for Nuclear Waste Immobilization. Wiley (2014). https://doi.org/10.1002/9781118511992

Cost Modeling of a Land Plot for Private Housing Construction

Olga Borovskikh(✉), Elvira Shagiakhmetova, Adilya Nizamova, and Tatiana Kazymova

Kazan State University of Architecture and Engineering, 420043 Kazan, Russia

Abstract. Construction or purchase of a detached house in the suburb, which is a more comfortable and demanded residential option for families, is really topical nowadays. According to this fact, large and small gated communities appear around large cities. They include modern infrastructure, protected territory and other amenities. The study analyzes 38 economy and business class gated communities around the city of Kazan. This analysis allowed us having the most complete understanding of events in this segment of residential real estate. The article selects pricing factors and reveals a mathematical pattern of changes in the market value of land plots intended for private houses construction in Kazan. As a result, the study reveals that three factors have the greatest impact on the value of land plots: distance from the city, transport accessibility and the current lifespan of the gated community. The materials of the article may be of practical importance for developer companies, as well as for potential buyers of land plots in gated communities in the Republic of Tatarstan.

Keywords: Private housing construction · Low-rise residential construction · Land plot · Gated community · Land plots value pricing factors

1 Introduction

The increased interest in the suburban real estate market arose not by chance: families dream of living in an ecologically clean area, minimizing city noise and stress from looking for parking spaces and noisy neighbors.

Despite the desire of metropolis residents to move out of town, few succeed in doing this, since prices for suburban real estate are quite high. A wide range of prices in the suburban real estate market is the subject of research by a large number of scientists and practitioners [1].

When evaluating a land plot, an extensive list of factors influencing the value of land is offered. Analysis of foreign studies has revealed the factors that receive increased attention when assessing the value of land: the number of building permits issued [2–4] and its cost the availability of loans for suburban real estate [5], location (remoteness from the city center) [6], transport accessibility and the availability of social infrastructure [7]. Swiss scientists identified the state of municipal planning as a resultant factor in assessing the value of land [8].

N. Vatin (Ed.): STCCE 2021, LNCE 169, pp. 47–57, 2021.
https://doi.org/10.1007/978-3-030-80103-8_6

Most of the factors considered are not suitable for quantitative analysis, so scientists try to apply the mathematical methods of expert assessments. It includes fuzzy methods, hierarchies analysis, which use subjective data obtained from the calculation of points [9–11]. When planning the construction of a property, subjective data can greatly distort the real situation [12–14]. The world scientific community is trying to visualize the process of building houses as much as possible, by calculating the necessary financial resources for the acquisition of a land plot and for the further implementation of construction [15–17].

The study is focused on the determination of the optimal cost of a land plot for private houses construction. To reveal this main topic of the study, we selected relevant indicators that can be quantified: the value of land, distance from the city centre, transport accessibility and the period of the existence of the gated community. Analysis of the assessment of the selected factors influence can be used to improve the efficiency of construction [17, 18] and increase the attractiveness of private investment [19].

2 Methods

Within the framework of the study, we used a method of indicators observation, which involves the study of data from the Federal State Statistics Service, the Central Bank of the Russian Federation and materials from various analytical centres and services. Also, to obtain data on gated communities we carried out some field studies and considered the information on the official websites of the communities.

The most widespread method of statistical analysis is correlation-regression modelling, which makes it possible to operate with objects of a complex multifactor nature. As a result of statistical analysis, it is possible to obtain a mathematical formula for the relationship between the dependent variable (the value of the land plot) and independent values (pricing factors). For the implementation of correlation-regression modelling, it is necessary to have a dataset for all investigated factors and predictions for the dependent variable, which will not change significantly when new information is added [20].

The process of modelling the equation begins with the choice of the form: linear, multiplicative power-law or exponential dependence [20]. All calculations are performed in Microsoft Excel using the supplementary add-in «Data Analysis». The coefficients in the model that have maximum values have the greatest influence on the resulting indicator, therefore, special attention should be paid to them [20]. The resulting statistical model must be statistically explicable and reliable in terms of market pricing.

3 Results and Discussion

To identify the characteristic features and trends of the private houses real estate market in the Republic of Tatarstan (Russian Federation) and in the city of Kazan, in particular, we studied and analysed 38 existing gated communities. They represent projects of both economy class and business segment, which have already taken place and are just starting. This study allowed us to draw up the most complete picture of events taking place in this market segment [21].

The gated communities were analysed according to the following criteria: the location of the gated community; land status; stage of project implementation; the number and average size of plots in the gated community; cost of 100 m^2 of land; availability of services at the land plots; the quality of the road surface inside the gated community; social infrastructure of the gated community; ensuring the safety of the gated community residents; environmental conditions at the location of the gated community; developer (investor) of the project [22].

In the course of this study, we revealed that the gated communities of Tatarstan are concentrated around the city of Kazan. The largest number of gated communities is located in the Laishevsky district (39%). Almost equal number of projects was implemented in Pestrechinsky, Verkhneuslonsky districts and in Kazan, the share of which is 33%. Vysokogorsky district accounts for 18% and Zelenodolsky and Rybnoslobodsky districts (11%) are at the bottom of the list. The percentage ratio of the number of gated communities in different districts of the Republic of Tatarstan is presented in Fig. 1.

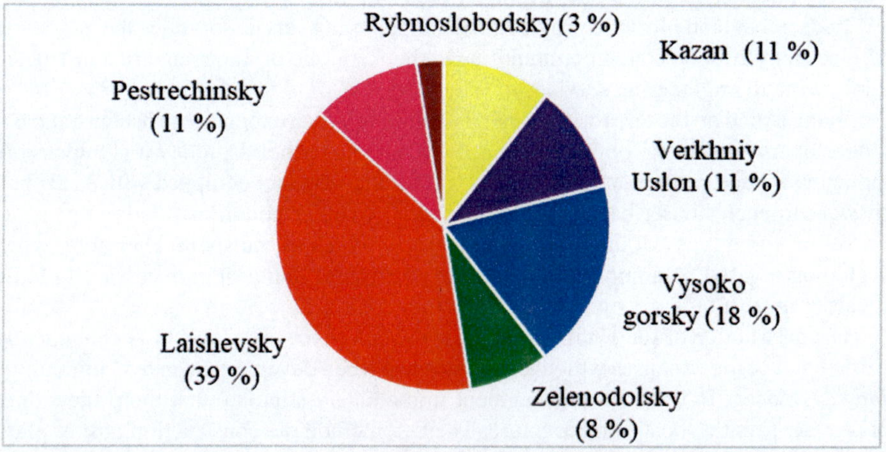

Fig. 1. The percentage ratio of the number of gated communities in different districts of the Republic of Tatarstan (Russia). Source: calculated by the authors.

The vast majority of gated communities (90%) is built on land intended for private housing construction. However, there are projects implemented on land intended for private subsidiary plots and suburban non-profit partnerships. This situation is connected with a shortage of land intended for private housing construction, as well as with an expensive and lengthy procedure for transferring the status of land to the required category [23]. Nevertheless, applying a competent approach, this factor does not prevent these projects from developing well.

In the suburban real estate market there are both completed projects, leading their history since 2006, and starting ones. This fact indicates that there is already a certain experience in the republic in the construction of organized gated communities, as well as this kind of activity is promising for investment.

The cost of 100 m^2 of land varies greatly from 40 to 280 thousand roubles. The main factor affecting the value of a land plot is undoubtedly its proximity to the capital. So, plots costing from 45 to 85 thousand roubles per 100 m^2 are located on average 60 km from the centre of Kazan, plots costing from 100 to 160 thousand roubles per 100 m^2 are 30 km from the centre of Kazan, and plots costing from 180–280 thousand roubles per 100 m^2 are 15 km from the centre of Kazan, that is, for the most part are located within the city borders.

At the same time, most of the gated communities (15 settlements) are located in the Laishevsky district of the Republic of Tatarstan (Russia). They were built relatively recently - the average life span is 6 years. There are few modern gated communities in Rybnoslobodsky and Zelenodolsky districts of the Republic of Tatarstan (Russia) (3 and 1 settlements, respectively). The average lifespan of gated communities in Zelenodolsky district is 11 years, that is, it began to be built up earlier. The active development of the Laishevsky district can be explained by its better transport accessibility: with the greater remoteness of the gated communities of this district from Kazan city centre, the way time does not exceed 60 min, the average time is 45 min.

The fact that land plots are provided with engineering services implies the gas, water and electricity supply. Gated communities in the Republic of Tatarstan are constructed mainly with all engineering services.

Laying a road on the territory of a gated community is an expensive procedure and not all developers take on the obligation to carry it out. Nevertheless, today it is impossible to imagine a gated community in business or elite segment, not equipped with an asphalt driveway to each house, because in case of bad weather conditions, it is impossible to drive on some types of unpaved roads by many types of transport. Therefore, when developing a gated community intended for year-round living, it is necessary to build asphalt or at least crushed stone roads [24].

The construction of social infrastructure is the most expensive and least economically justified part of the expenses which an investor has when developing a gated community. Many developers in the economy segment immediately stipulate that the project does not provide for any social infrastructure. The rest include the construction cost of playgrounds, kindergartens, shops, cafes, and the arrangement of beaches into the land plots value. But in fact they either do not implement it at all, or wait for a sufficient number of families living on a permanent basis to appear in the gated community. And some developers assume that the residents of the gated community will use the infrastructure of the nearest settlements.

Security means the presence of a fenced area, qualified security, a pass system for transport entering the gated community, video surveillance. A fenced area is one of the factors that distinguish an organized gated community from chaotic low-rise buildings, ensuring the closeness of residents.

Another feature of the gated communities of the Republic of Tatarstan is that the greatest part of the offer is made up of plots without a contract, that is, not ready-made houses are sold, but free land on which the buyer can build a house on his own. There are gated communities when the developer offers several options of houses to choose from and builds them on the order of the buyer after purchasing the land.

One of the main factors which the viability of the gated community project will depend on is the cost per 100 m^2 of land. The value of land plots should not be overstated. It must be consistent with the set of services provided by the developer, as well as provide the investor with an acceptable rate of return. In other words, the developer must be economically interested in the honest and timely implementation of his project, and the consumer must clearly understand what he is paying for and receive all the services promised by the developer on time and in full amount.

As a result of numerous studies, the authors have revealed that the most significant factors determining the choice of a land plot in a gated community and affecting its final value (Y) are the distance of the community from the centre of Kazan in km (X_1), transport accessibility (travel time to Kazan - X_2); the lifespan of the community in years (X_3). Information on the analyzed gated communities of the Republic of Tatarstan (Russia) is given in Table 1 (Source: compiled based upon Portal about suburban real estate in Kazan https://poselki116.ru/index/volzhskaya-gavan.html).

To obtain correct results of the model, it is necessary to determine the minimum sample size, which depends on the number of factors included in the model, taking into account the intercept. The minimum sample size for obtaining a statistically significant model can be determined using the formula:

$$N_{\text{min}} = 5 \times (m + n) \tag{1}$$

m – number of factors included in the model;

n – number of intercepts in the equation.

For our research $N_{\text{min}} = 5 \times (3 + 1) = 20$. To increase the significance of the study we analyzed 38 different gated communities when creating a multivariate model.

The selected factors (X_1, X_2, X_3) have different units of measurement, so it is necessary to bring them to dimensionless values by normalization. To perform normalization, it is necessary to find out the limits of change of the corresponding variables, that is, the maximum and minimum theoretically possible values. In our case, it is impossible to accurately establish the limits of change of variables, so they were set taking into account the minimum and maximum values in the available data sample [25].

At the next stage, pricing factors are analyzed by compiling a matrix of pairwise correlation coefficients. Each element from the correlation matrix is a paired coefficient of the relationship between factors and the resulting indicator. The coefficients take values from –1 to 1, while positive values indicate a direct relationship, and negative values indicate an inverse relationship: the greater is the value of the coefficient for a given factor, the smaller is the resulting value.

The resulting matrix of paired correlation coefficients of the main pricing factors for land plots in gated communities of the Republic of Tatarstan (Russia) is given in Table 2 (calculated by the authors).

An analysis of this table shows that the greatest inverse effect on the value of land plots is exerted by the distance from the city centre (X_1), two other factors have, respectively, an average inverse and direct effect on the cost (Y). Graphs of changes in the value of land plots for 100 m^2 (Y) and distance from the city centre of Kazan (Russia) (X_1) are presented in Fig. 2.

As can be seen in Fig. 2, there is an inverse relationship between the value of a land plot per 100 m^2 (Y) and distance from the city centre (X_1).

Table 1. Initial information about the gated communities of the Republic of Tatarstan (Russia).

№	Name of the gated community	District of the Republic of Tatarstan	Price, per 100 m^2, rub., Y*	Distance from the centre of Kazan in km, X_1	Transport accessibility, min., X_2	Lifespan of the community, years, X_3
1	Goluboe ozero	Kazan	280 000	14	45	6
2	Kazanskaya usad'ba	Kazan	250 000	15	25	11
3	Zagorodny'j klub	Zelenodolsky	250 000	25	34	15
4	Zoloty'e klyuchi	Zelenodolsky	190 000	26	35	9
5	Orekhovka	Zelenodolsky	190 000	26	34	13
6	Pyatidvor'e	Verkhniy Uslon	185 000	32	40	10
7	Sultan Ij	Kazan	180 000	17	25	9
8	Orlovsky	Laishevsky	160 000	36	30	8
9	Novy'e Salmachi	Pestrechinsky	155 000	16	20	5
10	Barvikha	Vysokogorsky	130 000	25	25	6
11	Novy'e Kuyuki	Pestrechinsky	130 000	19	20	5
12	Sem'ozer	Vysokogorsky	125 000	25	25	8
13	Udachny'j	Vysokogorsky	120 000	35	40	8
14	Shigali	Vysokogorsky	120 000	25	45	8
15	Bahchisaraj	Kazan	120 000	27	30	6
16	Rodny'e prostory'	Laishevsky	120 000	30	25	7
17	Semiozerskaya usad'ba	Vysokogorsky	105 000	25	25	5
18	Pallandiya	Vysokogorsky	100 000	25	25	12
19	Familiya	Laishevsky	99 900	44	29	4
20	Serebryanoe ozero	Laishevsky	92 000	35	30	4
21	Schastlivy'e dachi	Laishevsky	77 000	36	40	5
22	Zaprudny'j	Pestrechinsky	75 000	44	50	5
23	Bima	Laishevsky	70 000	46	60	3
24	Zimnyaya gorka	Laishevsky	70 000	32	30	8
25	Meshinskie usad'by'	Pestrechinsky	61 500	50	35	7
26	Teteevo	Laishevsky	60 000	50	60	8
27	Novo-Nikol'skoye	Laishevsky	55 000	37	30	6

(*continued*)

Table 1. (*continued*)

№	Name of the gated community	District of the Republic of Tatarstan	Price, per 100 m^2, rub.,Y*	Distance from the centre of Kazan in km, X_1	Transport accessibility, min., X_2	Lifespan of the community, years, X_3
28	Raduga	Laishevsky	55 000	40	50	5
29	Zeleniy bereg	Laishevsky	50 000	36	35	5
31	Solnechnaya dolina	Verkhniy Uslon	50 000	50	60	7
32	Studency	Verkhniy Uslon	50 000	51	50	7
33	Derzhavino	Laishevsky	48 000	50	60	9
34	Derzhavinskie prostory'	Laishevsky	47 000	44	50	6
35	Umy'rzaya	Laishevsky	46 000	52	70	10
36	Novy'e Tangachi	Laishevsky	45 000	52	60	7
37	Novy'j	Verkhniy Uslon	45 000	52	60	7
38	Serebryanaya Buhta	Rybnoslobodsky	40 000	85	110	6

*1$ = 74.82 rub.

Table 2. Matrix of paired correlation coefficients of the main pricing factors for land plots in gated communities of the Republic of Tatarstan (Russia).

Indicators	Y	X_1	X_2	X_3
Y	1			
X_1	−0.74344	1		
X_2	−0.49591	0.847463	1	
X_3	0.469783	−0.22445	−0.09379	1

The resulting regression equation for the value of land plots in gated community in the Republic of Tatarstan (Russia) is as follows:

$$Y = 0.451 - 1.337X_1 + 0.508X_2 + 0.352X_3 \tag{2}$$

Y – value of a land plot;
X_1 – distance from the city centre of Kazan, km;
X_2 – Transport accessibility (travel time to Kazan);
X_3 – lifespan of the gated community, years.

The regression equation for the market value of land plots is checked for compliance with the requirements of statistical adequacy. One of the most effective ways to test the

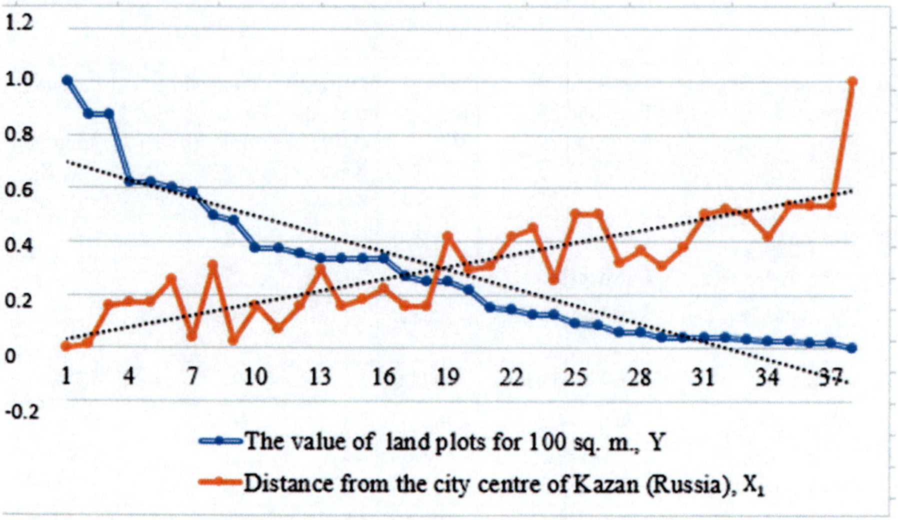

Fig. 2. Graphs of changes in the value of land plots for 100 sq. (Y) and distance from the city centre of Kazan (Russia) (X_1).

regression equation is to calculate the correlation coefficient (multiple R). Its values can vary from 0 to 1 and characterize the degree of connection between the variables (Table 3) (calculated by the authors).

Table 3. Regression statistics.

№	Indicators	Value
1	Multiple R	0.829827463
2	R-squared	0.688613618
3	Standardized R-squared	0.661138349

The correlation coefficient in our case was R = 0.829, which confirms the close relationship and accuracy of the model. The results indicate that the hypothesis of the relationship between the value of land plots, distance from the city centre, transport accessibility and the lifespan of the gated community was quantitatively confirmed.

The conducted research can be useful for development companies. So at the stage of the concept of the future gated community already they will be able to determine the approximate selling price per 100 m^2 of land and adjust their value proposition by choosing one or another set of benefits that the gated community will have. The potential buyer will also be interested in the results obtained. They will help determining the business proposal price correctness and understanding what the developer is trying to save on by offering to purchase a plot in the particular gated community.

4 Conclusions

The construction of houses in gated communities has become widespread around large urban agglomerations. A study of 38 gated communities near the city of Kazan showed that most of them are located either in Kazan itself or in the Laishevsky municipal district.

For land buyers, the value of the land is a fundamental element. The analysis of scientific publications revealed a high interest in this issue on the part of Russian and foreign scientists.

At the suburban real estate market in Kazan, there is a wide range of prices from 45 to 280 thousand rubles per 100 m^2. The most significant factors affecting the value of land plots are: distance from the city, transport accessibility and the current lifespan of the gated community. To confirm the obtained conclusion, a multifactor model was created, taking into account the degree of dependence of the land plots value on the identified factors.

The greatest influence on the value of land plots is exerted by the distance of the gated community from the centre of Kazan in kilometers, for which the coefficient value in the model is maximum (1.337), the second place is taken by the indicator of transport accessibility (0.508). The resulting indicator is inversely related to the most significant factor. The test showed that the created model is statistically adequate, since the correlation coefficient is 0.830. This value confirms the close relationship and accuracy of the model.

Further research can expand the possibilities for modeling the value of land plots. It is advisable to analyze the sensitivity of each price segment of gated communities for various categories of buyers and increase the potential of development projects marketing policy.

References

1. Mostafa, M.M.: A spatial econometric analysis of residential land prices in Kuwait Reg. Stud. Reg. Sci. **5**, 290 (2018). https://doi.org/10.1080/21681376.2018.1518154
2. Ay, J.S., Cavailhès, J., Hilal, M., Le Gallo, J.: Does issuing building permits reduce the cost of land? An estimation based on the demand for building land in France. Econ. Stat. (2018). https://doi.org/10.24187/ECOSTAT.2018.500T.1945
3. Chauvin, V., Muellbauer, J.: Consumption, household portfolios and the housing market in France Econ. Stat. (2018). https://doi.org/10.24187/ECOSTAT.2018.500T.1950
4. Muchová, Z., Konc, Ľ, Petrovič, F.: Land plots valuation in land consolidation in slovakia: a need for a new approach. Int. J. Strateg. Property Manag. **22**(5), 372–380 (2018). https://doi.org/10.3846/ijspm.2018.5221
5. Asiama, K.O., Bennett, R., Zevenbergen, J., Land, S.O.: Valuation in support of responsible land consolidation on Ghana's rural customar Asiama. Surv. Rev. **50**, 288 (2018). https://doi.org/10.1080/00396265.2018.1467672
6. Morales, J.A., Flacke, J., Zevenbergen, J.: Modelling residential land values using geographic and geometric accessibility in Guatemala City Environ. Plan. B Urban Anal. City Sci. (2019). https://doi.org/10.1177/2399808317726332
7. Syahar, F., Ramadhan, R., Putri, S.: Statistical interpretation of land value patterns in padang, Indonesia. Int. J. Geomate (2020). https://doi.org/10.21660/2020.72.ICGEO14

8. Rodas, J., Gómez, J., Castanho, R., Cabezas, J.: Land valuation sustainable model of urban planning development: a case study in Badajoz, Spain. Sustainability **10**(5), 1450 (2018). https://doi.org/10.3390/su10051450
9. Asiama, K., Bennett, R., Zevenbergen, J., Asiama, S.: Land valuation in support of responsible land consolidation on Ghana's rural customary lands. Surv. Rev. **50**(361), 288–300 (2018). https://doi.org/10.1080/00396265.2018.1467672
10. Bencure, J., Tripathi, N., Miyazaki, H., Ninsawat, S., Kim, S.: Development of an innovative land valuation model (iLVM) for mass appraisal application in sub-urban areas using AHP: an integration of theoretical and practical approaches. Sustainability **11**(13), 3731 (2019). https://doi.org/10.3390/su11133731
11. Oksaping, A.F.: Land valuation with analytical hierarchy process and regression method in Grogol Sub-District, Sukoharjo District. JGISE J. Geospatial Inf. Sci. Eng. **2** (2019). https://doi.org/10.22146/jgise.40845
12. Zawistowski, J., Kulejewski, J.: Influence of the contractor's payment method on the economic effectiveness of the construction project from the contractor's point of view. Open Eng. **8**, 562 (2018). https://doi.org/10.1515/eng-2018-0055
13. Bilal, M., et al.: Investigating profitability performance of construction projects using big data: a project analytics approach. J. Build. Eng. **26**, 100850 (2019). https://doi.org/10.1016/j.jobe.2019.100850
14. Ylmén, P., Peñaloza, D., Mjörnell, K.: Life cycle assessment of an office building based on site-specific data. Energies **12**(13), 2588 (2019). https://doi.org/10.3390/en12132588
15. Salem, D., Bakr, A., Sayad, Z.E.: Post-construction stages cost management: sustainable design approach. Alexandria Eng. J. **57**, 3429 (2018). https://doi.org/10.1016/j.aej.2018.07.014
16. Anysz, H., Rogala, W.: Sensitivity analysis of the contractor's financial effects achieved on a single building site. Sci. Rev. Eng. Environ. Sci. **28**, 183 (2019). https://doi.org/10.22630/PNIKS.2019.28.2.17
17. Marchi, B., Zanoni, S., Ferretti, I., Zavanella, L.E.: Stimulating investments in energy efficiency through supply chain integration. Energies **11**(4), 858 (2018). https://doi.org/10.3390/en11040858
18. Ratajczak, J., Riedl, M., Matt, D.: BIM-based and AR application combined with location-based management system for the improvement of the construction performance. Buildings **9**(5), 118 (2019). https://doi.org/10.3390/buildings9050118
19. Siniak, N.: Building information modeling for housing renovation - example for Ukraine. **27**, 97 (2019). https://doi.org/10.2478/remav-2019-0018
20. Borovskikh, O.: The main factors determining the cost of land for individual housing construction on the example of Kazan. Russ. J. Hous. Res. **5**, 367 (2018). https://doi.org/10.18334/zhs.5.3.39437
21. Podryadchikova, E.D., Gilyova, L.N., Dubrovsky, A.V.: Correlation-regression analysis of cadastre value properties and price forming factors (on the example of land plots of the city of Tyumen, designed for individual residential buildings). Vestn. SSUGT Siberian State Univ. Geosystems Technol. **25**, 274 (2020). https://doi.org/10.33764/2411-1759-2020-25-1-274-289
22. Daniilovich, R.G.: The development of the housing market in Russian regions by the example of Saint Petersburg. Mod. Manag. Technol. **10** (2014)
23. Gareev, F.F., Zagidullin, I.F., Khusnutdinov, A.R.: N low-rise construction around Kazan: scale, technologies, prospectso title. Ross. Predprin. **11**, 3815 (2017)
24. Koklyugina, L.A., Koklyugin, A.V., Gimranov, L.R., Nikiforov, G.A.: Modern technology of construction of multi-storey wooden houses. News Kazan State Univ. Archit. Eng. **1**(47), 231 (2019)

25. Saifullina, F., Loginovskaya, I.: Perspectives of low-rise buildings construction in Russia UPR. Econ. Syst. Sci. J. **5** (2015)
26. Shagiakhmetova, E., Borovskikh, O., Nizamova, A., Kazymova, T.: Multivariate model of construction project operational efficiency. In: IOP Conference Series: Materials Science and Engineering, vol. 890 (2020). https://doi.org/10.1088/1757-899X/890/1/012113

Locally Concrete Filled Reinforced Joints of RHS and SHS Trusses

Linur Gimranov(✉) and Ivan Kuznetsov

Kazan State University of Architecture and Engineering, Kazan 420043, Russia

Abstract. Today significant volume of roof structures comprises trusses made of rectangular and square hollow sections (RHS, SHS). Mentioned trusses have several advantages such as absence of gussets; cross sections optimized for compressing loads and sealed internal volumes of truss members. These advantages make them very cost effective and competitive on the market. However at the same time that particular type of trusses demands additional attention to the design calculations of its joints. There are several limit states determining strength of joints of the RHS and SHS members. One of the most frequently dominating limit states is chord face plastification in the area adjacent to the brace member. There are different but confirmed approaches in calculating strength of the joint determined by described limit state. There are several ways to address this limit state. One of them is filling with concrete chords of the trusses. This study is focused on the development of new approaches of calculating strength of locally concrete filled reinforced K type joints of RHS and SHS trusses. Local reinforcement of K joints is achieved by employing a special bag with cross sectional shape repeating cross section of the chord, installed in place through a hole prepared beforehand and filled with concrete.

Keywords: Rectangular hollow sections (RHS) · Square hollow sections (SHS) · Concrete reinforced joint · Chord face plastification

1 Introduction

Today structures from hollow sections are very common and took dominating positions especially among structural roof systems such as trusses and lattice girders [16]. This fact can be explained by the economic competitiveness and the architectural soundness of structural shapes comprised of rectangular, square or circlar hollow sections (RHS, SHS, CHS). Economic competitiveness is achieved by a low producing cost and by the structural effectiveness of the hollow sections in the work under compression loads [11]. Disadvantages of hollow sections include a lowered flexural stiffness [15, 19] and a special attention and stricter requirements to the weld execution. Connections of hollow sections transfer a load in a complex manner. The mechanism of that transfer depends on several parameters such as dimensions of RHS or SHS members, thickness, angle and a ratio of widths between adjacent members. If we consider a partial plastification of some local areas in the connection and buckling effects in a compressed part of plates, load

N. Vatin (Ed.): STCCE 2021, LNCE 169, pp. 58–71, 2021.
https://doi.org/10.1007/978-3-030-80103-8_7

transfer, in such kind of conditions, can be described as non-linear. The non-linearity leads to several modes of failure that can occur in connections of RHS or SHS members. According [1, 2] these are:

a) Chord face plastification
b) Punching shear failure of the chord
c) Uneven load distribution, in the tension brace
d) Uneven load distribution, in the compression brace
e) Shear yielding of the chord, in the gap
f) Chord side wall failure
g) Local buckling of the chord face

The most frequently occurring limit state of them all is a chord face plastification [3, 4]. This is due to a fact that a width of brace members is almost always less or equal than a width of the chord [12]. The accepted terminology and notations for EuroCodes and Russian Codes are shown in Fig. 1

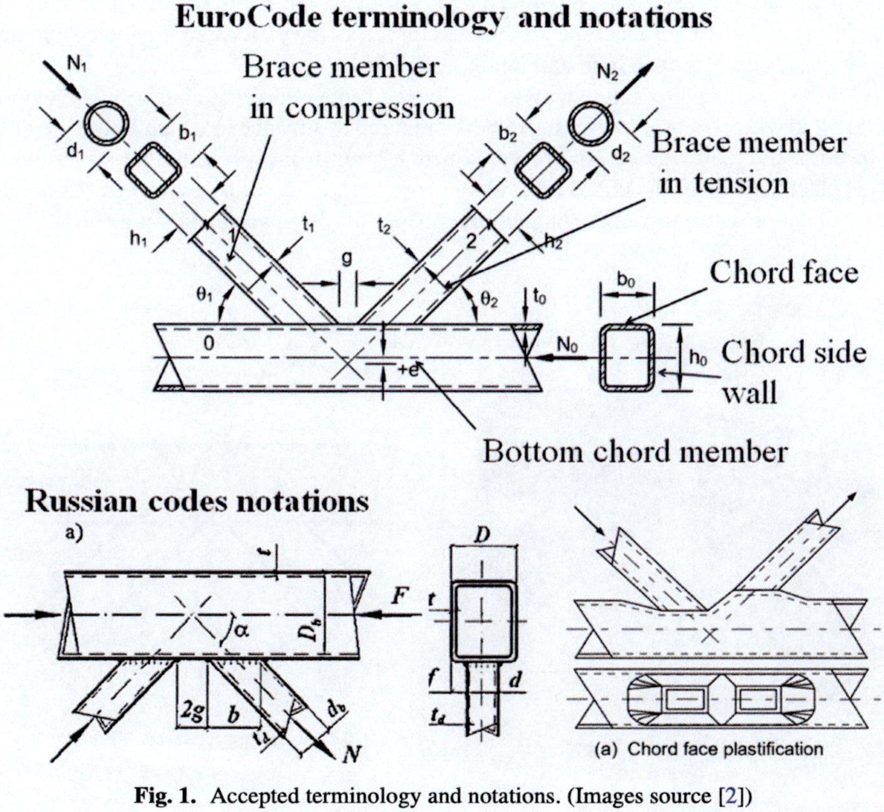

Fig. 1. Accepted terminology and notations. (Images source [2])

The mentioned limit state can be addressed by several ways. First, a chord face plate can be reinforced by the steel plate [9, 10] placed directly between a chord and the brace member or by filling a chord completely or in the vicinity of the joint with the concrete grout.

Studies [6] conducted by (Packer, 1995) and also other authors [5, 7, 8] have recently shown a significant increase in the yield strength of K-type joints with a chord filled with concrete. The same study also argues that brace members in tension and in compression should be calculated separately and that the governing limit state for a brace member in tension will be the premature (local) yielding of the tension brace, and (or) punching shear of the chord face around the brace.

2 Methods

2.1 Object of Research

The subject of this study is locally concrete filled reinforced joints of K-type. The weakest part of this type of joints is an area of a chord where the brace member in tension is adjacent to it. Concrete filling of chords can be done in different ways. In a fabrication shop truss can be tilted for filling procedure [2], for cases of entire chord replenishment. The filling of entire chord member of the truss increases the yield strength of the structure but also adds up to a dead load and increases the labor cost.

This study introduces new way to locally reinforce chord member (Fig. 2). Chord (1) is filled with concrete (4) through a bore prepared in advance in which a pipe plug (3) is inserted and then welded. To prevent concrete grout from spreading out a fabric bag (4) is placed inside of the chord through the bore before the filling procedure. Concrete grout under pressure forces the bag to unwrap and fill the space inside of a chord.

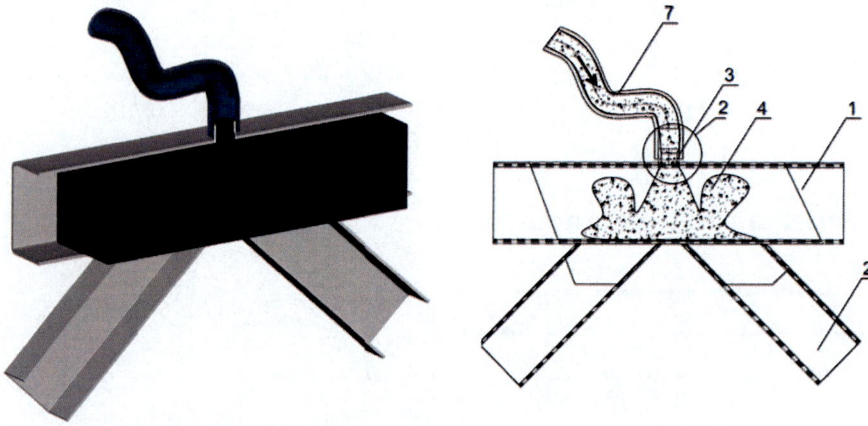

a) Overall view of a reinforced connection.

b) Filling procedure.

Fig. 2. Locally reinforced with concrete K-type connection.

Ends of the bag are sewed with a fabric (5) which shape is in a form of the chord's cross-section. The rubber hosepipe is connected to a pipe plug (Fig. 3 a). Concrete grout is pumped through a rubber hosepipe.

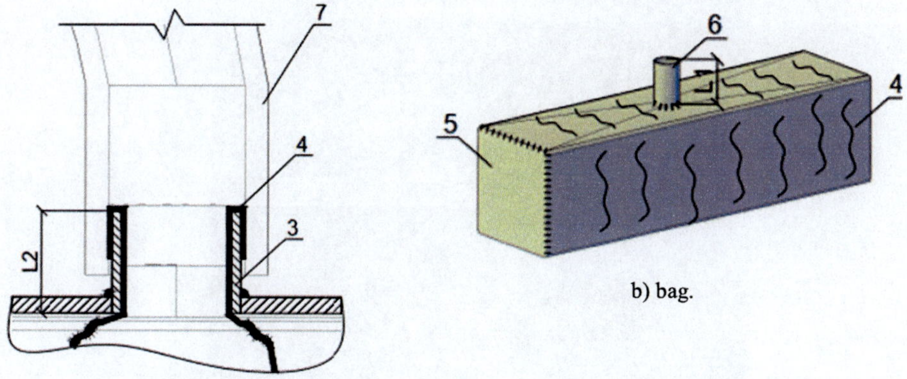

b) bag.

a) Overall view of a reinforced connection.

Fig. 3. Locally reinforced with concrete K-type connection(details).

2.2 Numerical Studies

Finite element model analysis (FEMA) is a known and widely acclaimed tool in assessment of strain-stress state of connections [10, 17, 18] This study was conducted with the help of Ansys Mechanical. Modeled connection is presented on Fig. 4

In order to effectively asses all relevant factors influencing the load bearing capacity of a reinforced connection following variable parameters were considered:

- a thickness of chord face - denoted as (t) [mm];
- an angle between brace member and the chord - denoted as (α) [^{0}];
- half the difference between chord width and brace in tension width - denoted as (f) [mm];
- the length of brace in tension cross-section height projection on chord member-denoted as (b) [mm];
- the gap between toes of brace members in tension and compression - denoted as (2g) [mm];

Nonlinear analysis consisted of nonlinear behavior of material and nonlinear contacts. Yield strength of steel elements was set at 350 MPa. Model was constrained by supports in such a way that an acting compressing force on the brace member in compression recreated the force distribution inside of the K-type connection as it supposed to be inside of whole truss.

Fig. 4. Modeled K-type connection with concrete element and the gap between it and chord.

Considering the focus of this study on the one particular limit state the strength of a reinforced connection is a function of several variable parameters and can be described as:

$$[N] = \psi\{R_s, f, 2g, F, N_i, \alpha, t\} \quad (1)$$

Where R_s - yield strength of the steel [MPa], f - half the difference between chord width and brace in tension width [mm], α - an angle between brace member and the chord [^{0}], t - a thickness of chord face [mm], F- acting force in the chord [N], N_i - acting force in the brace member in tension [N].

Finite element model also includes the gap between a concrete block and the inner surface of a chord Fig. 4 (bottom part). This gap takes into account possible concrete shrinkage during hydration process as also the fabric of the bag. Figure 5 and Fig. 6 present result of calculations by Ansys Mechanical.

Results of calculation point at following:

- limit state for the brace member in tension stays the same i.e. chord face plastification;
- chord face and side walls in the area of an adjacent brace member in tension are effectively reinforced by concrete;
- chord side walls in the area of an adjacent brace member in tension are partially reinforced by a concrete block.

2.3 Machine Learning Techniques

Machine learning techniques with the help of Sci-Kit-Learn module in Python [21] were implemented in particular Multiple linear regression [22] and Random Forest algorithm [20].

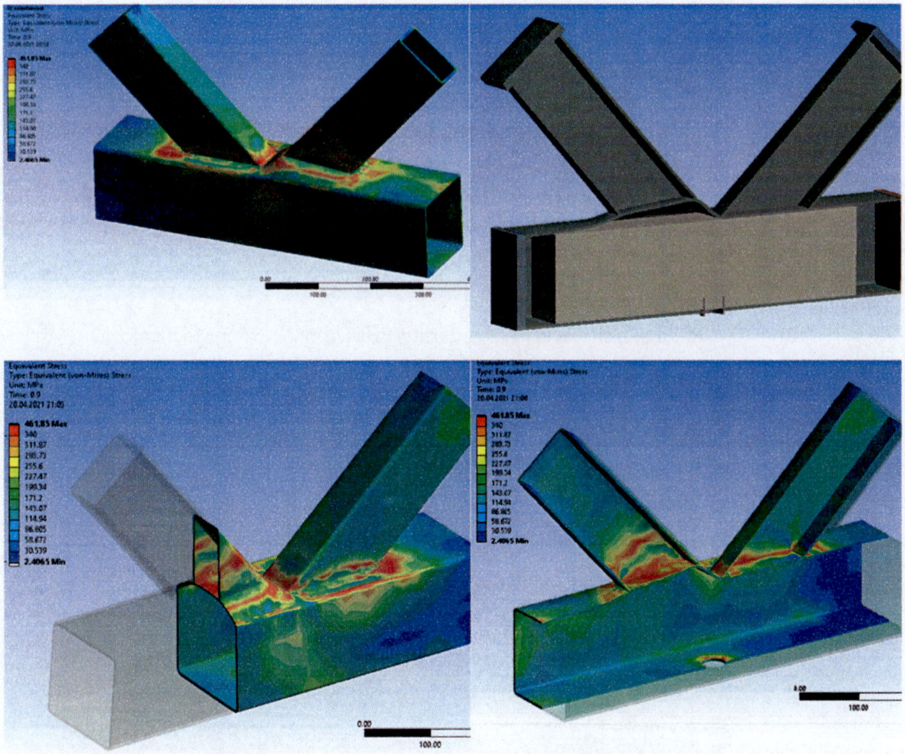

Fig. 5. Stress-strain state of reinforced K type connection

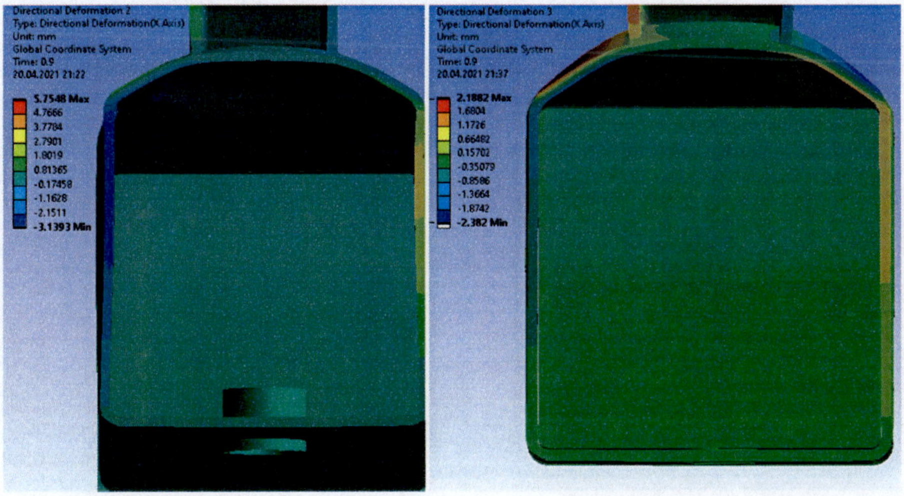

Fig. 6. Stress-strain state of reinforced K type connection (Sections and details)

Multiple linear regression according [22] described as:

$$y = \beta_0 + \beta_1 x_1 + \beta_2 x_2 + \ldots + \beta_i x_i + \epsilon \tag{2}$$

Random Forest algorithm can be implemented as a classifier or as a regression finding tool. And mathematical description of the algorithm is presented in [20].

Any machine learning technique to give a result needed to have corresponded data. In order to accumulate data for mentioned algorithms a set of variable parameters was generated Table 1.

Table 1. Model variations analyzed during FEMA in ANSYS Mechanical.

№	*d(mm)*	*α(deg)*	*f(mm)*	*b(mm)*	*t(mm)*	*gap 2g (mm)*
1	68.87	55.56	45.56	145.50	4.44	40.56
2	68.87	55.56	45.56	145.50	4.44	19.44
3	90.00	45.00	35.00	169.71	4.00	30.00
4	68.87	34.44	45.56	212.20	4.44	19.44
5	111.1	55.56	24.44	145.50	4.44	19.44
6	111.1	34.44	24.44	212.20	4.44	19.44
7	111.1	55.56	24.44	145.50	4.44	40.56
8	90.00	45.00	35.00	169.71	5.50	15.00
9	111.1	34.44	24.44	212.20	4.44	40.56
10	68.87	34.44	45.56	212.20	4.44	40.56
11	60.00	45.00	50.00	169.71	5.50	30.00
12	68.87	34.44	45.56	212.20	6.56	19.44
13	68.87	55.56	45.56	145.50	6.56	19.44
14	90.00	60.00	35.00	138.56	5.50	30.00
15	90.00	30.00	35.00	240.00	5.50	30.00
16	120.00	45.00	20.00	169.71	5.50	30.00
17	90.00	45.00	35.00	169.71	5.50	30.00
18	90.00	45.00	35.00	169.71	5.50	45.00
19	111.13	34.44	24.44	212.20	6.56	19.44
20	111.13	55.56	24.44	145.50	6.56	19.44
21	68.87	55.56	45.56	145.50	6.56	40.56
22	90.00	45.00	35.00	169.71	7.00	30.00

(continued)

Table 1. (*continued*)

№	*d(mm)*	*α(deg)*	*f(mm)*	*b(mm)*	*t(mm)*	*gap 2g (mm)*
23	111.13	55.56	24.44	145.50	6.56	40.56
24	111.13	34.44	24.44	212.20	6.56	40.56
25	68.87	34.44	45.56	212.20	6.56	40.56
26	100	235	30.00	146.49	4	30

Output parameter was chosen considering non-linear character of load transfer and occurring plastification of some areas inside of the connection. Considering a prevailing limit state and a stress-strain state of the connection in general, line located from a toe of the brace member in tension, at the distance of 15 mm and located under brace, was chosen. Average equivalent Von-Misses stresses along that line were calculated, which in turn allowed to take into account plastic strains inside of a chord face. Figure 7

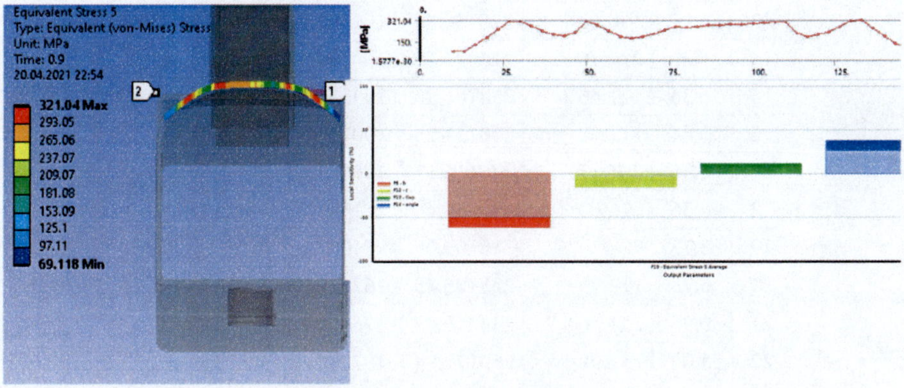

Fig. 7. Average equivalent Von-Misses stresses as output parameter.

3 Results and Discussion

In general 25 models were calculated each with its own set of variable parameters. The output parameter varied from 70 MPa to 250 MPa. In fact for the output parameter to reach to 250 MPa a significant amount of plastic strains had to take place.

Table 2 shows main output parameter and also additional parameters such as force in brace member in tension Ni [N], force in the chord F[N], vertical component of the force in a brace member in tension Ni * Sin(α) [N] and ratio of f[mm] to t[mm]. To get sense of the data it presented also in a chart form. Scatter plot in the Fig. 8 shows dependence of the main output parameter (equivalent Von-Misses stress[MPa]) from

Table 2. Output parameters.

№	f/t	Eq.Stress (MPa)	N_{chrd}(N)	N_i(N)	N_i*Sinα(N)
1	10.25	247.15	233590.48	188946.00	155833.09
2	10.25	226.77	227407.97	169541.11	139828.92
3	8.75	202.86	271015.77	195939.26	138549.98
4	10.25	182.49	320613.45	228119.17	129001.04
5	5.50	160.67	228644.81	164997.53	136081.61
6	5.50	107.18	320572.09	224449.14	126925.63
7	5.50	192.96	234114.99	181085.52	149350.17
8	6.36	132.69	265448.61	181180.20	128113.75
9	5.50	120.12	330887.24	238792.02	135036.51
10	10.25	238.98	332651.63	247463.90	139940.45
11	9.09	252.27	269803.80	194315.61	137401.88
12	6.95	119.86	318273.64	219652.22	124212.99
13	6.95	190.03	228156.84	161881.99	133512.07
14	6.36	234.10	216931.65	164096.42	142111.66
15	6.36	121.66	354761.85	253036.22	126518.11
16	3.64	113.68	270131.25	190723.91	134862.17
17	6.36	201.17	270116.17	190865.62	134962.38
18	6.36	226.03	275956.43	201577.97	142537.15
19	3.73	78.33	318197.05	221057.15	125007.47
20	3.73	122.79	228106.45	162028.75	133633.11
21	6.95	251.56	233210.97	176444.79	145522.73
22	5.00	152.98	269179.84	187834.60	132819.12
23	3.73	170.34	233604.32	173919.05	143439.63
24	3.73	110.88	327767.49	233060.76	131795.50
25	6.95	210.00	327833.34	233538.66	132065.75
26	7.50	195	233050	174530	142966.61

parameter ($2g$)[mm] and parameter (b)[mm] for different ratio of (f)[mm] to (t)[mm]. Therefore all variable parameters are aggregated in on chart.

As can be seen in the scatter plot, data in such form even with additional modification can hardly fit for regression analysis. In order to fit data for regression analysis better aggregated parameters were introduced:

$$\sigma_{\alpha} = \frac{1.5 * N_i * Sin\alpha * \mathrm{f}}{\mathrm{b}^2\mathrm{t}} * (1 + \mathrm{e}^{-\frac{\mathrm{f}}{2\mathrm{g}}}) \quad (3)$$

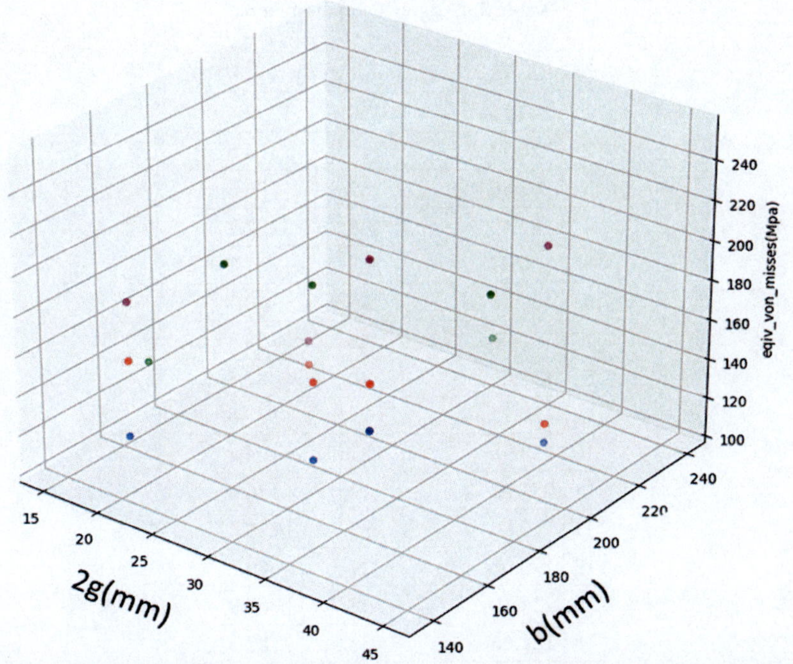

Fig. 8. Scatter plot of Eq. Stress dependence from *b* and *2g* for different ratio of *f/t*

$$\sigma_{\beta} = \frac{0.3 * N_i * Sin\alpha * 2g * A_{chrd} R_y}{d^2 tF} \tag{4}$$

Calculated aggregated parameters σ_α and σ_β are shown in Table 3 and dependence of the main output parameter from them is shown as a scatter plot in Fig. 9. Also by running sci-kit learn Multiple regression algorithm on data represented in Table 3, coefficients can be obtained for surface which approximate data shown in Fig. 9. Coefficients in turn can be used to introduce Eq. (5) for checking the yield strength of a connection corresponding to the limit state described above. The accuracy of multiple linear regression method (MLRM) is lower comparatively to the Random Forest regressor algorithm (RFRA), so Eq. (5) takes into account this fact and also instead of an average equivalent stress parameter utilizes the design yield stress of the steel R_y (Table 4).

Negative accuracy means predicted values lower than ones calculated explicitly.

Thus Eq. (5) also accommodates for lower predicted values at expense of overvaluing stresses for some combination of input parameters.

$$\frac{1.1}{R_y \gamma_c}\sigma_{\alpha} + \frac{0.536}{R_y \gamma_c}\sigma_{\beta} + \frac{131.4}{R_y \gamma_c} < 1 \tag{5}$$

Table 3. Aggregated parameters.

№	*Eq.Stress (MPa)*	$\sigma_\alpha(MPa)$	$\sigma_\beta(MPa)$
1	247.15	150.1	372.4
2	226.77	111.4	164.5
3	202.86	82.8	123.6
4	182.49	48.3	107.7
5	160.67	68.2	61.2
6	107.18	29.9	40.7
7	192.96	46.6	58.3
8	132.69	38.3	87.5
9	120.12	77.3	277.0
10	238.98	31.5	104.4
11	252.27	72.0	156.6
12	119.86	92.7	158.4
13	190.03	27.5	86.2
14	234.10	38.7	67.9
15	121.66	68.9	187.3
16	113.68	19.9	40.4
17	201.17	45.3	60.2
18	226.03	94.9	348.3
19	78.33	45.4	119.3
20	122.79	58.6	131.6
21	251.56	25.3	86.2
22	152.98	40.5	224.9
23	170.34	150.1	372.4
24	110.88	111.4	164.5
25	210.00	82.8	123.6
26	195	48.3	107.7

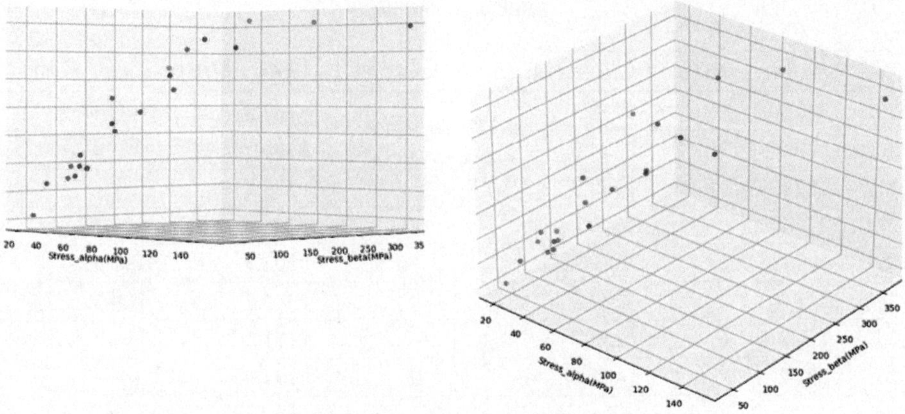

Fig. 9. Scatter plot of Eq. Stress dependence from σ_{α} and σ_{β}

Table 4. Comparison of ML techniques.

№	*Eq.Stress (MPa)*	*MLRM (MPa)*	*Accur. (%)*	*RFRA(MPa)*	*Accur. (%)*
1	201.2	165.6	−18	194.9	−3
2	252.3	224.0	−11	244.1	−3
3	113.7	134.0	18	116.5	2
4	202.9	200.0	−1	206.9	2
5	153.0	146.7	−4	146.2	−4
6	132.7	138.0	4	138.0	4
7	226.0	193.6	−14	219.6	−3
8	121.7	136.6	12	127.2	5
9	234.1	196.8	−16	219.2	−6
10	182.5	166.8	−9	181.6	0
11	107.2	124.2	16	110.4	3
12	119.9	137.3	15	120.3	0
13	78.3	110.2	41	91.9	17
14	239.0	234.0	−2	235.4	−1
15	120.1	151.2	26	139.6	16
16	210.0	179.1	−15	188.0	−10
17	110.9	127.8	15	112.4	1
18	226.8	225.0	−1	224.4	−1

(continued)

Table 4. (*continued*)

№	*Eq.Stress (MPa)*	*MLRM (MPa)*	*Accur. (%)*	*RFRA(MPa)*	*Accur. (%)*
19	160.7	158.5	−1	167.3	4
20	190.0	174.5	−8	189.7	0
21	122.8	132.9	8	128.2	4
22	247.2	320.1	30	245.9	−1
23	193.0	199.9	4	203.5	5
24	251.6	232.5	−8	244.7	−3
25	170.3	158.7	−7	171.2	0
26	201.2	165.6	−18	194.9	−3

4 Conclusions

Several conclusions can be made from result of this study:

- Finite element analysis tools in conjunction with machine learning techniques can be used to develop approximation models that allow to asses load bearing capacity of different structural elements and connections;
- Results significantly depend on what criteria were chosen as targeted need-to-predict value. ML tools allow using several need-to-predict values and offer means to analyze clustering in generated data underlying important variable parameters;
- Non-linear algorithms of machine learning are more accurate than linear regression methods;
- Accuracy of linear regression methods in case of current study can be improved by testing other hypothesis in formulating of σ_α and σ_β.

References

1. Wardenier, J., Packer, J.A., Zhao, X.-L., van der Vegte, G.J.: Hollow Sections in Structural Applications (2010)
2. Packer, J.A., Wardenier, J., Zhao, X.-L., van der Vegte, G.J., Kurobane Y.: Design Guide For Rectangular Hollow Section (RHS) Joints Under Predominantly Static Loading. Second Ed. (2009)
3. Hoadley, P.W., Yura, J.A.: Ultimate strength of tubular joints subjected to combined loads. In: Proceedings Offshore Technology Conference, OTC 4854 (1985)
4. Björk, T., Marquis, G., Kemppainen, R., Ilvonen, R.: The capacity of cold-formed rectangular hollow section K gap joints. In: Proceedings 10th International Symposium on Tubular Structures, pp. 227–234 (2003)
5. Huang, W., Fenu, L., Chen, B., Briseghella, B.: Experimental study on K-joints of concrete-filled steel tubular truss structures. J. Constr. Steel Res. **107**, 182–193 (2015). https://doi.org/10.1016/j.jcsr.2015.01.023
6. Packer, J.A.: Concrete-filled HSS connections. J. Struct. Eng. **121**(3), 458–467 (1995)

7. Xie, K., Wang, H., Pang, J., Zhou, J.: Study of the ultimate bearing capacity of concrete-filled steel tube K-Joints. KSCE J. Civ. Eng. **23**(5), 2254–2262 (2019). https://doi.org/10.1007/s12205-019-1268-7
8. Jiang, L., Liu, Y., Fam, A.: Stress concentration factors in joints of square hollow section (SHS) brace and concrete-filled SHS chord under axial tension in brace. Thin-Walled Struct. **132**, 79–92 (2018). https://doi.org/10.1016/j.tws.2018.08.014
9. Feng, R., Chen, Y., Chen, D.: Experimental and numerical investigations on collar plate and doubler plate reinforced SHS T-joints under axial compression. Thin-Walled Struct. **110**, 75–87 (2017). https://doi.org/10.1016/j.tws.2016.10.017
10. Khodaie, S., Mohamadi-shooreh, M.R., Mofid, M.: Parametric analyses on the initial stiffness of the SHS column base plate connections using FEM. Eng. Struct. **34**, 363–370 (2012). https://doi.org/10.1016/j.engstruct.2011.09.026
11. Broniewicz, M., Broniewicz, F.: Welds assessment in K-type joints of hollow section trusses with I or H section chords. Buildings (2020). https://doi.org/10.3390/buildings10030043
12. Qu, S., Wu, X., Sun, Q.: Experimental study and theoretical analysis on the ultimate strength of high-strength-steel tubular K-Joints. Thin-Walled Struct. **123**, 244–254 (2018). https://doi.org/10.1016/j.tws.2017.11.014
13. Cao, J.J., Packer, J.A., Yang, G.J.: Yield line analysis of RHS connections with axial loads. J. Constr. Steel Res. **48**(1), 1–25 (1998)
14. Driver, R.G., Grondin, G.Y., Kulak, G.L.: Unified block shear equation for achieving consistent reliability. J. Constr. Steel Res. **62**(3), 210–222 (2006)
15. Dutta, D.: Structures with hollow sections (1996)
16. Tubular structures in architecture. CIDECT and Delft University of Technology. Delft, The Netherlands
17. Chen, K., Huang, H., Qingxiong, W., Nakamura, S., Chen, B.: Experimental and finite element analysis research on the fatigue performance of CHS K-joints. Eng. Struct. **197**, 109365 (2019). https://doi.org/10.1016/j.engstruct.2019.109365
18. Grishin, I., Kayumov, R., Ivanov, G., Petropavlovckikh, O.: Computational model of Rib-reinforced plate. IOP Conf. Ser.: Mater. Sci. Eng. **890**(1), 012036 (2020). https://doi.org/10.1088/1757-899X/890/1/012036
19. Kayumov, R.A., Tazyukov, B.F., Shakirzyanov, F.R., Mukhamedova, I.Z.: Large deflections of beams, arches and panels in an elastic medium with regard to deformation shifts. Lobachevskii J. Math. **40**(3), 321–327 (2019). https://doi.org/10.1134/S1995080219030119
20. Louppe G.: Understanding random forests: from theory to practice. Ph.D. thesis, U. of Liege (2014)
21. Pedregosa, F., et al.: Scikit-learn: machine learning in python. JMLR **12**, 2825–2830 (2011)
22. McCullagh, P., Nelder, J.: Generalized Linear Models, Second Ed. (1989)

Parametric Vibrations of Viscoelastic Rectangular Plates with Concentrated Masses

Mirziyod Mirsaidov[1], Nikolay Vatin[2], Rustamkhan Abdikarimov[3], Dadakhan Khodzhaev[1](✉), and Bakhodir Normuminov[1]

[1] Tashkent Institute of Irrigation and Agricultural Mechanization Engineers, 100000 Tashkent, Uzbekistan

[2] Peter the Great St. Petersburg Polytechnic University, 195251 St. Petersburg, Russia

[3] Tashkent Institute of Finance, 100000 Tashkent, Uzbekistan

Abstract. In modern technology, construction, and mechanical engineering, thin-walled composite structures such as plates, panels, and shells are widely used. Often, various devices, assemblies, and overlays are attached to such structures. The structures can be subjected to various loads, both static and dynamic. Among such loads, periodic loads take a special place. Therefore, the study of the dynamic behavior of plates, panels, and shells with concentrated mass is of great importance. Parametric vibrations of isotropic viscoelastic rectangular plates with concentrated masses were considered. The equation of motion was obtained on the basis of the Kirchhoff-Love hypothesis, taking into account geometric nonlinearity. The hereditary Boltzmann-Volterra theory was used to describe the viscoelastic properties of the plate material. The effect of concentrated mass was taken into account using the Dirac delta function. The discretization of the system of equations was realized using the Bubnov-Galerkin method and was reduced to a system of ordinary nonlinear integro-differential equations of the Volterra type. In this case, the weakly singular Koltunov-Rzhanitsyn kernel was used as a relaxation kernel. To solve the resulting system, a numerical method based on the use of quadrature formulas was applied. The graphs of the frequency-amplitude response of the oscillations were constructed. The influence of the viscoelastic properties of the plate material, concentrated mass, geometric nonlinearity on the eigen frequencies was investigated.

Keywords: Viscoelastic rectangular plate · Concentrated masses · Geometric nonlinearity · Periodic load · Parametric oscillations · Bubnov-Galerkin method numerical method

1 Introduction

Thin-walled plates, panels, and shells made of composite materials are the basic elements of engineering structures and buildings. The importance of their strength and stability is due to the fact that they are the load-bearing elements of structures. Small masses in the form of various loads, engines, and equipment are attached to them due to a constructive or operational necessity. In the calculations, such attached masses are considered as

N. Vatin (Ed.): STCCE 2021, LNCE 169, pp. 72–79, 2021.
https://doi.org/10.1007/978-3-030-80103-8_8

concentrated ones at certain points and rigidly fixed. The "bearing surface-attached mass" system can be subjected to various static and dynamic loads, resulting in the initiation of the processes undesirable from the point of view of the dynamic strength and reliability of structures. Among dynamic loads, periodic loads are of particular interest. Such loads induce forced vibrations, which, together with the natural vibrations of structures, can cause the phenomenon of resonance and destruction of structures. Therefore, the study of the stability of bearing surfaces with attached elements under the action of periodic loads is an urgent problem.

Problems of dynamic stability of elastic systems under various dynamic loads and the methods for their solution were presented in the classical studies by V. Bolotin [1] and A. Volmir [2].

Recent advances in the study of nonlinear vibrations and stability of shells and plates were considered by M. Amabili [3].

The reference [4] is devoted to the study of nonlinear vibrations and stability of plates under static and dynamic loads. To solve the problems, a method based on the theory of R-functions was proposed.

In [5], the dynamic stability of composite plates under periodic loads was investigated. The problem was reduced to solving a system of ordinary differential equations of the Mathieu-Hill type using the Bolotin method.

The problem of the dynamic stability of thin composite cylindrical shells under the action of a periodic load was studied in [6]. Donnel's theory and Karman's nonlinearity were used to derive the equation of motion. Using the Galerkin method, the problem was reduced to solving the Mathieu-Hill equation.

A description of mathematical models and algorithms for solving problems on strength and stability of shells, taking into account various material properties and geometric nonlinearity, is given in [7].

In [8], a square plate rigidly fixed along all edges was investigated for stability; the case of complex bending was considered under the simultaneous action of transverse and compressive loads acting on the edges.

The study in [9] is devoted to the estimation of the dynamic stability of plates under the loads rapidly increasing along the axis. Hamilton's principle was used to derive the equation of motion. Analytical and numerical solutions were obtained.

In [10], the dynamic instability of rectangular plates under the action of periodic loads was considered. The equation of motion of the relative deflection was obtained according to the Hamilton principle, solved by the Galerkin method.

[11] is devoted to the study of vibrations of plates with concentrated masses. Numerical results were obtained using ANSYS software and were compared with the results found in literature sources.

In [12], under various boundary conditions, the effect of concentrated masses on the dynamic behavior of composite plates was studied. The results were obtained numerically using ANSYS software and were compared with the data obtained experimentally.

The results of studying the dynamic stability of rectangular plates with concentrated masses under the action of periodic loads are given in [13, 14]. The problem by the Galerkin method was reduced to solving the Mathieu-Hill equation.

The dynamic behavior of plates with concentrated masses under the action of a load applied along the edges of the plate was investigated in [15]. Numerical and analytical solutions to the problem were obtained.

In [16], free vibrations of multilayer composite plates and shells with a mass concentrated in the center were considered. The problem was solved by the finite element method under various boundary conditions.

Earlier, the dynamic stability of orthotropic viscoelastic plates and shells with concentrated masses under the action of rapidly growing compressive loads was investigated in [17, 18].

Parametric vibrations of orthotropic viscoelastic rectangular plates in a geometric nonlinear formulation are considered in the article.

2 Materials and Methods

Let us consider an orthotropic viscoelastic rectangular plate with sides a and b, constant thickness h, carrying concentrated masses M_i at points with coordinates (x_i, y_i). We assume that the plate has initial imperfections. Let the plate be subjected to the action of a periodic load along the side a in the following form

$$P(t) = P_0 + P_1 \cos \Theta t \tag{1}$$

where $P_0, P_1 = const$; Θ is the frequency of external periodic load.

Using the results given in [17], we obtain the equation of motion for an orthotropic viscoelastic rectangular plate with concentrated masses relative to deflection $w = w(x, y, t)$ and displacements $u = u(x, y, t)$, $v = v(x, y, t)$, where P(t), has the form (1).

A weakly singular Koltunov-Rzhanitsyn kernel of the form [19] is used as a relaxation kernel:

$$\Gamma(t) = Ae^{-\beta t}t^{\alpha-1}, \ (0 < \alpha < 1) \tag{2}$$

To solve the resulting system, the Bubnov-Galerkin method is used, based on a polynomial approximation of the deflection and displacements. Further, discretization with respect to spatial variables is performed at each point in time and the problem is reduced to solving a system of nonlinear integro-differential equations with variable coefficients and with weakly singular kernels. A numerical method based on the use of quadrature formulas [20] was applied to solve the resulting non-decaying system.

3 Results and Discussion

Figure 1 shows the results of studying the influence of the inhomogeneity parameter Δ (curve 1 - $\Delta = 1$; curve 2 - $\Delta = 1.5$ and curve 3 - $\Delta = 2$). Here: $\lambda = 1$; $\delta = 25$; $M_1 = 0.1$; $w_0 = 0.01$; $A_{ij} = 0.025$; $\delta_0 = 0.3$; $\delta_1 = 0.5$; $\theta = 1.1$; $\mu_1 = 0.32$; $\mu_2 = 0.27$; $g = 0.3833$. The figure shows that an increase in this parameter leads to a decrease in the vibration amplitude.

Figure 2 shows the vibration modes of a viscoelastic rectangular plate with concentrated mass without and taking into account the inhomogeneous properties of the plate material.

The effect of the concentrated mass on plate behavior is shown in Fig. 3. Here: $\lambda = 1$; $\delta = 25$; $w_0 = 0.01$; $A_{ij} = 0.025$; $\Delta = 1.5$; $\delta_0 = 0.3$; $\delta_1 = 0.5$; $\theta = 1.1$; $\mu_1 = 0.32$; $\mu_2 = 0.27$; $g = 0.3833$. The results obtained show that an increase in the concentrated mass leads to an increase in the vibration amplitude and a phase shift to the right.

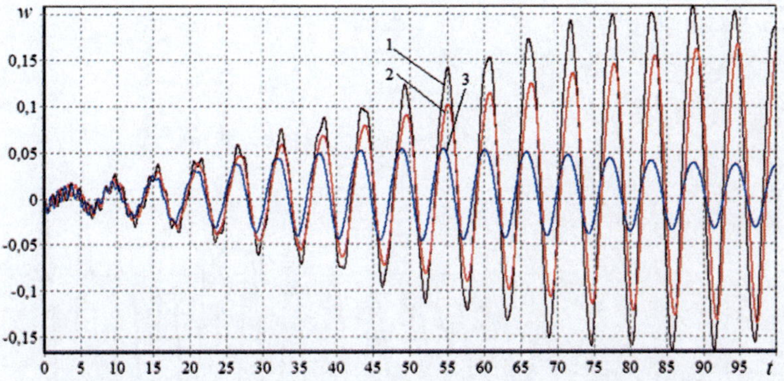

Fig. 1. Dependence of the deflection on time at $\Delta = 1$ (1); 1.5 (2); 2 (3)

Figure 4 shows the vibration modes of a viscoelastic rectangular plate without and taking into account concentrated mass.

The influence of other geometric parameters of a rectangular plate is also investigated. Here: $\delta = 25$; $w_0 = 0.01$; $A_{ij} = 0.025$; $\Delta = 1.5$; $\delta_0 = 0.3$; $\delta_1 = 0.5$; $\theta = 1.1$; $\mu_1 = 0.32$; $\mu_2 = 0.27$; $g = 0.3833$. Figure 5 shows the influence of the parameter λ. It is seen that an increase in this parameter leads to a decrease in the amplitude and frequency of vibrations.

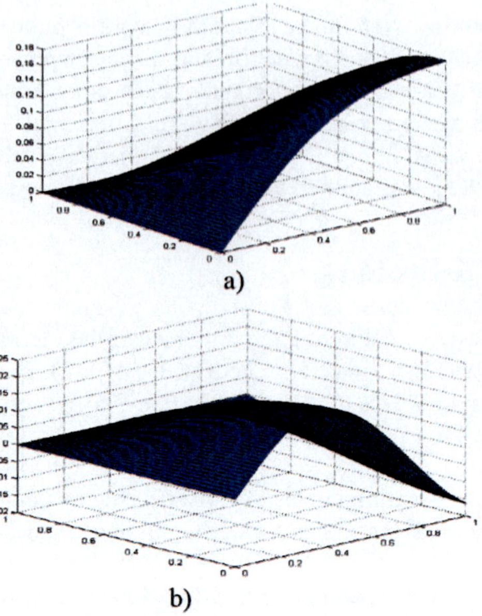

Fig. 2. Vibration form of a rectangular plate with concentrated mass at $t = 88$: a) A, $A_{ij} = 0$; b) A, $A_{ij} = 0.025$

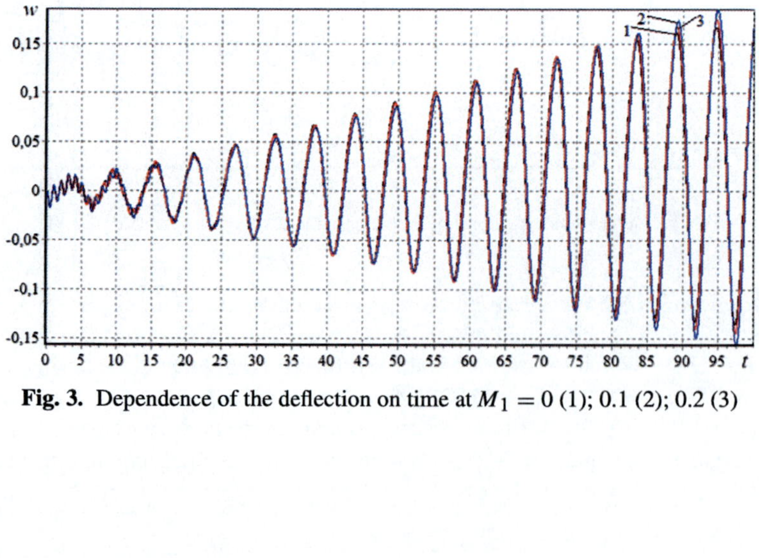

Fig. 3. Dependence of the deflection on time at $M_1 = 0$ (1); 0.1 (2); 0.2 (3)

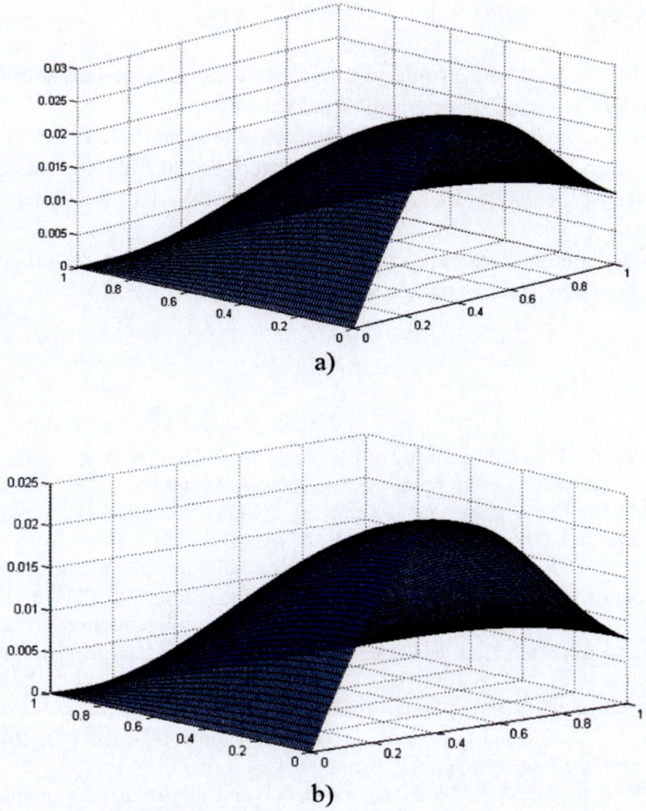

Fig. 4. Vibration form of a rectangular plate at $t = 88$: a) $M_1 = 0$; b) $M_1 = 0.1$

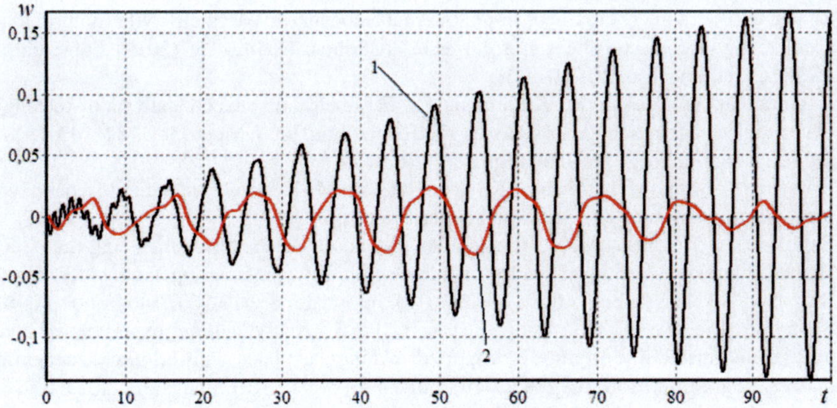

Fig. 5. Dependence of the deflection on time at $\lambda = 1$ (1); 2 (2)

4 Conclusion

1. Dynamic behavior of orthotropic viscoelastic rectangular plates with concentrated masses under the action of a periodic load was investigated.
2. The influence of the inhomogeneous properties of the material, the concentrated mass and its location on the amplitude-time characteristics of the plate was estimated.
3. It is shown that the concentrated masses significantly affect the dynamic instability of the plate.
4. The proposed method and algorithm make it possible to study other thin-walled structural elements such as panels and shells.

References

1. Bolotin, V.V.: The Dynamic Stability of Elastic Systems. Holden-Day, San Francisco (1964)
2. Volmir, A.S.: Stability of deformable systems. NASA AD (1965)
3. Amabili, M.: Nonlinear vibrations and stability of shells and plates. Cambridge University Press, Cambridge (2008). https://doi.org/10.1017/CBO9780511619694
4. Kurpa, L., Mazur, O., Tkachenko, V.: Dynamical stability and parametrical vibrations of the laminated plates with complex shape. Lat. Am. J. Solids Struct. **10**, 175–188 (2013)
5. Chen, W.-R., Chen, C.-S., Shyu, J.-H.: Stability of parametric vibrations of laminated composite plates. Appl. Math. Comput. **223**, 127–138 (2013). https://doi.org/10.1016/j.amc.2013.07.095
6. Darabi, M., Ganesan, R.: Non-linear dynamic instability analysis of laminated composite cylindrical shells subjected to periodic axial loads. Compos. Struct. **147**, 168–184 (2016). https://doi.org/10.1016/j.compstruct.2016.02.064
7. Karpov, V.V., Semenov, A.A.: Mathematical models and algorithms for studying strength and stability of shell structures. J. Appl. Ind. Math. **11**, 70–81 (2017). https://doi.org/10.1134/S1990478917010082
8. Sukhoterin, M., Baryshnikov, S., Knysh, T., Pizhurina, N.: Influence of transverse loading on the stability of a clamped rectangular plate. In: MATEC Web of Conferences (2018). https://doi.org/10.1051/matecconf/201823901022
9. Zhang, D.-B., Tang, Y.-Q., Chen, L.-Q.: Internal resonance in parametric vibrations of axially accelerating viscoelastic plates. Eur. J. Mech. - A/Solids. **75**, 142–155 (2019). https://doi.org/10.1016/j.euromechsol.2019.01.021
10. Pirmoradian, M., Torkan, E., Karimpour, H.: Parametric resonance analysis of rectangular plates subjected to moving inertial loads via IHB method. Int. J. Mech. Sci. **142–143**, 191–215 (2018). https://doi.org/10.1016/j.ijmecsci.2018.04.047
11. Hossain, N., Islam, M.S., Ahshan, K.H.N., Hossain, M.Z.: Effects on natural frequency of a plate due to distributed and positional concentrated mass (2015)
12. Rahmane, A., Benmansour, T., Bouakba, M., Meddour, I.: Factors influencing the effect of attached mass on dynamic behavior of laminate composite plates using Taguchi technique. J. Braz. Soc. Mech. Sci. Eng. **40**(2), 1–13 (2018). https://doi.org/10.1007/s40430-018-0987-x
13. Zhong, Z., Liu, A., Pi, Y.L., Deng, J., Lu, H., Li, S.: Analytical and experimental studies on dynamic instability of simply supported rectangular plates with arbitrary concentrated masses. Eng. Struct. **196**, 109288 (2019). https://doi.org/10.1016/j.engstruct.2019.109288
14. Zhong, Z.-L., Liu, A.-R.: Analytical study on the out-of-plane nonlinear dynamic instability of a thin rectangular plate with a concentrated mass. Gongcheng Lixue/Eng. Mech. **37**, 6–12 (2020). https://doi.org/10.6052/j.issn.1000-4750.2019.04.S018

15. Martin, P.A., Hull, A.J.: Dynamic response of an infinite thin plate loaded with concentrated masses (2020). https://doi.org/10.1016/j.wavemoti.2020.102643
16. Mandal, A., Ray, C., Haldar, S.: Free vibration analysis of laminated composite plates and shells subjected to concentrated mass at the centre. In: Singh, B.N., Roy, A., Maiti, D.K. (eds.) Recent Advances in Theoretical, Applied, Computational and Experimental Mechanics: Proceedings of ICTACEM 2017, pp. 49–57. Springer Singapore, Singapore (2020). https://doi.org/10.1007/978-981-15-1189-9_5
17. Khodzhaev, D., Abdikarimov, R., Normuminov, B.: Dynamic stability of viscoelastic rectangular plates with concentrated masses. IOP Conf. Ser. Mater. Sci. Eng. **896**, 012030 (2020). https://doi.org/10.1088/1757-899X/896/1/012030
18. Khodzhaev, D., Vatin, N., Abdikarimov, R., Normuminov, B., Mirzaev, B.: Dynamic stability of viscoelastic orthotropic shells with concentrated mass. IOP Conf. Ser. Mater. Sci. Eng. **890**, 012042 (2020). https://doi.org/10.1088/1757-899X/890/1/012042
19. Mal'tsev, L.E.: The analytical determination of the Rzhanitsyn-Koltunov nucleus. Mech. Compos. Mater. **15**, 131–133 (1979)
20. Abdikarimov, R.A., Khodzhaev, D.A.: Computer modeling of tasks in dynamics of viscoelastic thinwalled elements in structures of variable thickness. Mag. Civ. Eng. **49**, 83–94 (2014). https://doi.org/10.5862/MCE.49.9

Analysis of Light Comfort and Thermal Protection of a Building Taking into Account Changes in the Geometry of the Window Slope

Aleksey Ivantsov(✉) and Vadil Sirazitdinov

Kazan State University of Architecture and Engineering, Kazan 420043, Russia

Abstract. The article presents an analysis of the influence of the geometry of the window slope on the parameters of light comfort in the premises of buildings and the general heat loss of the building. It is shown that an increase in the opening angle of the outer slopes increases the parameters of daylighting, insolation and visual comfort without increasing the area of the window glazing. Changing the angle of the outer slope at the same time increases additional heat loss through this fragment of the facade. Combined consideration of the parameters of light comfort and thermal protection shows that this solution is effective at any level of thermal protection. The most total effect is observed in the wall with the largest.

Keywords: Thermal protection · Daylighting · Visual comfort · Thermal protection · Windows slope

1 Introduction

1.1 Relevance

Windows are an important part of any building in which people are constantly present. Windows are necessary to create light comfort in rooms as it provides positive psychological, mental and physiological effects on building occupants [1–5]. Solar radiation has a healing effect on the inhabitants of the building due to the penetration of UV radiation [6, 7]. In design practice, the daylight factor and the duration of insolation determines the light comfort in the rooms. To improve these indicators, it is necessary to increase the area of windows or select structures with high light transmission coefficients [8–11].

On the other hand, windows also affect the thermal and acoustic comfort of the room [12–15]. A large area of windows generally worsens the thermal balance of a building due to the low resistance to heat transfer of windows [14, 15]. Sound comfort is also reduced, as the sound insulation of the windows is much less than the sound insulation of the blank walls. The higher the area of the windows, the higher the performance characteristics must be applied to maintain thermal and acoustic comfort. The above parameters are typical for the design of any building.

With the current approach to the design of buildings, it is necessary to take into account other parameters that affect the quality of architecture and comfort in the rooms.

N. Vatin (Ed.): STCCE 2021, LNCE 169, pp. 80–89, 2021.
https://doi.org/10.1007/978-3-030-80103-8_9

One of these parameters is visual comfort, which is also determined by the area of the windows. The BREEAM building certification system, widely used in the world, takes into account the view from the windows along with other components. The view from the windows in this system means the relationship between the distance from the window to the workplace and the window area as % of the surrounding wall area. The further the workplace is located, the larger the window area should be [16]. Thus, the design of windows is a rather complex task, which must be considered taken together of various parameters [17–23].

When designing the area of windows, a large number of researchers pay attention to the optimal window-to-wall ratio in buildings for various purposes [24, 25]. As a rule, an increase in the coefficient of glazing of the facade leads to an increase in heat loss, if at the same time the thermal performance of the glazed part is not proportionally increased. However, the thermotechnical qualities of glazed structures have their own limitations both in technological terms and from an economic point of view.

1.2 Aims and Objectives

However, it is worth noting that only the window area and the window-to-wall ratio affect heat loss and sound insulation. In addition to the window area, light and visual comfort is also influenced by the viewing angle from the window. With a large wall thickness (in northern regions with large thermal insulation), the viewing angle can be significantly reduced, which reduces the flow of daylight and sunlight into the room. In this regard, it is possible to hypothesize that changing the angle of visibility through the windows without changing their area can increase the flow of daylight and solar radiation without deteriorating the quality of thermal protection and sound insulation of the entire facade as a whole. The main solution to this problem is to change the angle of the window slope in relation to the plane of the facade.

In world practice, there are examples of increasing the angle of visibility from windows with a large wall thickness by changing the angle of the window slope to the plane of the facade. This is mainly typical for historical buildings (Fig. 1a) [26]. In modern architecture, this technique of facade plastics is also known (Fig. 1b) [27]. However, the scientific rationale for this solution is often not presented: there is no method for calculating the intake of solar radiation with different geometry of window slopes. The effect of this solution on the change of heat loss of the building facade has not been sufficiently studied.

Based on the above, this article analyzes the design approach that allows you to increase the angle of visibility through the windows by changing the geometry of the outer window slopes without changing the area of the windows.

Fig. 1. Beveled outer slope in historical and modern architecture.

2 Materials and Methods

The analysis of the influence of the geometry of the outer window slope on the light comfort and thermal performance of the wall structure was carried out for a fragment of the outer wall with a window (Fig. 2) for a typical living space. The angle of the outer slope in relation to the plane of the facade was chosen as the main changing parameter of the structure under study. Figure 2 shows the design scheme of the window.

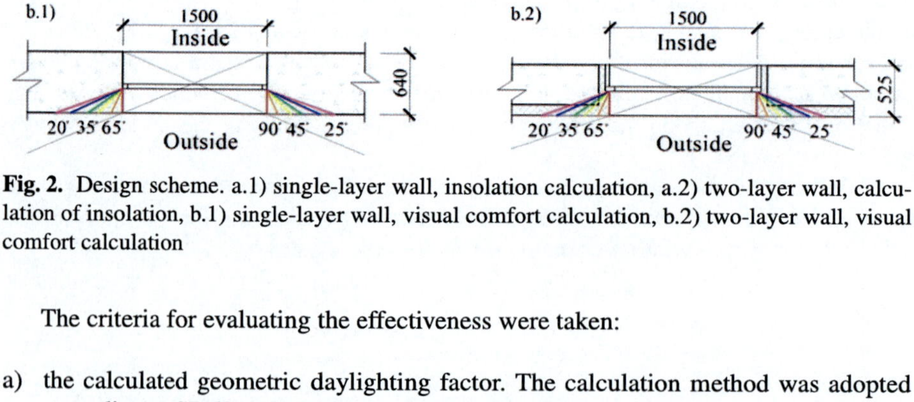

Fig. 2. Design scheme. a.1) single-layer wall, insolation calculation, a.2) two-layer wall, calculation of insolation, b.1) single-layer wall, visual comfort calculation, b.2) two-layer wall, visual comfort calculation

The criteria for evaluating the effectiveness were taken:

a) the calculated geometric daylighting factor. The calculation method was adopted according to SP 52.13330.2016 and SP 23-102-2003 "Daylighting of residential and public buildings". The calculation did not take into account the light transmission characteristics of the window and the reflection parameters of the room finish. The analysis was carried out for three points of the room: point 2 - in the centre, point 3 - at the outer and point 1 - at the inner walls.

b) the duration of the insolation of the room. The calculation method is based on the tracing of a cartogram for the shading of a window opening and its alignment with a solar map of a certain latitude of the area (48.5N by the example of Volgograd)
c) visual comfort of the room. This parameter was determined through the angle of visibility through the window based on the tracing of the cartogram of the window opening shading. The difference in the shading cartograms of the window opening in points b) and c) consists in different design schemes. When determining insolation, the horizontal shading angle is determined by the outer edge of the slope and the plane of the window (Fig. 1, a.1 and a.2). When determining the angle of visibility, the horizontal shading angle is determined by the outer and inner edges of the slopes (Fig. 1, b.1 and b.2).
d) reduced resistance to heat transfer of a wall section with a window slope. The calculation method is adopted according to SP 50.13330.2012 "Thermal protection of buildings". To compare the effectiveness of the design solution, the analysis of heat loss was carried out for two types of external wall: a uniform 600 mm thick (heat transfer resistance 2.15 $m^2 \cdot °C$ / W) and a two-layer wall with heat-insulating material with a total thickness of 525 mm (4 $m^2 \cdot °C$ / W). The heat transfer resistance of the window is taken as 0.8 $m^2 \cdot °C$ / W. When calculating heat fluxes, the following environmental parameters were taken: indoor air temperature +20° C, outdoor air temperature −22° C.

3 Results and Discussion

3.1 Daylighting

The geometric daylighting factor according to the standard method of SP 23–102-2003 is determined by the product of the rays of diffuse sky light determined by the section (n_1) and plan (n_2) of the room with a window according to Danilyuk's standard graphs:

$$\varepsilon = 0.01 \cdot n_1 \cdot n_2, \ \% \tag{1}$$

Table 1 shows the calculation of the geometric daylighting factor for one room with a constant window area and different angles of the window slope in relation to the plane of the facade. The value of Δ, % shows an increase in the calculated value of the coefficient ε, % in relation to a window with a slope of 90°. The calculation was carried out for three points: point 1 is located at the far wall of the room (at a distance of 1.2 m), point 2 - in the center of the room, point 3 - near the window at a distance of 1.2 m.

As can be seen from Table 1, a decrease in the angle of the window slope relative to the plane of the facade increases the natural illumination of the room along the entire depth (up to 42.2% for point 3 closest to the window and up to 11.8% for point 1 farthest from the window). However, the increase is uneven. Near the window, due to the greater change in the sky visibility angle, the increment is more significant and ranges from 22.2 to 42.2%. At the far point of the room, due to a slight change in the angle, the increase in the coefficient is 11.8%. In addition, for point 3, the coefficient increases until the slope

Table 1. The daylight coefficient of the room at different angles of the window slope

Angle of the window slope	Point 1 (far from the window)				Point 2 (center of the room)				Point 3 (near the window)			
	n_1	n_2	ε, %	Δ, %	n_1	n_2	ε, %	Δ, %	n_1	n_2	ε, %	Δ, %
90°	4.8	16	0.76	–	6	23,5	1.41	–	10	36	3.6	–
60°	5	17	0.85	+11.8	6.6	25	1.65	+17.0	11	40	4.4	+22.2
45°	5	17	0.85	+11.8	7	25	1.75	+24.1	11.8	40	4.72	+31.1
35°	5	17	0.85	+11.8	7	25	1.75	+24.1	12.8	40	5.12	+42.2
25°	5	17	0.85	+11.8	7	25	1.75	+24.1	12.8	40	5.12	+42.2
20°	5	17	0.85	+11.8	7	25	1.75	+24.1	12.8	40	5.12	+42.2

angle decreases to 35°; for point 1, after an angle of 60°, the illumination coefficient does not increase further.

An important conclusion of this analysis is that the change in the shape of the outer window slope insignificantly affects the illumination at the far point of the room. In accordance with this, for premises in which this particular point is normative (living quarters of residential buildings, preschool educational organizations, educational premises of general educational organizations, hospital wards), changing the geometry of the window slope in terms of increasing natural illumination is not an effective measure.

For rooms in which the point in the center of the room is normative (doctors' offices, administrative and office buildings, living quarters of hotels and hostels, etc.), the change of the geometry of the window slope can increase the value of natural illumination up to 24.1%.

It should be noted that in this article, the window is installed near the central plane of the outer wall, which determines the implementation of thermal protection standards. When the window is moved closer to the inner plane of the outer wall, the geometry of the outer slope is able to provide a greater increase in the coefficient of natural illumination over the entire area of the room.

3.2 Insolation and Visual Comfort

As noted in paragraph 2, the duration of insolation and the visual comfort of the room in this article were assessed by the cartograms of the shading of the window opening. Figure 3 shows cartograms with marked shading angles for windows with different slopes. Visually, the cartograms demonstrate that, depending on the differences in the design schemes (Fig. 2, a and b), the influence of the slope geometry is significantly different. From the point of view of the duration of insolation, the increase in the insolation time is not significant. The exact calculation values are presented in Table 2. Table 2 shows that changing the slope angle of the window from 90 to 20° increases the insolation time to 7.6% (for the southern orientation of the window). The spatial angle of visibility with the same changes in the geometry of the window slope increases to 14.9%. On the one hand, these percentages are not large values, but on the other hand, it is important to note that the increase in data by the parameter is obtained with a constant window area.

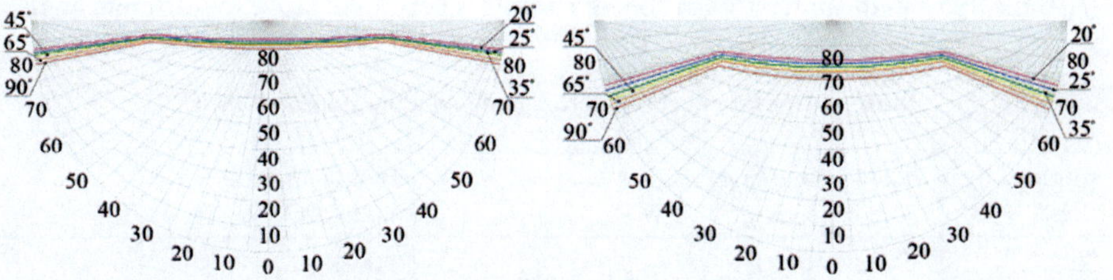

Fig. 3. Window opening shading cartograms a) for calculating the duration of insolation, b) for calculating the angle of visibility (visual comfort)

The last column of Table 2 shows the values of the area of the equivalent window. An equivalent window is a 90° slope window that, for a given wall thickness, produces the same percentage increment in sun duration and angle of view as a beveled slope window. The analysis shows that in order to achieve an increase in insolation time by 7.6% or a total angle of visibility by 15.0%, it is necessary to increase the window area by 120% or change the window slope angle to 20° relative to the facade plane. It is important to note that an increase in the area of the window, and hence the coefficient of glazing of the facade, will certainly lead to an increase in heat loss through the outer shell of the building.

In addition, the ratio between the area of the original window and the thickness of the outer wall should have a great influence on the relative increase in the time of insolation and the angle of visibility. This relationship is not considered in this article.

Table 2. Insolation duration and visual comfort at different angles of the window slope

Angle of the window slope	Insolation duration (increase %)	Visual comfort, increase %	Equivalent window area, increase, %
90°	8 h 20 min (–)	0	0
60°	8 h 34 min (+2.8%)	+4.3%	+24.4%
45°	8 h 40 min (+4.0%)	+6.8%	+40.0%
35°	8 h 45 min (+5.0%)	+9.3%	+59.1%
25°	8 h 52 min (+6.4%)	+12.2%	+90.2%
20°	8 h 58 min (+7.6%)	+15.0%	+120.0%

3.3 Thermal Performance

The window slope (the junction of the window and wall construction) is one of the significant areas of additional heat loss in relation to the heat loss of a flat outer wall.

This is due to the deviation of heat fluxes from a stationary position and an increase in the heat transfer area. With a decrease in the angle of the slope in relation to the facade, the thickness of the outer wall decreases locally, and, consequently, heat loss increases.

Table 3 shows the indicators of specific heat loss per 1 m of slope for different angles.

Table 3. Specific heat loss through the window slope, W/ m·°C

Angle of the window slope	One-layer wall, R = 2.15 m^2·°C / W		Two-layer wall, R = 4.0 m^2·°C / W	
90°	0.090		0.008	
60°	0.110	+22.2%	0.010	+25.0%
45°	0.130	+44.4%	0.014	+75.0%
35°	0.145	+61.1%	0.021	+162.5%
25°	0.172	+91.1%	0.040	+400.0%
20°	0.194	+115.5%	0.057	+612.5%

The values show that with a decrease in the angle of the slope in relation to the facade, additional heat loss increases. Moreover, for a homogeneous wall with low resistance to heat transfer, an increase in additional heat loss when the slope angle changes from 90° to 20° is slightly more than doubled. For a two-layer wall, the same increase occurs almost sevenfold. This difference is due to the difference in the heat transfer resistance of the window (R = 0.8 m^2·°C/W and of the analyzed walls): the higher the difference in the resistances of the elements of the assembly, the greater the contribution to thermal protection is made by the element with high resistance. Thus, if we consider the design method of changing the angle of the window slope from the point of view of thermal protection, then any change in the slope angle below 90° gives an increase in heat losses, and therefore worsens the overall thermal protection of the building. At the same time, the higher the thermal protection of the outer wall is, the more the overall thermal protection will deteriorate.

However, as noted above, the windows should be considered in a combination of different parameters. For this purpose, the calculation of the total heat losses through the windows with the changed geometry of the slope and the windows equivalent in area from the point of view of insolation and visual comfort, indicated in Table 2, was carried out. The calculation was carried out for a fragment of a facade with an area of 25 m^2 with a window with dimensions of 1.5 × 1.5 m. Table 4 shows the result of this calculation.

Analysis of Table 4 shows that, regardless of the thermal protection of the outer wall, changing the geometry of the window slope is a more effective measure for heat loss compared to a straight-sloped window with an equivalent area. The efficiency of a chamfered window slope increases with decreasing angle: from 1.8% to 9.1% in a single-layer wall, from 5% to 22% in a two-layer wall when changing the angle of slope from 60° to 20°. At the same time, the greater the resistance to heat transfer of the outer wall, the greater the energy saving effect is achieved when bevelling the outer slope and

Table 4. Heat flow through a fragment of the facade with different geometry of the window slope, W

Angle of the window slope	One-layer wall, R = 2.15 $м^2$*C/Вт			Two-layer wall, R = 4.0 $м^2$*C/Вт		
	Chamfered window	Equivalent window	Increase, %	Chamfered window	Equivalent window	Increase, %
90°	558	558	–	338.48	338.48	–
60°	568.19	578.82	+1.8%	339.93	356.98	+5%
45°	573.29	589.13	+2.76%	340.66	366.24	+7.5%
35°	576.13	600.84	+4.28%	341.98	380.07	+11.1%
25°	581.23	620.54	+6.76%	345.5	403.64	+16.8%
20°	585.39	638.68	+9.1%	348.7	425.52	+22.0%

increasing the light comfort in the room: at an angle of 20° the heat flow saving in a single-layer wall is 9.1%, in a two-layer wall - 22.0%.

This means that when the parameters of light comfort and heat loss through a section of a wall with a window are taken into account together, beveled slopes are an effective constructive means for increasing the flow of light at different levels of thermal protection.

4 Conclusion

The article shows that the geometry of the window slope affects various sanitary and hygienic parameters of the room and the thermal characteristics of the external structure as a whole.

Firstly, the influence of the geometry of the window slope on the daylighting of the room is manifested to a greater extent for areas of the room near the light opening. Moreover, the smaller the angle of inclination of the window slope to the plane of the façade is, the greater the value of illumination will be near the window (up to 40% increase). In the distant points of the room from the outer wall and window, the influence of the angle of inclination of the window slope is not so significant (no more than 12% increase). Moreover, the angle of the window slope is less than 60° to the plane of the facade practically does not affect the increase in illumination.

Secondly, the influence of the geometry of the window slope on the duration of insolation and the spatial angle of visibility is manifested depending on the design scheme. Taking into account the opening angle of the window along the glazing plane, the increase in the angle of visibility when the geometry of the slope changes is insignificant. If we take into account the opening angle of the window as a whole, taking into account the internal slopes, the increase in the total angle of visibility can reach almost 15%.

Thirdly, it is important to note that any change in the angle of the window slope below 90° increases the level of light comfort in the room. The same change in geometry leads to a deterioration in thermal performance. The value of additional heat losses through the window slope unit when changing the angle can increase by seven times, depending on the initial level of thermal protection.

Fourthly, with the combined consideration of the parameters of light comfort and thermal protection, it is possible to achieve overall efficiency. To increase the luminous flux in the room, it is necessary to increase the viewing angle through the window. Increasing the viewing angle by changing the angle of the outer slope with a constant window area is more effective than increasing the window area. Depending on the thermal protection of the wall, this effect can reach 9–22%.

References

1. Wirz-Justice, A., Skene, D.J.: Münch, M: The relevance of daylight for humans. Bio. Pharm. **28**, 114304 (2020). https://doi.org/10.1016/j.bcp.2020.114304
2. Bellia, L., Acosta, I., Ángel Campano, M., Fragliasso, F.: Impact of daylight saving time on lighting energy consumption and on the biological clock for occupants in office buildings. Sol. En. **211**, 1347–1364 (2020). https://doi.org/10.1016/j.solener.2020.10.072
3. Hosseini, S., Mohammadi, M., Rosemann, A., Schröder, T., Lichtenberg, J.: A morphological approach for kinetic facade design process to improve visual and thermal comfort. Rev. Build. Environ. **153**, 186–204 (2019). https://doi.org/10.1016/j.buildenv.2019.02.040
4. Michael, A., Heracleous, C.: Assessment of natural lighting performance and visual comfort of ducational architecture in Southern Europe: the case of typical educational school premises in Cyprus. Energy Build. **140**, 443–457 (2017). https://doi.org/10.1016/j.enbuild.2016.12.087
5. Tzempelikos, A.: Advances on daylighting and visual comfort research. Build. Environ. **113**, 1–4 (2017). https://doi.org/10.1016/j.buildenv.2016.12.002
6. Darula, S., Christoffersen, J., Malikova, M.: Sunlight and insolation of building Interiors. Energy Proc. **78**, 1245–1250 (2015)
7. Kupriyanov, V., Khalikova, F.: Field studies of the energy parameters of the insolation of residential premises. Izv. KazGASU **4**(32), 139–147 (2012)
8. Kim, S., Zadeh, P., Staub-French, S., Froese, T., Cavka, B.: Assessment of the impact of window size, position and orientation on building energy load using BIM. Procedia Eng. **145**, 1424–1431 (2016). https://doi.org/10.1016/j.proeng.2016.04.179
9. de Rubeis, T., Nardi, I., Muttillo, M., Ranieri, S., Ambrosini, D.: Room and window geometry influence for daylight harvesting maximization – effects on energy savings in an academic classroom. Energy Procedia **148**, 1090–1097 (2018). https://doi.org/10.1016/j.egypro.2018.08.050
10. Mangkuto, R., Feradi, F., Putra, R., Triyogo Atmodipoero, R., Favero, F.: Optimisation of daylight admission based on modifications of light shelf design parameters. J. Build. Eng. **18**, 195–209 (2018). https://doi.org/10.1016/j.jobe.2018.03.007
11. Rizki, A., Mangkuto, R., Rohmah, M., Dian, A.A.: Design optimisation for window size, orientation, and wall reflectance with regard to various daylight metrics and lighting energy demand: a case study of buildings in the tropics. Appl. Energy **164**, 211–219 (2016). https://doi.org/10.1016/j.apenergy.2015.11.046
12. Sierra, F., Gething, B., Bai, J., Maksoud, T.: Impact of the position of the window in the reveal of a cavity wall on the heat loss and the internal surface temperature of the head of an opening with a steel lintel. Energy Build. **142**, 23–30 (2017)
13. Adamus, J., Pomada, M.: Analysis of heat flow in composite structures used in window installation. Comput. Struct. **202**, 127–135 (2018). https://doi.org/10.1016/j.compstruct.2017.12.077

14. Liu, M., Heiselberg, P.K., Ivanov, Y., Mikkelsen, F.S.: Parametric analysis on the heat transfer, daylight and thermal comfort for a sustainable roof window with triple glazing and external shutter. Energy Build. **183**, 209–221 (2019). https://doi.org/10.1016/j.enbuild.2018.11.001
15. Borowska, J., Jezierski, W.: Heat parameters of multi-sash windows in residential buildings. Constr. Build. Mat. **176**, 112–117 (2018). https://doi.org/10.1016/j.conbuildmat.2018.04.231
16. https://www.breeam.com/internationalRFO2015/content/05_health/hea_01_rfrb.htm
17. Koohsari, A., Fayaz, R., Kari, B.: The influence of window dimensions and location on residential building energy consumption by integrating thermal and lighting analysis in a mild and humid climate. BRIS J. Adv. Sci. Tech. **3**, 187–194 (2015)
18. Vanhoutteghem, L., Skarning, G., Hviid, C., Svendsen, S.: Impact of facade window design on energy daylighting and thermal comfort in nearly zero-energy houses. Energy Build. **102**, 149–156 (2015). https://doi.org/10.1016/j.enbuild.2015.05.018
19. Altan, H., Mohelnikova, J., Hofman, P.: Thermal and daylight evaluation of building zones. Energy Proc. **78**, 2784–2789 (2015). https://doi.org/10.1016/j.egypro.2015.11.626
20. Zhu, L., Wang, B., Sun, Y.: Multi-objective optimization for energy consumption, daylighting and thermal comfort performance of rural tourism buildings in north China. Build. Environ. **176**, 106841 (2020). https://doi.org/10.1016/j.buildenv.2020.106841
21. Bakmohammadi, P., Noorzai, E.: Optimization of the design of the primary school classrooms in terms of energy and daylight performance considering occupants' thermal and visual comfort. Energy Rep. **6**, 1590–1607 (2020). https://doi.org/10.1016/j.egyr.2020.06.008
22. Al-Khatatbeh, B., Ma'bdeh, S.N.: Improving visual comfort and energy efficiency in existing classrooms using passive daylighting techniques. Energy Procedia **136**, 102–108 (2017). https://doi.org/10.1016/j.egypro.2017.10.294
23. Hosseini, M., Mohammadi, M., Guerra-Santin, O.: Interactive kinetic facade: improving visual comfort based on dynamic daylight and occupant's positions by 2D and 3D shape changes. Build. Environ. **165**, 106396 (2019). https://doi.org/10.1016/j.buildenv.2019.106396
24. Gioia, F.: Search for the optimal window-to-wall ratio in office buildings in different European climates and the implications on total energy saving potential. Sol. Energy **132**, 467–492 (2016). https://doi.org/10.1016/j.solener.2016.03.031
25. Bian, Y., Ma, Y.: Analysis of daylight metrics of side-lit room in Canton, south China: a comparison between daylight autonomy and daylight factor. Energy Build. **138**, 347–354 (2017). https://doi.org/10.1016/j.enbuild.2016.12.059
26. https://zurichguide.ru/switzerland-travel-guide/scuol
27. Tribus, M.: Architecture, Ex-Post. Refurbishment, Bolzano (IT), EURAC research
28. Petrov, A., Ivantsov, A.: Design and calculation of the internal roof drain system structure in terms of thermal protection and moisture condensation. IOP Conf. Ser.: Mater. Sci. Eng. **890**, 012141 (2020). https://doi.org/10.1088/1757-899X/890/1/012141
29. Ivantsov, A., Petrov, A.: The influence of architectural and construction parameters of residential buildings on energy efficiency in Russian Federation. IOP Conf. Ser.: Mater. Sci. Eng. **890**, 012142 (2020). https://doi.org/10.1088/1757-899X/890/1/012142
30. Kraynov, D., Maslennikov, I.: Calculation of the reduced heat transfer resistance of translucent structures made of aluminum profiles. Izv. KazGASU **4**(54), 35–43 (2020)
31. Kraynov, D.: Accounting of glazing of the facade when designing the heat protection of a building. Build. Mat. **6**, 32–34 (2017)

Improvements of Techniques for Determining Hazardous Area Near Facilities Under Construction

Almaz Shaehov(✉), Lyudmila Koklugina, Ruslan Ibragimov, and Yuliya Evstigneeva

Kazan State University of Architecture and Engineering, 420043 Kazan, Russia

Abstract. In this work, existing methods for determining the hazardous areas of tower crane operation were considered and analyzed, the behavior of cargo deflection, cargo weight at full wind load were revealed. The study investigates the impact of wind on lifting operations and their safety, therefore, on the hazardous area on the construction site. Improved methods for determination of the hazardous area of tower crane operation in the vicinity of facilities under construction were offered with wind velocity variations. It has been established that the deviation of the cargo caused by the wind directly affects the hazardous area. Analytical dependencies between additional wind velocity and the hazardous area of tower crane were obtained. Further development in this direction is associated with necessity of air flows impact research on the safety of erecting works, especially lifting operations with high parameters of cargo resistance to the wind.

Keywords: Hazardous area · Wind louds · Cargo resistance to the wind

1 Introduction

Construction in close city development is common, for the reason that site area is limited and there is a downward trend of free territory. There is a high probability of forming hazardous area outside the construction site [1, 2].

It is known that the border of hazardous areas, in places of cargo transfer by tower cranes [3], is taken from the extreme point of the horizontal projection of the outline smallest dimension of a moveable object or wall of a building [4] with the addition of the largest boundary dimension of a transported (falling) cargo and minimum cargo dropping distance (SP 49.13330.2010 appendix G) according to Table 1 and the diagram for determining the minimum cargo dropping distance:

$$R_h = R^{max}_{working} + L^{max} + \frac{B}{2} + r, \tag{1}$$

where $R^{max}_{working}$ – maximum working radius of operation of crane; L^{max} – largest boundary dimension of mounted construction; B – smallest boundary dimension of mounted construction; r – dispersion radius of cargo dropping with a crane.

N. Vatin (Ed.): STCCE 2021, LNCE 169, pp. 90–100, 2021.
https://doi.org/10.1007/978-3-030-80103-8_10

There are no recommendations for wind velocity accounting and its effect on the lifted cargo while determining the hazardous area [5]. Therefore, it is necessary to take into account that a powerful air flow is quite capable directly affected on the determination of the hazardous area of tower crane [6]. In this regard, the purpose of this work is to determine the dependence of wind velocity on the hazardous area of tower crane operation during construction and erection work.

Table 1. Main types of cargo.

Naming unit	Windward area, m^2	Weight, kg	Air resistance coefficient, c
Timbering MAXIMOMX (3300 × 2400 mm)	7,92	408	1,5
Aluminum form (1200 × 3000 mm)	3,6	110	1,5
Wall panel V1.3-65.27*12-1	17,28	5150	1,5
Ceiling panel PK 30-2-8	0,66	1080	1,5
Concrete bucket Zitrek BN-2.0 021-1067	3,41	330	1

During research we considered that this direction was pioneer examined in the scientific work of Chanyshev R.O. [7]. In this work, the author determined the effect of the wind load on the lifted cargo during operations of the light cranes and string lifts. Having regard to the horizontal outreach of light cranes (at string lifts – consoles) outside the construction hesitates from 0,6 to 1 m, in research we take into account that cargo in case of deviating from the vertical position must not hurt or hit the building under construction. The formula is established in the research and determines the value dependence of the cargo deviation from the lifting height, full wind load and mass of cargo.

In the science work [8] Ian Skelton considered the impact of airflow on tower cranes and their down-time due to high wind velocity. It was found out that low-powered magnification of wind velocity exercised a significant influence on the safe lifting operations of the tower crane. As consequence that led to the accelerated difficulties for the operator during lifting operations on an overloaded construction site. This posed a significant security risk not only for the crane operator, but also for all workers near the crane and cargo. In the course of research Ian Skelton proposed a new solution by new type of equipment called «Lifting Wing» [9].

In the research [10] it was noted that the wind increased the risk of delayed high-rise construction. It was also revealed that the wind could increase working radius of operation of crane, while the distance from the center of the crane grew and, therefore, the overturning moment increased, which led to instability of tower crane [11].

Formation of dangerous dynamic wind loads not only exists, which is confirmed by long-term observations, but also demonstrates a tendency of growth ability [12, 13].

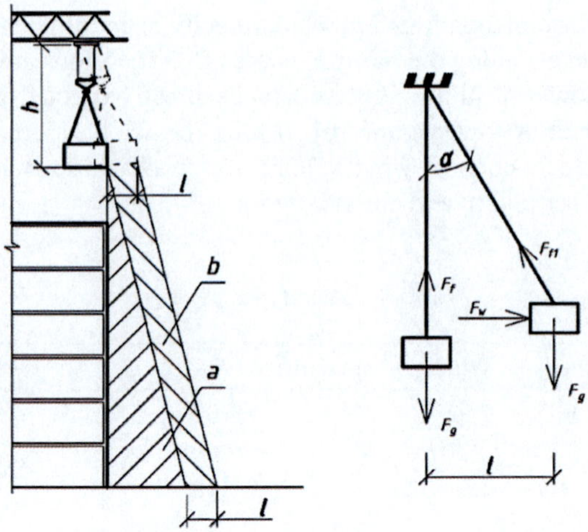

Fig. 1. Wind load impact on lifted cargo: a – hazardous area of potential cargo dropping with a crane; b – increase of hazardous area.

The effect of wind on the lifting load is evident in the form of load, the value of which depends on the wind velocity [14]. Depending on the wind force intensity, the area of cargo resistance to the wind and its direction, the radius of the hazardous area will increase [15–17]. From Fig. 1 it is clear that the radius of the hazardous area increases by value *l*.

We introduce the *l* value in the formula for determining the hazardous area for high cargo resistance to the wind:

$$R_h = R_{working}^{max} + L^{max} + \frac{B}{2} + r + l, \quad (2)$$

l – Increase of hazardous area due to wind pressure.

2 Materials and Methods

For illustration, we compare the main materials and elements, used during construction (Table 1).

When analyzing the data from the table, it may be said that some elements have a large area and low weight, while others, on the contrary, have a large weight and small area. It is necessary to compare their deviations in the same conditions. For this reason, we compile a table, where we compare cargo under investigation as far as they deviate vertically at different wind velocities. Moreover, all elements will have a low point on one level. That means the total distance from the lowest point of cargo to the crane trolley will be the same. Wind velocities assume 1–10 m/s. In our research we disassembled the following example: the process of lifting operations with the selected hook of the tower crane Giraffe TDK-8.180. For all cargos dimension *L* (distance from the crane trolley to the bottom of the cargo) we take 10,3 m.

These passport specifications are required for subsequent calculation Giraffe TDK-8.180:

- maximum radius of operation of horizontal crane arm – 30 m;
- maximum velocity of the cargo trolley – 0,66 m/s;
- velocity of rotation – 0,0133 p/s (1/s).

3 Results

It is known that the suspended cargo, being at rest, is in an upright position.

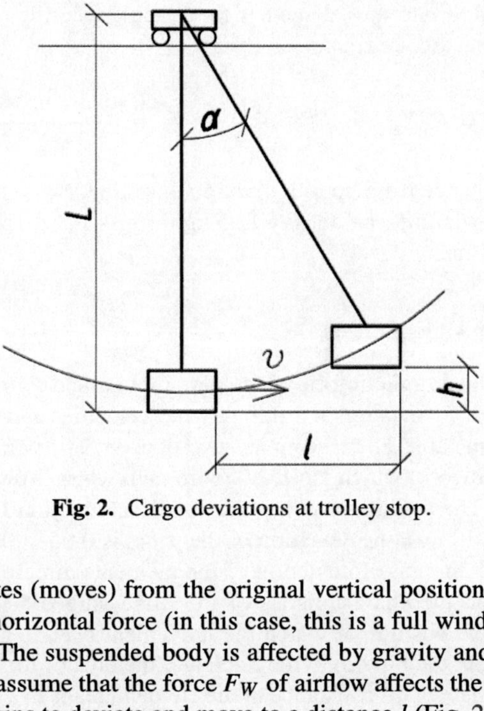

Fig. 2. Cargo deviations at trolley stop.

The cargo deviates (moves) from the original vertical position to a certain angle α under the action of horizontal force (in this case, this is a full wind load). It determines the deviation value. The suspended body is affected by gravity and the funicular force, so $F_G = F_F$. Let us assume that the force F_W of airflow affects the cargo and under this impact, the body begins to deviate and move to a distance l (Fig. 2).

3.1 Deviations of Lifted Load at Static Crane Arm

Ratio of deviation depends on wind pressure forces to gravity ratio, then:

$$tg\alpha = \frac{F_W}{F_G} = \frac{S_W \cdot p}{mg}, \tag{3}$$

where p– velocity air head.

Horizontal deviation under action of force F_W ignoring insignificant variation with altitude *h*, can be determined by:

$$l = L * tg\alpha. \tag{4}$$

After certain calculations, we get the behavior of the deviation distance on the physical characteristics of the cargo and on the air flow:

$$l = \frac{L * S_w * v^2 * \frac{\rho}{2}}{mg}. \tag{5}$$

It follows that the cargo, when deviating increases the hazardous area of tower crane. It determines the impact of material characteristics and airflow rates. As a result, it turns out there is an additional new element in the formula according to SP 49.13330.2010 for determining the hazardous area:

$$R_h = R_{working}^{max} + L^{max} + \frac{B}{2} + r + \frac{L * S_w * v^2 * \frac{\rho}{2}}{mg}, \tag{6}$$

where L – the distance from bottom of cargo to trolley; S_w – cargo resistance to the wind; v – velocity of wind; ρ – air density 1,25 kg/m^3; m – mass of lifting cargo; g – free fall acceleration 9,81 m/s^2.

3.2 Comparison of Main Cargos

Table 2 is completed using the information from Table 1 and Formula (5). Each cargo type is determined by its deviation at different wind velocities and the same conditions.

It can be seen from Table 2 that the greatest vertical deviation occurs due to lifting the aluminum panel formwork, which is 41,4% more than when lifting the MAXIMOMX 15 panel formwork. The ceiling panel has the smallest deviation from the vertical. So, at a wind velocity of 10 m/s, the deviation of the panel is 0,06 m.

To sum it up, the amount of deviation of the elements directly depends mainly on the wind velocity and the characteristics of the lifted cargos. These include weight, windward area and the position at which the movement occurs. Changing parameters such as the length from the bottom to the hook and the area of lifting cargo, as well as, increasing the coefficient of air resistance of the element, leads to a directly proportional increase in vertical deviation. In case of increase in wind velocity, it leads to the square expand [18–25].

Therefore, it is necessary to pay more attention to the airflow velocity on the construction site during work with large cargo resistance to the wind.

Table 2. Cargo deviations results.

Velocity of wind, m/s	Timbering MAXIMO MX 3.3 × 2.4 m	Aluminum form 1200 × 3000 mm	Wall panel V1.3-65.27-12-1	Ceiling panel PK 30-12-8	Concrete bucket Zitrek BN-2.0 021-1067
1	0,02	0,03	0,00	0,00	0,01
2	0,08	0,13	0,01	0,00	0,03
3	0,17 58,6%	0,29 100%	0,03 10,3%	0,01 1,9%	0,06 21,1%
4	0,31 58,6%	0,52 100%	0,05 10,3%	0,01 1,9%	0,11 21,1%
5	0,48 58,6%	0,81 100%	0,08 10,3%	0,02 1,9%	0,17 21,1%
6	0,69 58,6%	1,16 100%	0,12 10,3%	0,02 1,9%	0,24 21,1%
7	0,94 58,6%	1,58 100%	0,16 10,3%	0,03 1,9%	0,33 21,1%
8	1,22 58,6%	2,06 100%	0,21 10,3%	0,04 1,9%	0,43 21,1%
9	1,55 58,6%	2,61 100%	0,27 10,3%	0,05 1,9%	0,55 21,1%
10	1,91 58,6%	3,22 100%	0,33 10,3%	0,06 1,9%	0,68 21,1%

3.3 Consideration of Dynamic Impacts

During the tower crane operation, swinging of the cargo occurs during starting, braking, turning of the crane arm/turret, as well as during lifting and lowering of the cargo.

We investigate two types of cargo movement:

1) Movement of a crane trolley with cargo at static crane arm;
2) Rotation of a crane arm at maximum working radius and at static trolley.

The combination of two movement types will be inefficient. Firstly, when the crane arm is turned, the cargo increases the radius of work. Secondly, the movement of the trolley from the crane partially compensates for this deviation of the cargo. Therefore, it is necessary to consider these two types separately from each other.

3.4 Crane Cargo Trolley Movement

First, let us consider the case where the trolley moves with the fixed crane arm. Then the horizontal movement of the trolley from the crane axis will cause the cargo deviation for the crane. Consequently, the movement velocity of the trolley and the cargo will be identical. In the case of the cargo trolley stop, the cargo will continue moving along the circumference equal to the length of the cable and after the equilibrium point it will deviate in the opposite direction. As a result, it increases the working radius of the tower crane.

The maximum velocity of the cargo will be at the lower point – at the point of equilibrium. And at this velocity, the load will deviate from the extreme position. At the extreme position, the maximum potential energy is numerically equal to the kinetic energy at the equilibrium point.

This deviation is obtained during erection works and wind force must be taken into account. The wind pressure will only increase the deviation and the forces causing the deviation will be summed up.

Cargo deviation is determined by:

$$l = tg\alpha * (L - h). \tag{7}$$

The horizontal deviation l can be found according to the formula:

$$l = tg(\arccos\left(\frac{L - \frac{\upsilon_T^2}{2g}}{L}\right))(L - \frac{\upsilon_T^2}{2g}). \tag{8}$$

Taking into account the formulas (5) and (8), it is finally obtained that the deviation under the action of the wind load and the movement of the crane trolley is determined by:

$$l = \frac{L(S_w * \frac{\rho \cdot \upsilon^2}{2} + mg * sin(\arccos(\frac{L - \frac{\upsilon_T^2}{2g}}{L}))}{mg}, \tag{9}$$

where L – the distance from bottom of cargo to trolley; S_w – resistance to the wind of lifting cargo; v – wind velocity; ρ – air density 1,25 kg/m^3; m – mass of lifting cargo; g – free fall acceleration 9,81 m/s^2; v_T – velocity of trolley movement (passport specifications).

3.5 Rotation of Tower Crane Arm

When the arm of the tower crane rotates, a force occurs. This force applied to the suspended cargo and directed radially from the center is called centrifugal (Fig. 3).

Horizontal deviation by force F_{cf}, ignoring the change in height h, as insignificant, can be determined by:

$$l = L * tg\alpha = L * \frac{F_{cf}}{F_G}. \tag{10}$$

Horizontal deviation l under the action of centrifugal force:

$$l = \frac{4L * R * \pi^2 * \nu^2}{g}, \tag{11}$$

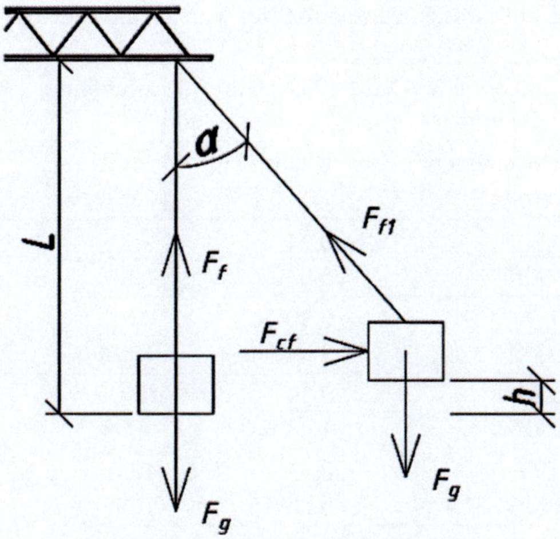

Fig. 3. Deviation under centrifugal force.

where L– length of cable; R– maximum operating radius of the crane; ν– crane velocity of rotation (passport data); g– free fall acceleration.

This deviation is obtained during erection works. Moreover, wind load must be taken into account. The wind pressure will only increase the deviation and the forces causing the deviation will be summed up. Thus, the force of the wind load F_W and the centrifugal force F_{cf} are summed. Considering the formulas (10–11), we obtain that the deviation under the action of wind load and crane arm rotation is determined by:

$$l = \frac{L(S_w \cdot \frac{\rho \cdot \upsilon^2}{2} + 4m \cdot R \cdot \pi^2 \cdot \nu^2)}{mg}, \quad (12)$$

where L– the distance from bottom of cargo to trolley; S_w – resistance to the wind of lifting cargo; v – wind velocity; ρ – air density 1,25 kg/m^3; m – mass of lifting cargo; g – free fall acceleration 9,81 m/s^2 (passport specifications).

Table 3 is compiled on the basis of foregoing formula. The dynamic effects arising during the operation of the tower crane and the air flow pressure are considered in Table 3. The most susceptible to cargo deviation is considered to be large-shield formwork. Calculations show that the maximum working radius of operation of crane is 30 m.

The results of the calculation show that the greatest danger is the deviation arising from the movement of the cargo trolley. In all investigated wind ranges, this deviation is 18% greater than in crane arm rotation when using MAXIMO formwork and 12% greater when in aluminum formwork.

Table 3. Results of deviation calculation (m) with an allowance of crane dynamics.

Velocity of wind, m/s	Timbering MAXIMO MX 3300 × 2400 mm		Aluminum form 1200 × 3000 mm	
	Movement of carriage	Boom swing	Movement of carriage	Boom swing
1	0,70 100%	0,24 34,3%	0,72 100%	0,25 34,7%
2	0,76 100%	0,30 39,5%	0,81 100%	0,35 43,2%
3	0,86 100%	0,39 45,3%	0,97 100%	0,51 52,6%
4	0,99 100%	0,53 53,5%	1,20 100%	0,74 61,7%
5	1,16 100%	0,70 60,3%	1,49 100%	1,03 69,1%
6	1,37 100%	0,91 66,4%	1,84 100%	1,38 75%
7	1,62 100%	1,16 71,6%	2,26 100%	1,80 79,6%
8	1,91 100%	1,44 75,4%	2,75 100%	2,28 82,9%
9	2,23 100%	1,77 79,4%	3,29 100%	2,83 86%
10	2,59 100%	2,13 82,2%	3,91 100%	3,44 87,9%

4 Conclusions

In this work the problem of determining the hazardous area increase of the tower crane, depending on the wind velocity, was solved. During operation, the crane will swing the lifted cargo due to the influence of the wind load on it and the dynamic impact during erection. The amount of deviation mainly depends on the parameters of the cargo, and this is necessary while calculating the hazardous area in order to increase the safety of construction and erection work. Dependencies recommended to determine the increase in the hazardous area of operation of tower crane, variation with the wind velocity were obtained. The main cargos used with high resistance to the wind are also considered and their possible deviations in the construction and erection works are determined theoretically.

The wind load is a dangerous factor that must be taken into account as an influence of increase the radius of operation of crane and, first of all, the hazardous area. It is very important to take into account wind velocity at the design stage of the construction general plan.

References

1. Shakhiev, A.D., Evlakhova E.Yu.: Features of the placement and operation of tower cranes during the construction of multi-storey buildings in dense urban areas, Science and innovation – modern concepts. Collection of scientific articles on the results of the work of the International Scientific Forum, Moscow, Russia (2019)
2. Sadeghi, S., Soltanmohammadlou, N., Rahnamayiezekavat, P.: A systematic review of scholarly works addressing crane safety requirements Saf. Sci. **133**, 105002 (2021). https://doi.org/10.1016/j.ssci.2020.105002
3. Raviv, G., Fishbain, B., Shapira, A.: Analyzing risk factors in crane-related near-miss and accident reports. Saf. Sci. **91**, 192–205 (2017). https://doi.org/10.1016/j.ssci.2016.08.022
4. Im, S., Park, D.: Crane safety standards: Problem analysis and safety assurance planning. Saf. Sci. **127**, 104686 (2020). https://doi.org/10.1016/j.ssci.2020.104686
5. Ramli, L., Mohamed, Z., Abdullahi, A.M., Jaafar, H.I., Lazim, I.M.: Academic Press, vol. 95 (2017)
6. Zhou, W., Zhao, T., Liu, W., Tang, J.: Tower crane safety on construction sites: a complex sociotechnical system perspective. Saf. Sci. **109**, 95–108 (2018). https://doi.org/10.1016/j.ssci.2018.05.001
7. Chanyshev, R.: Moscow, Russia, 288 (1975)
8. Skelton, I., Demian, P., Glass, J., Bouchlaghem, D., Anumba, C.: Buildings, **4**(2), 245–265 (2014). https://doi.org/10.3390/buildings4020245
9. Demian, P., Skelton, I., Bouchlaghem, D., Anumba, C.: Tall Building Boom – Now Bust? CRC Press, Boco Raton (2009)
10. Raviv, G., Shapira, A., Fishbain, B.: AHP-based analysis of the risk potential of safety incidents: case study of cranes in the construction industry. Saf. Sci. **91**, 298–309 (2017). https://doi.org/10.1016/j.ssci.2016.08.027
11. Swuste, P.: A 'normal accident' with a tower crane? An accident analysis conducted by the Dutch safety board. Saf. Sci. **57**, 276–282 (2013). https://doi.org/10.1016/j.ssci.2013.03.002
12. Stroganov, V., Sagadeev, E., Ibragimov, R., Potapova, L.: Mechanical activation effect on the biostability of modified cement compositions. Constr. Build. Mater. **246**, 118506 (2020). https://doi.org/10.1016/j.conbuildmat.2020.118506
13. Khuzin, A., Ibragimov, R.: Processes of structure formation and paste matrix hydration with multilayer carbon nanotubes additives. J. Build. Eng. **35**, 102030 (2021). https://doi.org/10.1016/j.jobe.2020.102030
14. Das, S., Dhalmahapatra, K., Maiti, J.: Z-number integrated weighted VIKOR technique for hazard prioritization and its application in virtual prototype based EOT crane operations. Appl. Soft Comput. J. **94**, 106419 (2020). https://doi.org/10.1016/j.asoc.2020.106419
15. Lesovik, V., et al.: Improving the behaviors of foam concrete through the use of composite binder. J. Build. Eng. **31**, 101414 (2020). https://doi.org/10.1016/j.jobe.2020.101414
16. Fediuk, R.S., Yevdokimova, Y.G., Smoliakov, A.K., Stoyushko, N.Y., Lesovik, V.S.: Use of geonics scientific positions for designing of building composites for protective (fortification) structures. IOP Conf. Ser.: Mater. Sci. Eng. **221**(1), 012011 (2017). https://doi.org/10.1088/1757-899x/221/1/012011
17. Murali, G., Fediuk, R.: A Taguchi approach for study on impact response of ultra-high-performance polypropylene fibrous cementitious composite. J. Build. Eng. **30**, 101301 (2020). https://doi.org/10.1016/j.jobe.2020.101301
18. Khuziakhmetov, R.A., Kashina, S.G., Khuziakhmetova, K.R.: Investigation of the circumstances of the fall of reinforced concrete wall panel during the dismantling of a panel house. Izvestiya KGASU **1**(47), 239–249 (2019)

19. Mudrov, A.G.: Reduction of peak loads when starting the mechanism of lifting the load in the cranes. Izvestiya KGASU **2**(44), 239–245 (2018)
20. Ibragimov, R.A., Korolev, E.V., Deberdeev, T.R., Leksin, V.V., Solovev, D.B.: Energy parameters of the binder during activation in the vortex layer apparatus. Mater. Sci. Forum. **945**, 98–103 (2018). https://doi.org/10.4028/www.scientific.net/MSF.945.98
21. Galeev, R.R., Nizamov, R.K., Abdrakhmanova, L.A.: Filling of epoxy polymers with chemically precipitated chalk from chemical water treatment sludge. In: Klyuev, S.V., Klyuev, A.V. (eds.) ICICC 2021. LNCE, vol. 147, pp. 93–97. Springer, Cham (2021). https://doi.org/10.1007/978-3-030-68984-1_14
22. Khozin, V., Khokhryakov, O., Nizamov, R.: A «carbon footprint» of low water demand cements and cement-based concrete. IOP Conf. Ser. Mater. Sci. Eng. **890**(1) (2020). https://doi.org/10.1088/1757-899X/890/1/012105
23. Galeev, R., Nizamov, R., Abdrakhmanova, L., Khozin, V.: Resource-saving polymer compositions for construction purposes. IOP Conf. Ser. Mater. Sci. Eng. **890**(1) (2020)
24. Vdovin, E.A., Stroganov, V.F.: Properties of cement-bound mixes depending on technological factors. Mag. Civ. Eng. **93**, 147 (2020). https://doi.org/10.18720/MCE.93.12
25. Bulanov, P.E., Vdovin, E.A., Mavliev, L.F., Kuznetsov, D.A.: Development of road soil cement compositions modified with complex additive based on polycarboxylic ether. IOP Conf. Ser. Mater. Sci. Eng. **327**(3) (2018). https://doi.org/10.1088/1757-899X/327/3/032014

Analysis of the Functional Planning Development of Cities in the USSR on the Example of Kazan

Maria Grishina(✉)

Kazan State University of Architecture and Engineering, 420043 Kazan, Russia

Abstract. Today, planning theory must meet the challenges of reality. Success in the rational solution of urban planning problems and sustainable planning based on experience in this activity. In this publication, the author makes an objective assessment of the accumulated experience of planning socialist industrial cities. In the course of the analysis of functional planning development, the article compares two chronological periods. This comparison is used to assess the rate of development of the city. In addition, the publication presents conceptual schemes of territorial and functional planning for the development of a socialist city. As a result of the study of the possibilities of solutions for the urbanization of territories. One of the results of the study is the revealed relationship between the population of the city and urban infrastructure organizations. The author suggests in professional use some terms that more accurately express aspects of the planning of socialist cities in English.

Keywords: Industrial socialist city · Industrial center · Planning area · Street network · Stalinist type of building · Khrushchev type of building · Panel type of building

1 Introduction

The theory of urban planning in the USSR draws on great interest in the modern professional and scientific community. The publications in this area of scientific research are often associated with a comparison of various historical eras with the period of industrialization in Russia. One of the aspects studied is the analysis of the effectiveness of urban development management. For example, S.O. Kuznetsov, in his study of the management of the city of Moscow, examined three historical periods. As a result, the author revealed a cyclical nature in the dynamics of city development associated with variable success and the rate of accumulation of fundamental knowledge and empirical experience [1]. The study of the biographies of Soviet urban planners allows us to understand the ideas of architects and the implementation of their plans. For example, having studied the publications and biography of M.D. Zagryatskov, the author of the publications revealed the decisions he had not implemented about the effective urbanization of territories. It is interesting for modern researchers to compare the functional planning development of an

N. Vatin (Ed.): STCCE 2021, LNCE 169, pp. 101–112, 2021.
https://doi.org/10.1007/978-3-030-80103-8_11

industrial city in capitalist and socialist countries. In the publication of this orientation, it is possible to objectively determine the reasons and the level of efficiency of urban planning management [2, 3]. These publications identify the main trends that have led to certain consequences in the development of cities. The authors of these publications cite critical factors that were taken into account or not taken into account. In the study of objects (socialist cities), the identified factors of urban development make it possible to assess the effectiveness of the methods used and use the experience gained in solving modern issues of sustainable development of post-industrial cities in Russia.

Researchers are interested in the study of closed socialist cities. For example, research by L.G. Mikhailova [4], N.N. Makarova [5], Yu.A. Mylova, N.P. Kradin [6], A.A. Makhnina [7] and G.A. Reuta [8] are devoted to the planning of socialist autonomous cities of the closed type. In these publications, both the advantages and disadvantages of urban planning of the Soviet period are sharply expressed. In the study of Komsomolsk-on-Amur, Novouralsk, Magnitogorsk and Zheleznogorsk, a similar periodization of the tendencies of capital construction in socialist cities is given. The first half of the 20th century was characterized by the formation of free-line buildings and a gradual transition to perimeter-frontal ones. In the second half of the 20th century, the emergence and gradual development of typical block buildings is observed [9].

M.S. Ilchenko in his publication considers the incompleteness of functional planning solutions of the first half of the 20th century as an ambiguous factor that interferes with objectively assessing their prospects [10].

Studies devoted to the analysis of various infrastructures of industrial socialist cities give an idea of the peculiarities of the development of individual territories in the general urban infrastructure [11–13].

Studies of industrial cities in Europe and compilation with Soviet cities in the 20th century show the relative pace of urban development. This allows us to identify the key mistakes that have negative consequences in the sustainable development of cities [14–16]. Such mistakes in development planning took place both in capitalist industrial cities and in socialist industrial cities. For example, at the local level, in the management of housing estates, the community of urban users turned out to be more effective than municipal socialism. But the collapse of the city as an integral socio-territorial system gave a strong impetus for the decentralization of urban infrastructures. This contributed to the large-scale urbanization of suburban areas [2, 17, 18]. The choice of the priority «to provide residential apartments in socialist cities» or «to preserve the psychological climate and intellectual atmosphere of the capitalist city» also had an ambiguous impact on the development of socialist and capitalist cities [19, 20].

Urban planning in the USSR every year of the bygone era arouses more and more interest in the scientific community. This is confirmed by the growth of research on the architectural heritage of the Soviet period and publications on various aspects of that era. But the contemporaries of the USSR are not able to abstract themselves well enough and objectively assess the experience of their era. And the post-Soviet generation does not possess sufficiently complete reliable knowledge about the planning of industrial socialist cities.

The aim of the study is to reveal the objective experience of planning an industrial socialist city.

The hypothesis for studying changes in the dynamics of the city's population is associated with the development of urban infrastructure planning.

The task is to analyze and compare the territorial planning of a socialist city with the previous period. Transport infrastructure and capital construction are chosen for analysis. The accumulated domestic and foreign experience of industrial urban planning is described.

2 Materials and Methods

In the course of the research, cartographic materials of the Kazan city planning were used. The analysis of the urban development of the city was made on the basis of a comparison of the cartographic material of the 19th and 20th centuries. The study is associated with the Soviet period (20th century) and a comparable previous period of development (19th century). Such a comparison made it possible to determine the relative rates of the city's planning development in the Soviet period.

The study analyzed the zone of individual residential development and the zone of multi-apartment residential development. The division into categories of zones is determined by the number of storeys of buildings. When identifying the boundaries of the building, the obvious differences in the planning solutions of individual residential buildings and multi-apartment residential buildings were taken into account.

The analysis of the transport infrastructure was carried out in the following categories:

- railway transport (freight and passenger);
- river navigation (cargo and passenger);
- car roads.

The study took into account the annexation of the territory and the change in the administrative boundaries of the city. The rates of development of the indicated urban planning infrastructures have been studied.

Further, the figure shows the territorial and planning development of the city of Kazan for the historical period that preceded the period under study.

Territory. City maps in Fig. 1. illustrate the rate of urbanization of the territory in different periods of development. For a comparable period in the 19th century, the territory of the city of Kazan increased from 26.1 km^2 to 98.8 km^2. This means that the territory of the city has increased by 74%. In 1833, the territory of the Admiralty settlement was included in the city border. By 1950, a dam was erected across the mouth of the Kazanka River.

Transport in the 19th century. By 1830 the Kazan Admiralty was moved to Astrakhan. The Ust-Kazan pier continues to operate. Developed shipping ensured the prosperity of the exchange on Yarmorchnaya Square. River shipping was the most important mode of transport for Kazan until the end of the 19th century. At the end of the 19th century, plans for the construction of a large river port in Kazan were discussed. In the second half of the 19th century, most of the ships were steamers and motor ships.

A horse-drawn railway connected the Near and Far quays.

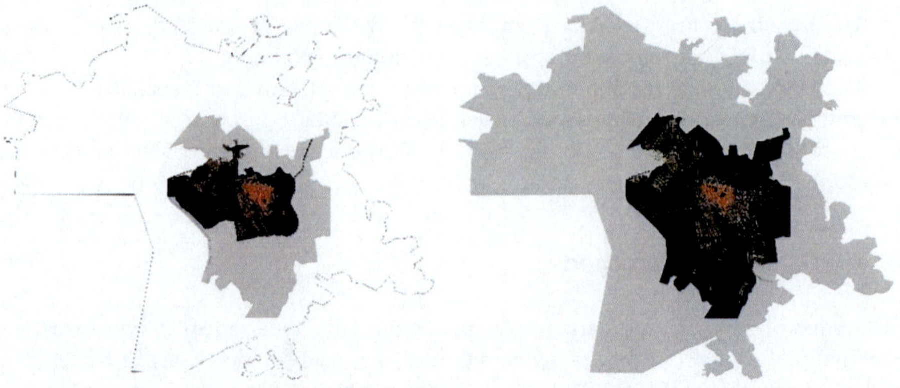

Fig. 1. Territorial development of the city of Kazan in the 19th century.

Railway. In 1880, the construction of the «Romanovsky Bridge» across the Volga was completed. In 1894 a railway station was opened in Kazan. The Krasnaya Gorka station (Yudino) was opened the same year. The main rolling stock is steam locomotives. Freight traffic prevails.

Roads. From the west, in the 19th century, it was possible to get to the city of Kazan through a ferry to Nizhnie Vyazovye. From the eastern side, the city of Kazan could be reached along the Siberian tract. From the southern side, the city could be reached along the Orenburg tract.

According to the decree of Alexander 1, since 1812, all river vessels brought cobblestones for paving the streets of the city. By the end of the 19th century, a network of streets in Kazan united the city with suburban settlements. The territory of the city was located on the left bank of the Kazanka. Suburban settlements on the right bank of the Kazanka. The city's street network at the end of the 19th century consisted of 270 streets, paths, lanes and city squares. In the central part of the city, the main streets were paved with cobblestones. The squares at the locations of the city gates were used as retail spaces.

Production. Due to the appearance of steam engines by the end of the 19th century, a significant number of industries developed on the territory of the city of Kazan and suburban settlements. The most developed were the Krestovnikov brothers' enterprise (Zakabanye), the Alafuzov factory (Admiralteyskaya Sloboda), the Powder Factory (Porokhovaya Sloboda), the Spartak shoe factory (Sukonnaya Sloboda). Some factories by the beginning of the 20th century used steam generators of electricity [13].

Capital construction. New stone houses and hotels were present and built in the city. In suburban settlements and within the city limits, a large number of one-story and two-story wooden residential buildings remained. The territory of the city itself was divided into an upper part and a lower part. The upper part was located on the upper terrace of the relief and was divided into aristocratic and merchant. In addition, the buildings of the Kazan State University, Gostiny Dvor, the building of the Alexander Passage, the Noble Assembly, the Varvara Church were on the upper terrace of the city. the Hermitage Garden, which served as a public center for walking the townspeople [12, 21, 22] was on the upper terrace. In the lower part, there were mainly trade, craft and residential quarters of the estate type. But there were stone buildings and places of worship (churches, mosques). Residential buildings were formed mainly near trade routes and squares. The type of residential development in Kazan in the 19th century was a lowercase estate. In the central part there were mainly administrative buildings. For capital buildings, a merchant style was characteristic. The height of the buildings was 2 or 3 floors. About 95% of the city's buildings were manor-based, without utilities. Landscape features (relief, water, plants) had a decisive influence on urbanization because construction technologies were not developed.

According to statistics, the population of the city in the 19th century increased from 41300 to 146495 people. This represents 71% of the original amount at the beginning of the 19th century (Fig. 2).

Fig. 2. Territorial development of the city of Kazan in the 19th century.

City maps in Fig. 1. illustrate the rate of urbanization of the territory in different periods of development. For a comparable period in the 20th century, the territory of the city of Kazan increased from 98.8 km^2 to 450 km^2. This means that the city's territory has increased by 78%. New types of transport are available in the city. The main route of trade is not shipping, but the railroad. Differentiated public transport was being developed within the city.

Transport of the 20th century.

River transport. From 1936 to 1956, the Blizhnee Ustye and Dalnee Ustye cargo and passenger docks were actively developing. However, due to the unstable water level in the Volga River and the peculiarities of the relief, the docks of the Dalniy Ustye with the adjacent trading village were flooded during the flood period. In 1957, the operating mechanized river port of the city of Kazan was put into operation. The construction costs of the new river port was fully paid off in 6 years. In 1963 the river port became one of the leading enterprises on the Volga. River shipping by 1990 was supplemented by suburban shipping, which connects the city with the right bank of the Volga River. This made it possible to urbanize the territory of the right bank of the Volga River as summer cottages for the townspeople. Suburban shipping traffic was formed from the Kamsky Ustye to the mouth of the Sviyaga River. For the transportation of goods, bulk liquid and self-propelled barges were used. The passenger ships used motor ships, high-speed ships («Raketa», «Meteor»), river trams («Om», «Moscow-74»). The use of modern technologies for that time made remote settlements and inaccessible islands (Kamskoye Ustye, Zaton, Sviyazhsk, Verkhniy Uslon, Matyushino, Tashevka) affordable for a large number of townspeople. The creation of a transport connection along the Volga contributed to the development of these settlements as industrial centers. It was planned to carry out cargo transportation of food and mineral resources from the settlements of the Trans-Volga region using river transport. In terms of cargo logistics, by the end of the 20th century, the Kazan river port was of great importance on the Volga River. This is due to the implementation of the largest hydrological and energy project on the Volga River. The implementation of this project made it possible to connect the city of Kazan with the Caspian and Baltic Sea through cascades of hydroelectric power stations and canals.

Railway. In 1912–1941 the construction of a freight railway through the Vosstanie station. Trains on the Sergach – Kazan – Sarapul section were fully opened in 1918. By the end of the 20th century, the rolling stock of the railway was represented by diesel locomotives, diesel engines and electric locomotives. Railway transport had a developed typology – it was freight and passenger transportation. Passenger transportation by rail was also divided into suburban and long-distance trains. Suburban railway communication made it possible to connect Kazan with Zelenodolsk and Izhevsk high-speed transport routes. This had a positive effect on the development of transport accessibility of the city's territory for external passenger traffic.

Roads. From the west until 1990 it was possible to get there only using the ferry crossing Nizhnie Vyazovye – Zelenodolsk. Highways to the Urals and Mamadysh were being developed in the north-east and east directions. In the southern overland direction in the settlement of Sorochy Gory, a crossing was being created across the Kama River. In 1981, the construction of a bypass road from the north of the city of Kazan began. Thus, by the end of the 20th century, the external relations of the city were being actively developed in three equivalent directions: river shipping, railways, and highways.

In the 20th century, capital concrete bridges appeared in the city of Kazan across the Kazanka River. The Kirov dam provided rail and road connections with the city. Leninskaya dam – road connection with the city and the Moscow region (Leninsky district). The third transport dam (Sovetskaya dam) was a road connection between Sovetsky and Novo-Savinovsky (Eastern District) districts. The pontoon bridge connected Amirkhan and Tolstoy streets.

The network of streets grew within a radius of 20 km. from the central part of the city. The urbanization of the territory of the right bank of the Kazanka River began. By the end of the 20th century, the network of streets in Kazan consisted of about 2000 streets, paths, lanes and city squares. In 1977, the construction of the first transport hub in the city began. The lower level of the transport hub was supposed to connect the network of streets on the right and left banks of the Kazanka River. At the top level, services were organized for passengers of all types of urban public transport (bus, trolleybus, tram, commuter train).

An asphalt concrete road surface appeared in the improvement. But by the end of the 20th century in Kazan, the highways of city and regional significance mainly had two lanes with asphalt concrete pavement. More than half of the city streets had an asphalt concrete road surface. But the improvement and maintenance of pedestrian links remained at a low level. Not most of the sidewalks were park pedestrian.

2.1 Capital Construction. Administrative Division of the City into Districts

In 1934 the eastern, northeastern region of the city was allocated to the Soviet district of the city. Since 1947, the industrial center of the region has been the Kazan Compressor Plant. In 1951, the industrial center of the Soviet region was supplemented with the «Electronic computers» plant and the «Typewriter» plant. Jobs at these industrial giants provided an increase in urban residents and the demand for new residential complexes. The diagram shows that the development was dominated by apartment buildings built in 1964–1980. This means that the construction of the area of the Soviet district of the city began in the second half of the 20th century. Village settlements, together with natural outskirts, which were located in the area of the Soviet district of the city, were preserved and were not rebuilt.

Since 1920, the Volga region has been formed in the southern part of the city. A vast area of the industrial center can be distinguished here. It includes a chain of different production targets. The industrial center of the Volga region began a cargo river port and completed a complex of urban treatment facilities. Development in its area began to form not much later than in the Soviet district. This was due to the stabilization of groundwater level differences due to the construction of protective structures of the Kuibyshev reservoir. This construction made it possible to eliminate a large area of unstable flooded landforms in the city.

Since 1994, sections of the Novo-Savinovsky district have been formed in the northern part of the city. This is the youngest and most compact district of the city on the right bank of the Kazanka River. Novo-Savinovsky district is the first district of the city, which is fully built according to the planning project. There is no clear industrial center in this area. There are more dense residential complexes, more spacious areas of parks and boulevards. But the layout and area of residential complexes does not correspond to the comfortable perception of the surrounding space by the townspeople.

Since 1960, the industrial center of the Moscow region began to form. It was based on the city-forming enterprise Kazanorgsintez. The location of this large chemical production plant was due to the presence of railway sidings, thermal power plants and the possibility of forming a buffer sanitary-protective area of the city forest park. In parallel with the construction of the plant, residential complexes were built in the Zhilploshchadka microdistrict.

Since 1940, the formation of an urban-type settlement for workers in the aircraft industry began along the freight track of the railway. The industrial center of the aircraft building region was the 22nd plant and engine building plant. Residential complexes here were built in the 50s. The buildings are represented by the «Stalinkas». Hypothetically, for planning districts of the Stalinist type, the construction of 3 or 5-storey residential buildings with 2 and 3 room apartments, 5 houses per quarter, is typical. On the outskirts of the urban area, dense development of individual residential buildings remained. In the second half of the 20th century, the Aviation District remained the largest industrial area of the city. But the railroad infrastructure has isolated the area from the rest of the city. From the city it was possible to get here only along one street of the Dekabristov [21].

In the western part of the city, the Kirovsky district is located (1920). There, the center of capital construction is the Powder Sloboda and the Fortieth Plant. Powder settlement – in the first half of the 20th century, a completely autonomous settlement on the outskirts of the city. By the end of the 20th century, this region was one of the largest industrial centers of the city of Kazan. In this industrial center of the Kirovsky district of the city, in addition to the gunpowder factory, there are also the Flax Factory, the Hammer and Sickle factory, and the Helicopter Factory. Admiralteyskaya Sloboda and Porokhovaya Sloboda are fully integrated into the network of city streets.

The historical center of the city is located in the Vakhitovsky district (Baumansky district). There, to the existing factories and plants were added: 1924 the Zarya confectionery factory in Sukonnaya Sloboda, 1905 the Petzold factory (later the Krasny Vostok), the Spartak shoe factory, 1928 The Fur Factory (later the Melita factory).

In the first half of the 20th century, socialist townships for workers of defense plants were built along the railway freight track. One of the largest Sotsgorod in the Aviastroitelny city district and the urban-type settlement Derbyshki in the Soviet city district. State educational, cultural, shopping and entertainment establishments (TSUM, Detsky Mir) were being built in the city. In the central part of the city, an opera and ballet theater, a conservatory, a philharmonic society, university complexes, and libraries were being erected. In the second half of the 20th century, the typical construction of residential buildings made it possible to build up vast annexed territories with high-rise buildings from 5 to 10 floors. A well-developed engineering infrastructure was being formed. With the development of construction technologies, the boggy areas of the right bank of the Kazanka River were urbanized.

According to statistics, the population of the city for the 20th century increased from 206562 to 1105000 people. This is 81% of the original amount at the beginning of the 20th century.

3 Results

In this study, the author confirms the cyclical nature of the consequences of city development management proposed by S.O. Kuznetsov [1].

In the 19th century, external transport links of the city of Kazan were represented mainly by river shipping. Not significant, but year-round external communication, in terms of freight and passenger turnover, was supported by roadways. In the 20th century, the external transport links of the city received a developed typology. The connection with other cities and regions was supplemented by rail transport, a full-fledged road infrastructure and air traffic.

Intra-city pedestrian and transport connections were not differentiated. The central shopping plazas and city streets leading to these plazas were paved with cobblestones. In the 20th century, new types of transport appeared in the city of Kazan: suburban electric trains, trams, trolleybuses. This development of public transport infrastructure significantly increased the accessibility of urban areas for citizens. Projects of street profiles made it possible to differentiate pedestrian, transport and engineering communications. An integrated approach to the development of a network of city streets made it possible to form their hierarchy and calibrate according to the individual characteristics of a city street in accordance with its purpose in the hierarchy.

In the 19th century, the residential buildings of the city were predominantly of an individual, private character. The construction of a new residential building was an urban event. Capital development was of a line type, that is, it was carried out along the existing streets. In the 20th century, an integrated approach and phased planning of urbanization of urban areas was used. This made it possible to efficiently use the city's territory. The development of technology increased the rate of urbanization and the efficiency of urbanization of the territory. In addition, technological progress made accessible areas that were considered uninhabitable until the 20th century. Rationing in urban planning of the 20th century made it possible to form a comfortable urban environment, taking into account many factors not related to the architecture of buildings and structures, but determining the comfort of the urban environment. These are sanitary standards, social aspects, climatic factors, and more. An integrated approach to planning urban areas became one of the fundamental prerequisites for the differentiation of their area of architecture in a separate scientific direction «Urban Planning» at the end of the 20th century.

The graph below clearly shows the growth rate of the population of the city of Kazan. The first row of pillars reflects the population dynamics in the 19th century. The second row of pillars reflects population dynamics in the 20th century. In percentage terms, the territory of the city in the 19th and 20th centuries increased by 74% and 78%. This means that the territorial development in both periods is almost the same. If we consider the percentage by population, then in the 19th and 20th centuries, the city's population increased by 71% and 81%. Thus, the relationship between the population of the city and the quality of life becomes obvious (Fig. 3).

To some extent, the development of all urban infrastructures was positively influenced by the disintegration of the city as a single integral socio-territorial system. This process of disintegration made possible the territorial development of the socialist city due to the annexation of already urbanized industrial centers (Sotsgorod, urban settlement Derbyshki and other «Workers' settlements») in the suburbs.

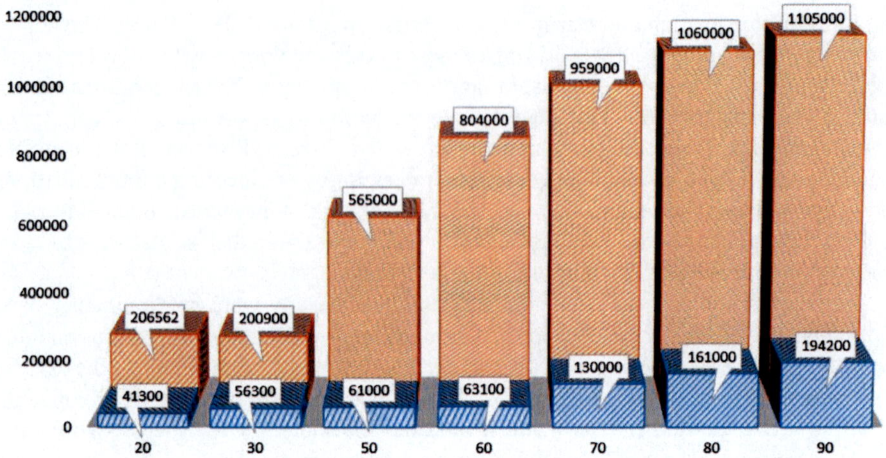

Fig. 3. Dynamics of the population of the city of Kazan in the 19th and 20th centuries.

4 Discussion

Of course, the scientific and technological revolution at the beginning of the 20th century influenced all areas of life. The rapid development of transport became a positive result. In the 20th century, transport accessibility determined the development of existing and the creation of new cities and settlements in locations previously considered inaccessible [10]. However, villages located not near transport routes (railways, highways) were on the verge of extinction.

The city-forming and city-forming industrial centers made it possible to decentralize the city as a single social-territorial center. This decentralization contributed to the development of intra-city connections and increased mobility of the urban population by the end of the 20th century. But the ambitions of providing the population with residential apartments and totalitarian government at all levels prevented the development of architectural identity in the socialist cities of the 20th century.

Forcing a systematic approach to urban planning made it possible to test tools and methods for managing agglomerations of autonomous suburban settlements for workers in a unified structure of functional planning of the city. However, insufficient attention to changing the paradigm of urban development with a predominance of road and high-speed transport made Soviet transport infrastructures unpromising.

Ambiguous in the Soviet urban planning and the use of landscape resources [2, 19] the formation of a typology of landscaping systems, innovative approaches to landscape construction and active borrowing of foreign successes, contributed to a rapid increase in the level of improvement of urban spaces, the formation of a normative technical base for landscape construction. But the war and post-war period of decline objectively reflects the relationship between the well-being of cities and the level of improvement in them.

Paradoxically, in the era of technological progress in the country where the first man visited space, high technical capabilities in the domestic and civil spheres were practically not used [5, 8]. For example, in the first half of the 20th century, most of the UK was already filmed by aerial photography for conceptual planning and land surveying [8]. While in the USSR, it was only in the second half of the 20th century that aerial photography began to be used in forest taxation.

5 Conclusion

More detailed studies on the development of socialist cities will help to identify the reliable reasons for the current problems of sustainable urban development, and give full-fledged characteristics of an industrial city. A detailed analysis of domestic and foreign experience will reveal «points of growth» and «pain points» in the urban planning of technologically advanced cities of the future. A correct assessment of the accumulated experience will exclude the repetition of «mistakes» in urban planning management with long-term negative consequences.

References

1. Kuznetsov, S.O.: Analysis of activities for the management of the urban planning process in Moscow early. In: XVIII-Late XX Century Academia, Architecture and Construction, vol. 1, pp. 29–38 (2020)
2. Kosenkova, Yu.L.: City as a collective dwelling: the mission of professor M.D. Zagryatskov 1930s. Acad. Archit. Constr. **4**, 40–49 (2020)
3. Kotova, E.S.: Peculiarities of representation of urban development problems in the USSR and Great Britain 1950-1970s. Mod. Hist. Russia **3**(14), 182–192 (2015)
4. Mikhailova, L.G.: Architecture of Novouralsk 1940–1950s. Acad. Bull. UralNIIproekt RAASN, **3**(46), 53–58 (2020)
5. Makarova, N.N.: Housing construction in Magnitogorsk in 1953-1964. Bull. Udmurt Univ. Ser. Hist. Philol. **1**(29), 37–45 (2019)
6. Mylova, Yu.A., Kradin, N.P.: The specifics of the formation of the architectural and urban planning environment of the Soviet city in the middle of the twentieth century (on the example of Komsomolsk-on-Amur, Khabarovsk Territory). Balandinsky Read. **1**(9), 359–367 (2014)
7. Makhnin, A.A.: Formation and development of the architectural environment of Rybinsk in the late 1950s-1980s. Bull. Volgograd State Univ. Archit. Civ. Eng. Ser.: Build. Archit. **22**(41), 138–148 (2011)
8. Reut, G.A.: Zheleznogorsk (Krasnoyarsk-26) is a secret urbanization in the closed city of historical courier, **4**(6), 143–152 (2019)
9. Veretennikov, D.B.: Structural and functional aspects of the formation of the system of Togliatti centers. In: The Collection: Post-Industrial Environment of Russian Megacities. Collection of Articles Based on the Materials of a Scientific and Technical Conference with International Participation, Samara, pp. 27–33 (2020)
10. Ilchenko, M.S.: An unfinished project as a form of perception of Soviet urban planning in the 1920s-1930s: the experience of socialist cities. Siberian Hist. Res. **2**, 56–79 (2017)
11. Sementsov, S.V., Akhmedova, E.A., Volkov, V.I.: Water spaces as the main public spaces of urban planning compositions and functional systems of the largest cities. Water Ecol. Prob. Solut. **4**(72), 86–94 (2017)
12. Zhandarova, A.A., Denisenko, E.V.: Historical and theoretical aspects of the development of bioarchitecture. News KSUAE **1**(47), 18–25 (2019)

13. Zakieva, L.F., Dembich, A.A.: Features of urban planning placement of medical institutions in large urban agglomerations. News KSUAE **2**(48), 127–133 (2019)
14. Suel, E., Polak, J.W., Bennet, J.E., Ezzati, M.: Measuring social, environmental and health inequalities using deep learning and street imagery. Sci. Rep. **9**, 6229–6239 (2019). https://doi.org/10.1038/s41598-019-42036-w.17
15. Saunders, P., Campbell, P., Webster, M., Thawe, M.: Analysis of small area environmental, socioeconomic and health data in collaboration with local communities to target and evaluate 'triple win' interventions in a deprived community in Birmingham, UK. Int. J. Environ. Res. Public Health **16**(22), 4331–4346 (2019). https://doi.org/10.3390/ijerph1622433.18
16. Adedeji, T., Proverbs, D., Xiao, H., Cobbing, P., Oladokun, V.: Making Birmingham a flood resilient city challenges and opportunities. Water **11**, 1699–1716 (2019). https://doi.org/10.3390/w11081699.19
17. Leach, J.M., Mulhall, R.A., Rogers, C.D.F., Bryson, J.R.: Reading cities: developing an urban diagnostics approach for identifying integrated urban problems with application to the city of Birmingham UK. Cities **86**, 136–144 (2019). https://doi.org/10.1016/j.cities.2018.09.012.20
18. Lee, S.E., Quinn, A.D., Rogers, C.D.F.: Advancing city sustainability via its systems of flows: the urban metabolism of Birmingham and its hinterland. Sustainability **8**(3), 220–244 (2016). https://doi.org/10.3390/su8030220.21
19. Azevedo, J.A., Chapman, L., Muller, C.L.: Quantifying the daytime and night-time urban heat island in Birmingham, UK: a comparison of satellite derived land surface temperature and high-resolution air temperature observations. Remote Sens. **8**, 153–170 (2016). https://doi.org/10.3390/rs8020153.22
20. Warren, S., Jones, P.: Local governance, disadvantaged communities and cultural intermediation in the creative urban economy. Environ. Plan. C: Gov. Policy **33**(6), 1738–1752 (2015). https://doi.org/10.1177/0263774X15614661
21. Gataullina, A.A., Novikov, S.V.: Architectural classification of the inner-quarter spaces of the Kazan social city. News KSUAE **4**(50), 89–99 (2019)
22. Grishina, M.: Testing the methods of reconstruction of the spatial solution of the pedestrian street. IOP Conf. Ser.: Mater. Sci. Eng. **753**, 042069 (2020). https://doi.org/10.1088/1757-899X/753/4/042069

Prediction of Creep for a Reinforced Concrete Beam Strengthened with an External Reinforcement System Using the Stepped Isothermal Method

Almaz Shakirov(✉) and Alfred Sulejmanov

Kazan State University of Architecture and Engineering, 420043 Kazan, Russia

Abstract. The principle of analogies is known to be true for most polymers. This principle makes it possible to shorten the duration of an experiment for durability prediction by changing the intensity of various factors without affecting rheological properties of a polymer. Specifically, temperature (temperature-time superposition) or humidity (humidity-time superposition) are changed. There is an impressive amount of studies about successful application of the principle of analogies for prediction of polymer durability. However, the absence of legal framework makes it legally impossible to use this method for prediction of durability of external reinforcement systems. In the accelerated weather test laboratory of Kazan State University of Architecture and Engineering, a number of tests was performed, which were based on the principle of analogies, to predict the durability of structures strengthened with an external reinforcement system. This method was studied in regard to its suitability for reinforced concrete beams strengthened with an external reinforcement system. The method was developed to predict the durability of these structures for the entirety of their operating period by performing short-term laboratory tests.

Keywords: CFRP · Epoxy · Concrete · Creep test · Time-temperature superposition principle · Stepped isotherm method

1 Introduction

One of the important factors that affect durability of reinforced concrete structures strengthened with external reinforcement systems with epoxy binders is creep [1–3]. The existing studies discuss development of creep under constant loading and register its stabilization within 120 days. However, the development of creep may be of a different nature under conditions of varying temperature factors. Thus, this raises the question about prediction of creep for structures strengthened with an external reinforcement system under specific temperature conditions [4, 5].

The principle of time-temperature superposition is known to be true for most polymers. According to this principle, varying temperature does not affect rheological properties of a polymer. However, it makes it possible to shorten a time interval, which

N. Vatin (Ed.): STCCE 2021, LNCE 169, pp. 113–121, 2021.
https://doi.org/10.1007/978-3-030-80103-8_12

is useful when applying this principle for investigation and prediction of stress-strain behaviour of polymer materials [6–8].

The principle of time–temperature superposition is a well-established method for such materials as polyurethane [9–11], asphalt [12–15] and bitumen [16–18], rubber [19], geopolymers [20]. Also, there are studies that prove the adequacy of the time–temperature superposition for epoxy binders [21, 22].

However, it is more convenient to predict creep of large structures using the stepped isothermal method (SIM) [23] based on the principle of time–temperature superposition. The papers [24, 25] provide the results of SIM application for prediction of creep of geoplolymers. The advantage of this method is that it makes it possible to predict creep of a polymer material for up to 100 years within a short time (24 h).

This paper investigates the possibility of using the stepped isothermal method to predict creep of reinforced concrete beams strengthened with an external reinforcement system.

2 Materials and Methods

2.1 Making Samples

The test beam samples were made from B15 concrete, section of 120x140 (h) mm, total length of L = 1000 mm. In the lower area, the beams were strengthened with two bars Ø8 A400 with a length of 1000 mm.

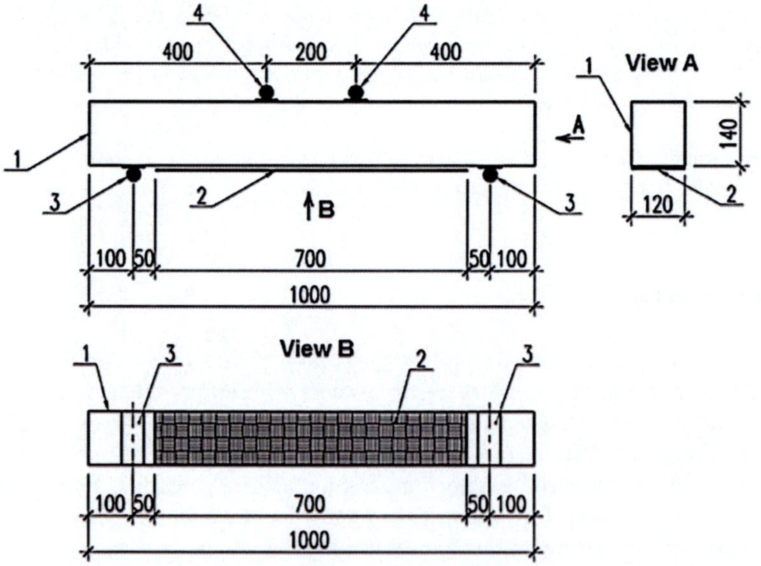

Fig. 1. Layout of beam reinforcement with carbon fiber cloths: (1) reinforced concrete beam; (2) polymer composite; (3) holding supports; (4) loading supports.

Longitudinally, the beams were strengthened with one layer of FibARM Tape-230 carbon fiber tape with a width of 120 mm, with the longitudinal cloths being 50 mm from the supports. Epoxy compound FibArm Resin 230 + was used for reinforcement. The layout of reinforcement and supports is shown in Fig. 1.

Load carrying capacity of the reinforced beams was determined by means of calculations according to the instructions specified in SP 164.1325800.2014 [26]. Breaking load of the beam is $P_{break} = 40$ kN.

2.2 Equipment and Instruments

Load frame 1 was made to apply load to beams (Fig. 2). The beams rested on cylindrical supports 3, 4 (Fig. 1) via the metal plates. Loading was performed with hydraulic jack 3 (Fig. 2) via the traverse beam with two-point distribution of forces in the beam. Figure 2 shows the loading layout for the beam in the test bench.

The load P on the traverse beam was measured with radial pressure gauge 9 (Fig. 2) on pump 6 (Fig. 2).

The deflections in the middle of the beam were measured with dial gauges ICh10–0.01 5 (Fig. 2).

Temperature exposure was simulated in thermal vacuum chamber STBV-1000-IV (ILKA, GDR) 2 (Fig. 2), directional infrared heater 4 (Fig. 2) was used to provide uniform heating of the composite. The temperature of the composite was monitored using thermocouples 7 (Fig. 2) which were installed between the surface of concrete and composite.

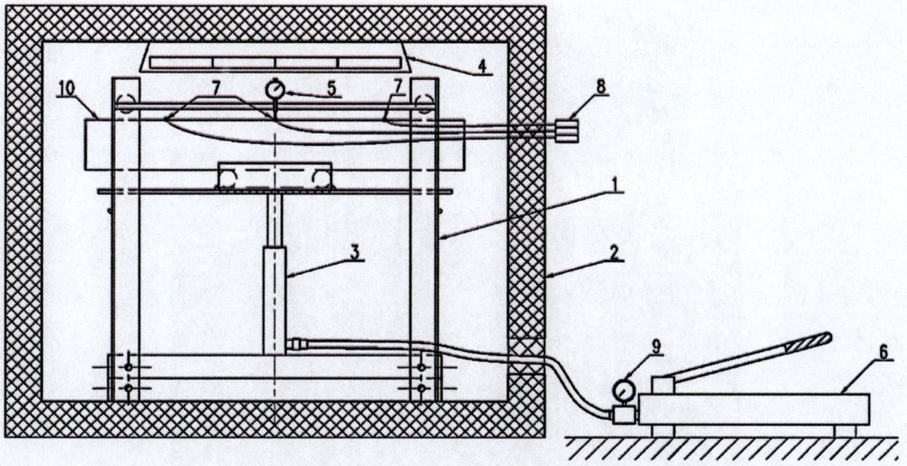

Fig. 2. Layout of the loading system: (1) load frame; (2) climate chamber; (3) hydraulic jack; (4) infrared heater; (5) dial gauge; (6) hydraulic pump; (7) thermocouple installation points; (8) bank of thermocouple indicators; (9) pressure gauge; (10) test sample.

2.3 Test Procedure

The time-temperature relationship for creep of the reinforced concrete beam strengthened with the external reinforcement system was studied under conditions of constant load at gradually changing temperature. The test load was 70% of the breaking load. The duration of one temperature cycle was 2 h. The initial temperature was 25 °C, the final temperature was 55 °C, the temperature interval was 5 °C. The total experiment duration was 14 h.

The experiment algorithm is as follows:

1. The test bench is put in the thermal vacuum chamber (Fig. 3), and the initial temperature of the experiment is set.
2. After the temperature reaches the constant level and the beam is heated the initial deflection values are recorded.
3. The sample is loaded until the test load is achieved, and the beam deflection is recorded.
4. Then every change in deflection during the temperature cycle is recorded.
5. After the time of the temperature cycle runs out the temperature is increased by 5 °C.
6. The sample is heated until the current cycle temperature is reached.
7. Up to the end of the experiment, the steps 4–6 are repeated.

The data of thermal creep at gradually changing temperature are used to plot a master curve to predict deflections of the structure for the required operating temperature over the long term.

Fig. 3. Test bench.

Master curve plotting order (Fig. 4):

1. Creep curves for different temperatures are plotted in one plot (Fig. 4a). Time (t, s) is measured on the horizontal axis, deflection (f, mm) is measured on the vertical axis at different temperatures.
2. Then the creep curve is divided into separate segments, where each of them corresponds to a certain temperature step, and is replotted on coordinates creep modulus *E*, kN/m – log time l_{nt}, s (Fig. 4b).

Creep modulus is calculated using the following formula:

$$E = P/f(t) \tag{1}$$

where *P* is the test load;
f(t) is the deflection at a given time.

3. The creep modulus curves for different temperatures are shifted along the time axis to align them with each other with partial overlap (Fig. 4c).
4. The creep modulus-log time curve is replotted on coordinates «deflection-log time» (Fig. 4d).

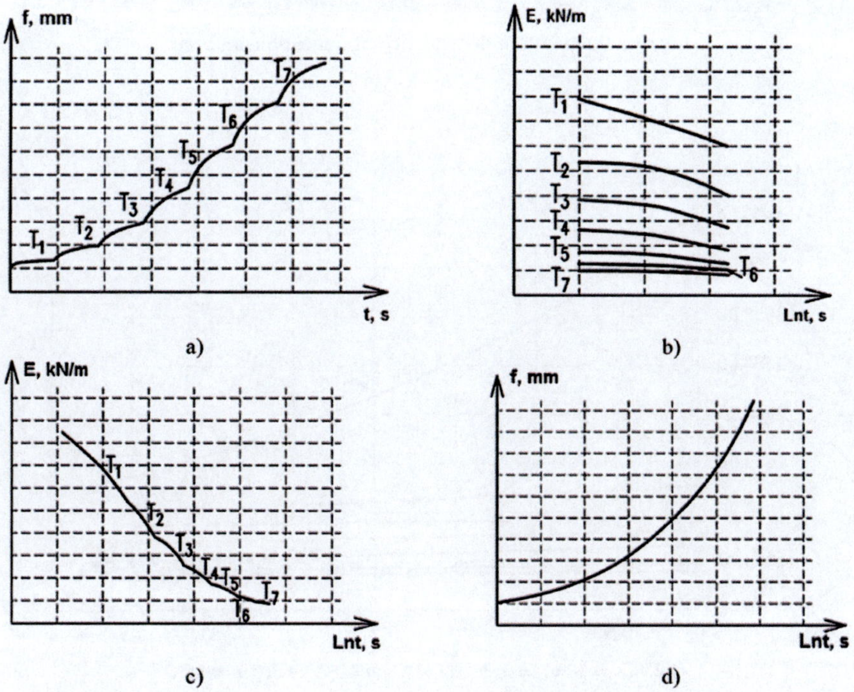

Fig. 4. Plotting of a master curve for the stepped isothermal method.

The resulting master curve is a predicted long-term creep curve for normal temperature.

3 Results and Discussions

Figure 5 shows the results of the experimental investigation of creep of the reinforced concrete beam strengthened with the external reinforcement system using the stepped isothermal method.

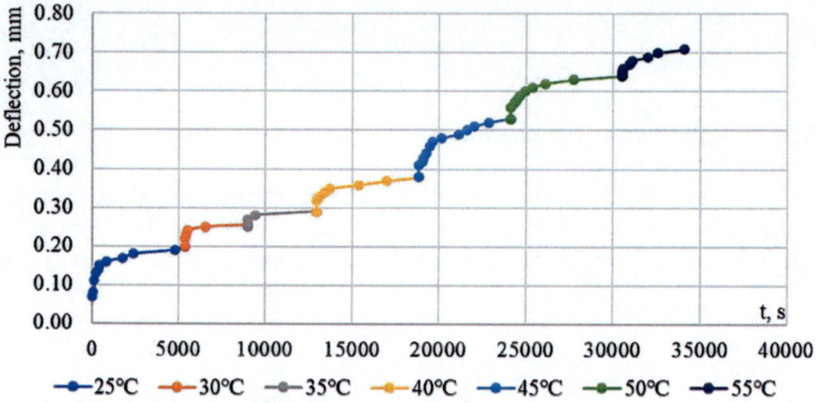

Fig. 5. Temperature dependence of deflections in time.

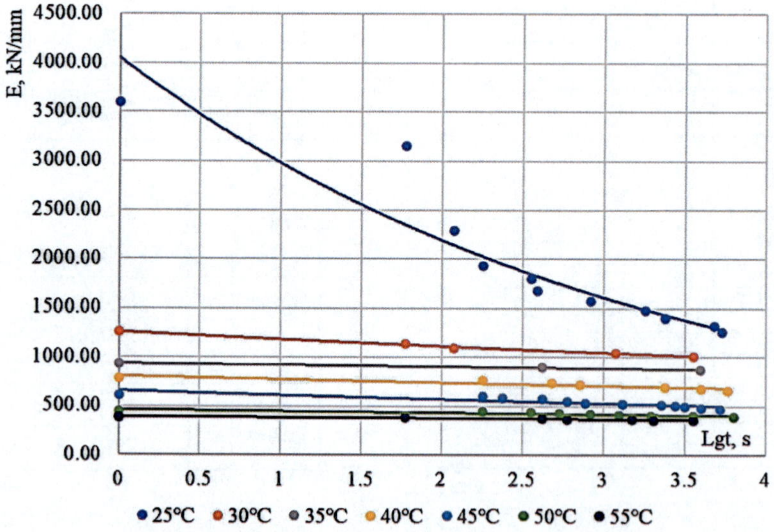

Fig. 6. Relationship between creep modulus and time.

Let us replot the resulting creep curves for each temperature cycle on coordinates «creep modulus-log time» (Fig. 6). It can be seen from the graph in Fig. 6 that the resulting curves are similar. It indicates that the test sample behaves as a thermorheologically simple system within the specified temperature ranges. Thus, the plots show that rheology of the materials is not affected within the studied temperature range.

To construct a prediction master curve for the creep modulus, we shift the curves of the creep modulus until they coincide one after another (Fig. 7). Next, we rearrange the creep modulus curves into the creep versus time curve, the resulting data were used to plot the master creep curve (Fig. 8).

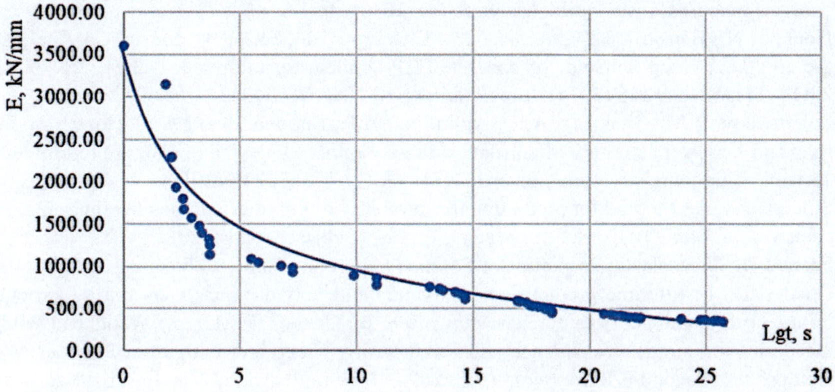

Fig. 7. Master creep modulus curve.

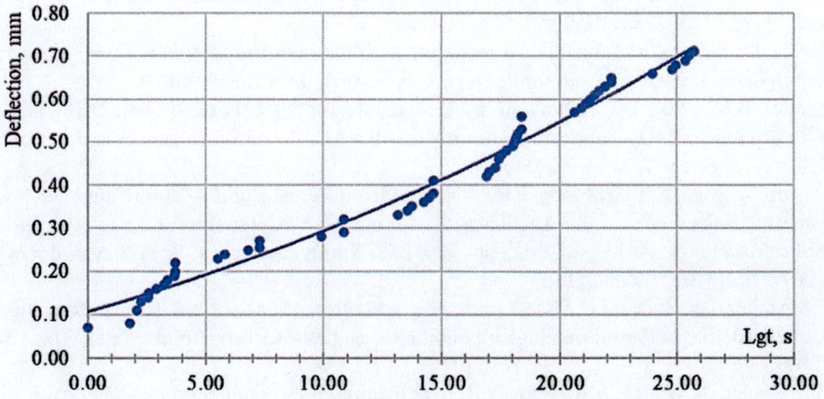

Fig. 8. Master creep curve.

4 Conclusions

The experimental analysis showed that the stepped isothermal method could be used to evaluate creep of a reinforced concrete beam strengthened with an external reinforcement system. In future, the designed method can be used for development of adhesives for external reinforcement systems. Importantly, this method takes into consideration the combined action of a system which consists of adhesive, concrete, and carbon fiber.

References

1. Hadjazi, K., Sereir, Z., Amziane, S.: Creep response of intermediate flexural cracking behavior of reinforced concrete beam strengthened with an externally bonded FRP plate. Int. J. Solids Struct. **94–95**, 196–205 (2016)
2. Al Chami, G., Thériault, M., Neale, K.W.: Creep behaviour of CFRP-strengthened reinforced concrete beams. Constr. Build. Mater. **23**(4), 1640–1652 (2009)
3. Houhou, N., Benzarti, K., Quiertant, M., Chataigner, S., Fléty, A., Marty, C.: Analysis of the nonlinear creep behavior of concrete/FRP-bonded assemblies. J. Adhes. Sci. Technol. **28**(14–15), 1345–1366 (2014). https://doi.org/10.1080/01694243.2012.697387
4. Suleimanov, A.M., Shakirov, A.R., Agliullina, A.F., Starovoitova, I.A.: Research of short-term and long-term strength of adhesive adhesive joints for device of external reinforcement systems of building structures. Izvestia KSUAE **4**(46), 309–318 (2018)
5. Suleimanov, A.: Method for predicting the durability of polymer-composite reinforcement in cement concrete. IOP Conf. Ser.: Mater. Sci. Eng. **890**, 012100 (2020)
6. Ferry, J.D.: Viscoelastic Properties of Polymers, 3rd edn. Wiley, Hoboken (1980)
7. Guedes, R.M.: Lifetime prediction of polymers and polymer matrix composite structures: failure criteria and accelerated characterization. In: Guedes, R.M. (ed.) Woodhead Publishing Series in Composites Science and Engineering, Creep and Fatigue in Polymer Matrix Composites (Second Edition), vol. 9, pp. 269–301. Woodhead Publishing (2019)
8. Nakada, M.: Accelerated testing methodology for long-term creep and fatigue strengths of polymer composites. In: Guedes, R.M. (ed.) Woodhead Publishing Series in Composites Science and Engineering, Creep and Fatigue in Polymer Matrix Composites (Second Edition), vol. 11, pp. 325–348 (2019)
9. Ionita, D., Cristea, M., Gaina, C.: Prediction of polyurethane behaviour via time-temperature superposition: meanings and limitations. Polym. Test. **83**, 106340 (2020)
10. Mahdi, E.M., Tan, J.-C.: Dynamic molecular interactions between polyurethane and ZIF-8 in a polymer-MOF nanocomposite: microstructural, thermo-mechanical and viscoelastic effects. Polymer **97**, 31–43 (2016)
11. Kayumov, R.A., Shakirzyanov, F.R.: Large deflections and stability of low-angle arches and panels during creep flow. In: Altenbach, H., Eremeyev, V.A., Igumnov, L.A. (eds.) Multiscale Solid Mechanics. ASM, vol. 141, pp. 237–248. Springer, Cham (2021). https://doi.org/10.1007/978-3-030-54928-2_18
12. Hernandez-Fernandez, N., Ossa-Lopez, A.: Validation of partial time-temperature superposition principle in thermorheologically complex asphalts. Constr. Build. Mater. **276**, 122224 (2021)
13. Sudarsanan, N., Fonte, B.R., Kim, Y.R.: Application of time-temperature superposition principle to pull-off tensile strength of asphalt tack coats. Constr. Build. Mater. **262**, 120798 (2020)
14. Luo, R., Hou, Q.: Development of time-temperature-humidity superposition principle for asphalt mixtures. Mech. Mater. **156**, 103792 (2021)
15. Chen, Z., Zhang, H., Duan, H., Wu, C.: Determination of time-temperature superposition relationship of SBS modified asphalt based on special rheological phenomenon caused by SBS-formed structure in asphalt matrix. Constr. Build. Mater. **260**, 119835 (2020)

16. Chen, Z., Zhang, H., Liu, X., Duan, H.: A novel method for determining the time-temperature superposition relationship of SBS modified bitumen: effects of bitumen source, modifier type and aging. Constr. Build. Mater. **280**, 122549 (2021)
17. Cuciniello, G., Leandri, P., Polacco, G., Airey, G., Losa, M.: Applicability of time-temperature superposition for laboratory-aged neat and SBS-modified bitumens. Constr. Build. Mater. **263**, 120964 (2020)
18. Miglietta, F., Tsantilis, L., Baglieri, O., Santagata, E.: A new approach for the evaluation of time–temperature superposition effects on the self-healing of bituminous binders. Constr. Build. Mater. **287**, 122987 (2021)
19. Peng, Q., Zhu, Z., Jiang, C., Jiang, H.: Effect of stress relaxation on accelerated physical aging of hydrogenated nitrile butadiene rubber using time-temperature-strain superposition principle. Adv. Ind. Eng. Polym. Res. **2**(2), 61–68 (2019)
20. Zornberg, J., Byler, B., Knudsen, J.: Creep of geotextiles using time–temperature superposition methods. J. Geotech. Geoenviron. Eng. **130**(11), 1158–1168 (2004)
21. Krauklis, A.E., Akulichev, A.G., Gagani, A.I., Echtermeyer, A.T.: Time – temperature – plasticization superposition principle: predicting creep of a plasticized epoxy. Polymers **11**, 1848 (2019)
22. Ghosh, S., Rajesh, P., Srikavya, B., Rathore, D., Prusty, R., Ray, B.: Creep behaviour prediction of multi-layer graphene embedded glass fiber/epoxy composites using time-temperature superposition principle. Compos. Part A: Appl. Sci. Manuf. **107**, 507–518 (2018). https://doi.org/10.1016/j.compositesa.2018.01.030
23. Achereiner, F., Engelsing, K., Bastian, M., Heidemeyer, P.: Accelerated creep testing of polymers using the stepped isothermal method. Polym. Test. **3**(32), 447–454 (2013)
24. Koerner, R.M., Koerner, G., Hsuan, Y.: Creep tension testing of geosynthetics. GSI White Paper 29, July 26, Grace (2014)
25. ASTM D6992 – 16: Standard test method for accelerated tensile creep and creep-rupture of geosynthetic materials based on time-temperature superposition using the stepped isothermal method
26. SP 164.1325800.2014: Strengthening of reinforced concrete structures by FRP composites (2015)

Economic Aspects of Infrastructure Projects Implementation in Towns and Medium-Sized Cities

Yulia Medyanik(✉), Elvira Shagiakhmetova, Liliya Gimadieva, and Dilyara Vakhitova

Kazan State University of Architecture and Engineering, 420043 Kazan, Russia

Abstract. The state of the infrastructure is a reflection of the socio-economic development and investment attractiveness of the city. However, the decision on attracting investments for the implementation of infrastructure projects should be made taking into account the indicators of the project's financial stability and possible negative scenarios. The aim of the research is to assess the investment attractiveness of the infrastructure project in the city of Leninogorsk (Russia, the Republic of Tatarstan) and to model the risks of its implementation. The following methods were used in the calculations: the method of calculating the net present value of the project (NPV), the method of calculating the discounted payback period (PBP), the method of calculating the internal rate of return (IRR). A two-parameter sensitivity analysis was used for risk simulation of the investment project. The net present value of the project (NPV) and the internal rate of return (IRR) were taken as the result indices. It was concluded that long payback and low financial sustainability make infrastructure projects in towns and medium-sized cities unattractive for private investors. The simultaneous increase in the discount rate and the investment costs of the project has the greatest influence on the NPV indicator. The results can be used in the strategy for the development of infrastructure projects in towns and medium-sized cities.

Keywords: Infrastructure facilities · Construction · Towns and medium-sized cities · Economic efficiency of investments

1 Introduction

Towns and medium-sized cities make up 84.5% of the total number of all urbanized areas in Russia. They are home to more than 26 million people (about 18% of the population). Such cities play an important role in the socio-economic development of the country and reflect not only a separate municipality, but also the country as a whole [1]. The development of social infrastructure is one of the most important problems faced by the majority of towns and medium-sized cities. The economy of small towns lacks the social focus. Furthermore, social infrastructure and urban environment have unsatisfactory quality which do not allow ensuring a decent standard of living for the

N. Vatin (Ed.): STCCE 2021, LNCE 169, pp. 122–134, 2021.
https://doi.org/10.1007/978-3-030-80103-8_13

population. Consequently, improving the comfort of the urban environment and creating the mechanisms for its formation is a priority area in the development of modern Russia.

Programmes for the development of technical infrastructure are being implemented, the innovative infrastructure for conducting business is being created [2], and the sports infrastructure is developing rapidly in the Republic of Tatarstan (Russia). At the same time, a distinguishing feature is the discrepancy between the rates of the social infrastructure development of towns and medium-sized cities and the rates of economic and socio-cultural changes taking place in the society. However, the level of infrastructure development varies in the cities, municipal districts and rural settlements of the Republic of Tatarstan [3].

The construction of multifunctional shopping centres has become very popular in large cities of Russia, which are also used as public spaces. These centres perform not only shopping functions but also combine entertainment, sports, recreational and other public functions. The transformation of shopping centres into public spaces contributes to the economic growth of cities and it is an important factor in improving standards of living.

However, the construction of multifunctional shopping centres in towns and medium-sized cities has not become widespread yet, therefore, it is an urgent task. We hypothesize that the construction of multifunctional shopping centres, hotels and other infrastructure facilities will provide employment opportunities in towns and cities. This fact will also contribute to the development of small and medium-sized businesses, domestic tourism and provide a competitive advantage of the city for the development of investment potential and business environment. The expected social effect of the construction of such centres involves increasing the availability, diversity and quality of cultural leisure activities as well as the attractiveness of living conditions by creating a modern social infrastructure and reducing the migration outflow of the population.

The purpose of this work is to assess the investment attractiveness of an infrastructure project on the example of the construction of a multifunctional shopping centre in Leninogorsk (Republic of Tatarstan, Russia).

Leninogorsk is a city with a population of 61974 people, located in the south-east of the Republic of Tatarstan. The city was founded in 1955 near the explored oil field. In accordance with the classification applied in Russia, Leninogorsk can be attributed to the group of medium-sized cities (with a population of 50 to 100 thousand people). The economy of the Leninogorsk municipal district and the city of Leninogorsk is highly dependent on the oil and gas producing industry; therefore, fluctuations in oil prices are the most important external factors that affect the socio-economic development of the region.

The previous studies [4] contributed to the identification the range of the most pressing problems for the city residents: low standards of living, unfavorable demographic trends, lack of jobs offering decent salaries, inconsistency between the development level of the urban environment and social infrastructure and the modern needs of the society.

Socio-economic problems of urban infrastructure development are the area of research conducted by many scientists around the world. These studies show that modern cities operate in a very dynamic, changeable environment and struggle with many economic, social and spatial problems, which make them seek an optimal development model [5, 6]. G. Gemenetzi [7], S.M.F. Costa [8] note that the specifics of life in towns and medium-sized cities differ significantly from life in large residential areas. However, each city is unique and has its own range of problems due to the historical features of its formation, geographical location, production structure and a number of other factors [6, 9–11].

E.V. Shcherbina [11], D.F. Meyer [12] state in their research that infrastructure is one of the most significant factors influencing the long-term economic growth of both cities and rural settlements. In addition, infrastructure significantly improves the quality and standard of living of the population [13].

Modern society is changing ideas about the quality of the environment [14]. The urban environment should meet the diversity of needs of the population [15] and be functional, physically and psychologically comfortable; the residents themselves should be active participants in its development [14]. A decrease in the quality and comfort of the human life environment leads to an aggravation of social problems in the society [12]. The creation of a comfortable urban environment not only contributes to a more efficient performance of labor activities of the population, cultural leisure activities, recreation and daily routine, but also determines largely the investment attractiveness of the city [16].

Currently, many towns in Russia are monotowns [14] with only one town-forming industry, and their economic situation is difficult [17]. The global financial and economic crisis of recent years has strongly affected the socio-economic development of such towns. The enterprises turned out to be uncompetitive, many of them ceased work.

There is an increase in migration processes among the economically active population in towns and medium-sized cities of Russia. According to M. Eremenko, I. Gareev, insufficient development of social infrastructure is one of the reasons for this phenomenon [18].

N. Zubarevich [19] notes in the study that the degradation of the social infrastructure of small towns aggravates the processes of depopulation and migration, which are one of the significant barriers to the country's spatial development.

Studying the distinguishing features of the territorial settlement of the population of Russia, R. Fattakhov, M. Nizamutdinov and V. Oreshnikov predict an increase in the gap between megalopolises and provincial cities [20], which will inevitably lead to an aggravation of economic, social and infrastructural problems. Therefore, when forming a socio-economic policy for the development of towns and medium-sized cities, it is necessary to take into account their cultural, historical, production, territorial and social potentials [4]. This will have a positive effect on the integrated development of territories and will reduce the level of differentiation between regions in terms of economic indicators.

One of the most effective ways to solve infrastructure problems is the transition to managed development [12, 20]. Y. Skorobogatova [21] notes in her study that the government should pursue an effective policy of support and development of small business, especially in terms of tax and financial policies. The state should contribute to the development of small business in monotowns by creating infrastructure support, providing financial and information assistance, legal regulation of small business activities

[21–23]. At the same time, the town should be not only attractive for business, but also comfortable for its residents [24].

Studying the relationship between infrastructure development and entrepreneurship, Daniel L. Bennett argues that investment in infrastructure can play both a stimulating and restrictive role [25]. Private investment in infrastructure provides benefits for business and job creation, while state investment has the opposite effect.

A number of studies deal with the problems of transport infrastructure and transport accessibility of cities as a driving force of urbanization [13, 26–28], but they do not focus on the issues of investment attractiveness of infrastructure projects.

The review showed that towns and medium-sized cities have significant economic, intellectual and cultural potential and can be considered as important engine of the national economy [29]. The development of social infrastructure is an important factor in ensuring the effective functioning of the economy of such cities, stimulating its growth and promoting employment. Thus, the task is to determine what factors affect the financial stability indicators of the investment project for the construction of a social infrastructure facility, namely, a multifunctional shopping centre.

2 Methods

2.1 Discounted Method of Evaluating the Effectiveness of Investment Projects

The following methods were used in the calculations:

- method of calculating the net present value of the project (NPV);
- method of calculating the discounted payback period (PBP);
- ymethod of calculating the internal rate of return (IRR).

The calculations were made using the software product «Alt-Invest. Construction».

Net present value method (NPV) is based on comparing discounted income and discounted expenses during the implementation of the investment project. This method possesses sufficient stability under various combinations of initial conditions, which allows finding an economically rational solution in all cases.

The NPV indicator characterizes the net cash flow from the project modified to the present value:

$$\mathrm{NPV} = \sum_{t=1}^{n} \frac{CF_t}{(1+d)^t} - \sum_{t=0}^{n} \frac{I_t}{(1+d)^t} \tag{1}$$

where CF_t – cash inflow at the end of the period t, d – discount rate, I_t – investment cost in the period t.

The project is effective if NPV > 0.

The discounted payback period (PBP) is the amount of time necessary for an investment to generate cash flow to recover, taking into account the time value of money. This indicator demonstrates accurately the risk degree of the project.

Internal Rate of Return (IRR) shows the maximum allowable level of interest expense related to the project financing:

$$\mathrm{IRR} = \mathrm{d}_1 + \frac{NPV(d_1)}{NPV(d_1) - NPV(d_2)} \cdot (d_2 - d_1) \quad (2)$$

where d_1 – discount rate when NPV (d_1) > 0, d_2 – discount rate when NPV (d_2) < 0.

The project is effective if the IRR exceeds the discount rate included in the calculation.

2.2 Risk Simulation of an Investment Project

This method is based on the change in the resulting performance indicators of the project in case of a possible increase or decrease in the initial project parameters. Two-parameter sensitivity analysis allows finding out how the net present value of the project (NPV) will change in two cases:

- a simultaneous increase in investment cost and the discount rate;
- a simultaneous increase in operating costs and the discount rate.

The simulation was carried out using the software product «Alt-Invest. Construction».

3 Results and Discussion

The research deals with a multifunctional shopping centre in the city of Leninogorsk with a total area of 1932.77 m^2 with further leasing of its retail space. The location of the shopping facility is one of the most important factors affecting the popularity and, as a result, its attendance. The main difficulty of choosing a location for a multifunctional shopping centre relies in the necessity to take into account a number of factors that show the attractiveness of this centre for the population not only for this area, but also for the surrounding areas. The key point is its location in relation to traffic flows, visitor profile, the profile of the prospective tenants, specifics of management [30].

In accordance with the general plan of the city of Leninogorsk, the site selected for construction refers to public and business areas intended for the construction of administrative, business, commercial facilities as well as consumer services for cultural and leisure facilities, sporting venues, religious building sand healthcare facilities.

The construction site is located near the city center along one of the largest streets in the city and has good transport links, which will simplify its search and save time for visitors from the neighbouring areas.

Other social infrastructure facilities are located in close proximity, which is a positive factor in the formation of groups of visitors.

The construction site is provided with the following engineering networks: high pressure gas pipeline; water supply, sewerage; telephone and radio communication systems;

electric power supply. The existing road network of this area has good traffic capacity, which allows carrying out the transportation necessary for the construction.

The criteria for choosing a site for the construction of a multifunctional shopping centre are given in Fig. 1.

When choosing a location, possible competitive advantages of the city for the development of investment potential and business environment were also taken into account: favorable economic, geographical and transport position of the region (at the intersection of the main types of transport routes); availability of free land resources and production sites for infrastructure development; promising opportunities for the development of certain areas of recreation and tourism within the unified tourist and recreation system of the Republic of Tatarstan (Russia) [4].

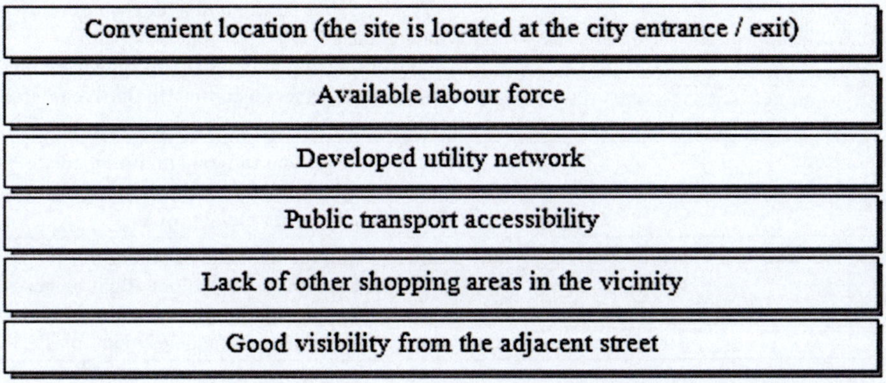

Fig. 1. Criteria for choosing a site for construction. (Proposed by the authors).

The following areas can be found on the territory of the multifunctional shopping centre: trade, consumer services, culture, leisure and entertainment zones. It is planned to build a food court (3 restaurant chains) having 120 seats, a cinema, an entertainment area for children and parents, more than 20 shops. It is also possible to equip the adjacent territory with entertainment zones.

The project provides the following service and utility areas of the multifunctional hall: a cloakroom, 2 washrooms for customers and staff, necessary rooms for the staff.

The plan of the site improvement includes its division into functional zones and the work on the development of the area, driveways and pedestrian ways. The area in front of the building, the sidewalk and pedestrian zones are made of paving slabs.

The blind area around the buildings, trash container enclosures, a utility site with an entrance for service and special vehicles, parking lots for cars have an asphalt concrete surface.

The territory of the centre has outdoor artificial lighting due to lamps placed on the facade of the building from the side of the utility site and three park lamps of the standing type from the side of the main entrance.

The description of structural elements of a multifunctional shopping centre in the city of Leninogorsk (Russia) is given in Table1.

Table 1. Structural elements of a multifunctional shopping centre.

№	Element name	Characteristic
1	Foundations	Foundations for external and internal walls made of concrete blocks
2	Framing	Frameless building
3	Floor slab	Precast concrete floor slab
4	External walls	Three-layer structure: 1stlayer – 380 mm thick inner bearing layer made of silicate bricks; 2ndlayer – 100 mm thick heat insulation made of non-combustible mineral wool boards with «Tyvek» vapor barrier and hydro-windproof membranes; 3rdlayer – facing with porcelain stoneware tile son a metal frame (according to the «ventilated facade» system)
5	Roof	Attic roof based on rafters with intermediate support and lathing. Pitched roof made from PK «Metalloprofil» metal profile
6	Architectural appearance of the facade	Exterior wall facing made of porcelain stoneware of contrasting colors «light beige», «terracotta» and «dark brown» on the metal frame. White windows made of metal-plastic

The total cost of construction of the building of a multifunctional shopping centre was calculated using the software product «Grand Smeta» and amounted to 85370.9 thousand rubles.

A construction schedule was developed on the basis of the cost estimate summary (Table 2).

Table 2. Schedule for the construction of a multifunctional shopping centre in the city of Leninogorsk.

№	Name	Start date	Completion date
1	Preparation of the construction site	01.09.2020	30.10.2020
2	Temporary construction facilities	15.09.2020	30.10.2020
3	Power facilities	25.09.2020	30.10.2020
4	Transport and communication facilities	17.07.2021	20.08.2021

(*continued*)

Table 2. (*continued*)

№	Name	Start date	Completion date
5	External engineering networks	01.09.2020	16.11.2020
6	Shopping and entertainment centre building	30.10.2020	01.09.2021
7	Landscaping	15.07.2021	01.09.2021
8	Other work and costs	01.09.2020	01.09.2021

Source: calculated by the authors.

Calculations to substantiate the construction efficiency were carried out using the software product «Alt-Invest. Construction». The chosen planning horizon is 12 years. When performing the calculations, the investment costs of the project were taken into account (Table 3).

Table 3. Investment costs.

№	Name	Cost, thousand rub, (1$ = 74.82 rub)
1	Land plot	1482
2	The cost of construction of a multifunctional shopping centre per sq. m	44.170
3	IP video surveillance kit for 4 indoor and 4 outdoor 2MP PST cameras IPK08BH	65.430
4	Sound equipment Rebox Street M	86.552
5	Snow removal machine Weima WXS0722A 21000100	114.980
6	Mini loader CDM307 Lonking	3940
7	Light box	36.720

Source: calculated by the authors.

The income of the multifunctional shopping centre at the operational stage will be generated by renting out space, placing advertisements on the facade of the building. The costs are given in Table 4.

Table 4. Total operating costs for 11.5 years of a multifunctional shopping centre operation in the city of Leninogorsk.

№	Name	Cost, thousand rub.
1	Salary of the employees	40 359
2	Utility payments	3 071
3	Facility insurance	12 285
4	Advertisement	575
5	Taxes	147 292

(Calculated by the authors taking into account 4% inflation rate).

The result efficiency indices of the construction and operation of a multifunctional shopping centre in the city of Leninogorsk are given in Table 5.

Table 5. Performance indicators of total investment costs.

№	Name	Values
1	Annual discount rate, %	15
2	Net present value (NPV), thousand rub	28 069
3	Internal rate of return (IRR), %	21.8
4	Profitability index (PI), times	1.32
5	Discounted pay-back period (PBP), years	8.22

Source: calculated by the authors.

When performing the calculations, a two-parameter risk simulation was carried out: the effect of simultaneous increase in the discount rate and investment costs of the project on NPV was studied (Fig. 2).

Calculations show that at the planned level of investment costs (91.942 million rubles), safety margin of the project based on IRR is 6%. In the case of 20% increase in the investment costs, the safety margin will be only 2%.

Similar simulation by operating costs shows a slightly lower NPV sensitivity to the simultaneous increase of operating costs and the project discount rate (Table 6).

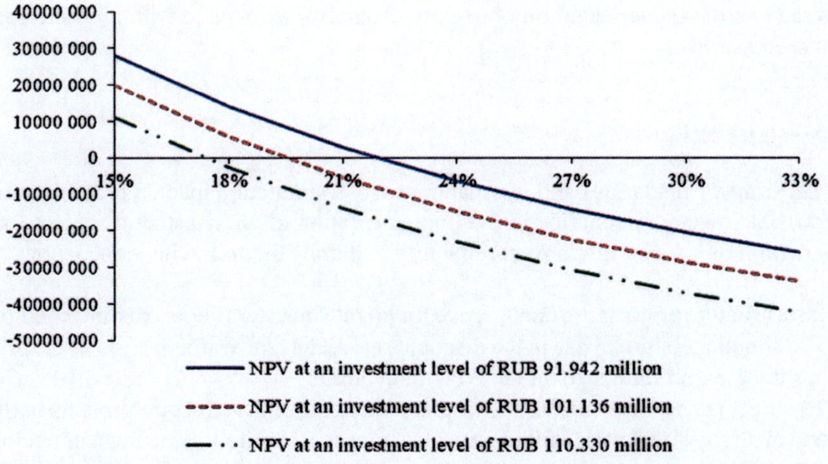

Fig. 2. NPV change in the case of a simultaneous increase in investment costs and discount rate. Source: calculation tables from the software product «Alt-Invest. Construction» (calculated by the authors).

Table 6. NPV change in the case of a simultaneous increase in operating costs and discount rate.

№	Discount rate, d, %	NPV, rub.		
		level of operating costs 16.965 million rubles a year	level of operating costs 18.661 million rubles a year	level of operating costs 20 358 0 million rubles a year
1	15%	28 069 554	21 437 870	13 787 459
2	18%	14 087 392	8 255 279	1 518 513
3	21%	2 773 451	−2 407 436	−8 399 980
4	24%	−6 475 073	−11 119 675	−16 499 261
5	27%	−14 107 460	−18 305 817	−23 175 272
6	30%	−20 462 217	−24 285 606	−28 726 331
7	33%	−25 797 055	−29 302 468	−33 379 548

Source: calculation tables from the software product «Alt-Invest. Construction» (calculated by the authors).

Table 6 shows that in the case of 20% increase in operating costs per year the safety margin of the project according to IRR will be 3%.

The simulation shows that the project does not have sufficient financial stability in the case of a possible negative 20% change in the initial indicators: investment and operating costs, discount rate for the chosen planning horizon is 12.5 years (1 year of construction and 11.5 years of operation). Long payback and low financial stability make infrastructure development projects unattractive for private investors. However, it should be noted that the implementation of negative scenarios associated with 20% increase in costs seems unlikely.

4 Conclusions

The performed calculations and two-parameter risk simulation made it possible to identify possible negative scenarios in the implementation of an investment project for the construction of a social infrastructure facility and draw the following conclusions:

1. Infrastructure projects are unattractive for private investors under existing conditions.
2. The simultaneous increase in the discount rate and the investment costs of the project has the greatest influence on the NPV indicator.
3. To attract private investors to the implementation of infrastructure projects in towns and medium-sized cities of Russia, state support is required in the form of tax incentives, compensation for pre-project preparation costs and the use of a public-private partnership scheme at the stage of infrastructure facilities operation. The inflow of private investment can be increased by means of the mechanism of infrastructure mortgage and the conclusion of special investment contracts to ensure a stable business environment.

Further research in the development of this topic will allow establishing the optimal ratio of state and private capital for the implementation of infrastructure investment projects. This will give the opportunity to use them as the basis of a strategy for the development of infrastructure projects in towns and medium-sized cities.

References

1. Yu, A., Laamarti, A.V.: Kofanov: small towns in contemporary Russia. Sotsiologiya Goroda **4**, 3–16 (2012)
2. Zagidullina, G.M., Klescheva, O.A.: The development of innovation infrastructure of investment-building complex. Izvestija KGASU **2**(16), 271–277 (2011)
3. Bakhareva, O.V., Ilina, E.V., Serova, A.S.: Methodology of formation of social infrastructure in the region investing in the quality of life. Manage. Econ. Syst. Sci. Eelectron. J. **12**(94), 32 (2016)
4. Yu, V.: Medyanik: trends in the socio-economic development of towns and cities (on the example of Leninogorsk in the Republic of Tatarstan). Ekonomika predprinimatelstvo i pravo **10**(10), 2489–2504 (2020). https://doi.org/10.18334/epp.10.10.111072
5. Gunko, M., Eremenko, Yu., Batunova, E.: Planning strategies in the context of urban shrinkage in Russia: evidence from small and medium-sized cities. Mir Rossii **29**(3), 121–141 (2020). https://doi.org/10.17323/1811-038X-2020-29-3-121-141
6. Wierzbicka, W.: Socio-economic potential of cities belonging to the Polish national cittaslow network. Oeconomia Copernicana **11**(1), 203–224 (2020). https://doi.org/10.24136/oc.2020.009
7. Gemenetzi, G.: Exploring the relationship between urban sprawl and the urban system. Evidence from Thessaloniki, 1991–2011. Spatium **37**, 1–11 (2017)

8. Costa, S.M.F., Carmo, M.B.S., Barja, P.R.: The urban hierarchy at the delta of the Amazon River and the importance of small cities, urbe. Revista Brasileira de Gestão Urbana **11**, e20180014 (2019). https://doi.org/10.1590/2175-3369.011.001.AO09
9. Zemtsov, S.P., Smelov, Y.A.: Factors of regional development in Russia: geography, human capital and regional policies. J. New Econ. Assoc. **4**(40), 84–108 (2018). https://doi.org/10.31737/2221-2264-2018-40-4-4
10. Yu, V., Medyanik, E.E.: Valeeva: actual problems of development of small towns of Russia (on the example of Leninogorsk of the Republic of Tatarstan). Int. J. Hum. Natural Sci. **11–1**(38), 101–104 (2019). https://doi.org/10.24411/2500-1000-2019-11707
11. Shcherbina, E.V., Gorbenkova, E.V.: IOP Conf. Ser.: Mater. Sci. Eng. **753**, 032086 (2020). https://doi.org/10.1088/1757-899X/753/3/032086
12. Meyer, D.F., Neethling, J.R.: Meaurement of the enabling developmental environment: a comparative study in a developing region. Econ. Soc. **10**(4), 67–82 (2017). https://doi.org/10.14254/2071-789X.2017/10-4/6
13. Thondoo, M., Marquet, O., Márquez, S., Nieuwenhuijsen, M.J.: Small cities, big needs: urban transport planning in cities of developing countries, J. Transp. Health **19**, 100944 (2020). https://doi.org/10.1016/j.jth.2020.100944
14. Antonova, N.L., Grunt, E.V.: IOP Conf. Ser.: Mater. Sci. Eng. **687**, 055053 (2019)
15. Palicki, S.: Housing resources and needs in light of modern urban transformations: the case of Poznan. Real Estate Manage. Valuation **25**(2), 72–78 (2017). https://doi.org/10.1515/remav-2017-0014
16. Shagiakhmetova, E.I., Medyanik, Y.: Assessment of investment in the development of recreational areas of the city of Kazan. Manage. Econ. Syst. Sci. Eelectron. J. **12**(130), 86 (2019)
17. Bakhareva, O.V., Romanova, A.I., Safina, R.S., Kursina, I.M., Shindina, T.A.: Infrastructure in the region: long-term investments of institutional investors in Russia. J. Adv. Res. Law Econ. **6**(3), 488–503 (2015). https://doi.org/10.14505/jarle
18. Eremenko, M.M., Gareev, I.F.: Research on the relationship between housing construction and sustainable development of territories. Zhilischnye Strategii **6**(3), 333–376 (2019). https://doi.org/10.18334/zhs.6.3.41188
19. Zubarevich, N.V.: Development of the Russian space: barriers and opportunities for regional policy. Mir novoj ekonomiki **2**, 46–57 (2017)
20. Fattakhov, R.V., Nizamutdinov, M.M., Oreshnikov, V.V.: Analysing and modelling of trends in the development of the territorial settlement system in Russia. Ekonomika Regiona **15**(2), 436–450 (2019)
21. Skorobogatova, Y.A., Bovkun, A.S., Ivanov, M.Yu., Shilova, O.S.: IOP Conf. Ser.: Mater. Sci. Eng. **880**, 012103 (2020)
22. Sharma, P.: Opportunities and struggles of decentralized governance reform for urban municipalities in India. World Dev. Perspect. **17**, 100174 (2020)
23. Gimadieva, L.S.: Strategy and methods of engaging of investment resources at municipal level. Sci. J. Kub Gau **74**(10), 378–387 (2011)
24. Oleinik, E.B., Zakharova, A.P.: City: Economic growth and social attractiveness issues. Entrepreneurship Sustain. Issues **7**(1), 454–470 (2019). https://doi.org/10.9770/jesi.2019.7.1(32)
25. Bennett, D.L.: J. Bus. Venturing **34**, 105907 (2019). https://doi.org/10.1016/j.jbusvent.2018.10.005
26. Kasraian, D., Maat, K.: Bert van Wee: the impact of urban proximity, transport accessibility and policy on urban growth: a longitudinal analysis over five decades. Urban Anal. City Sci. **46**(6), 1000–1017 (2019). https://doi.org/10.1177/2399808317740355

27. Angel, S., Blei, A.: The spatial structure of American cities: the great majority of workplaces are no longer in CBDs, employment sub-centers, or live-work communities. Cities **51**, 21–35 (2016). https://doi.org/10.1016/j.cities.2015.11.031
28. Engelfriet, L., Koomen, E.: The impact of urban form on commuting in large Chinese cities. Transportation **45**(5), 1269–1295 (2017). https://doi.org/10.1007/s11116-017-9762-6
29. McDonald, C., Buckley, P.J., Voss, H., Cross, A.R., Chen, L.: Place, space, and foreign direct investment into peripheral cities. Int. Bus. Rev. **27**(4), 803–813 (2018). https://doi.org/10.1016/j.ibusrev.2018.01.004
30. Latypova, T.A.: Spaces of urban activity in Kazan. Izvestija KGASU **4**(54), 161–171 (2020)

Carbonate Phase in the Formation of Binding Substance in Dolomite Cement

Nikolay Shelikhov and Ruslan Sagdiev(✉)

Kazan State University of Architecture and Engineering, 420043 Kazan, Russia

Abstract. According to different sources of information the durability of magnesia binders, depending on raw material and technology of production, is equal to 20–100 MPa. It is considered that thanks to the presence of passive phase $CaC0_3$ in the caustic dolomite which doesn't dissociate during the calcinations and which comprises 71% of the total mass of binder, the solidity of caustic dolomite must be lower than the durability of caustic magnesite. Judging by the quantitative correlation of active phrases consisting of MgO in caustic magnesite and caustic dolomite (100 and 29%), it is obvious that the durability of the first one must approximately be 3 times as much the durability of the second one. In fact, the correlation of durability is within the limits of 1:1.5 to 1:2. And what's more some articles as well as our articles dealing with magnesia binders, show the possibility of receiving the caustic dolomite, or as we call it further, dolomite cement with the durability which can be correlated with the durability of caustic magnesite. Besides the durability limit was reached up to 110 MPa when the condition of calcination (temperature, duration, size of the grain of raw material) was optimized.

Keywords: Dolomite · Dolomite cement · Calcination

1 Introduction

The role of alternative feedstocks for the production of building materials has been increasing in recent years. There are works, where the dolomite is used to get the natural cement and hydraulic lime [1–4] or as the filler for concrete [5–7]. The simplest way to use the dolomites in the binding substances consists in the preparation of mixed cements. The dolomite-based mixed cements are of wide range. There are complex additives both binary and many more [8–11]. The more updated way to get the dolomite-based binding components consists in the production of magnesia cement [12], which particularly facilitates the mitigation of carbon footprint [13, 14]. The magnesia cements are usually obtained by burning magnesium-containing carbonate rocks such as magnesites and dolomites. They are tempered by the water solutions of magnesium chloride. The magnesium phosphate cement is a kind of magnesia cements [15–17]. The magnesium phosphate cements feature the superior technical characteristics but require the higher burning temperature in comparison with the magnesia cement. The main trends in the magnesia cements enhancement are focused on the production of magnesia cement mixtures [18] as well as inclusion of crystallization centers [19] and other additives [20].

N. Vatin (Ed.): STCCE 2021, LNCE 169, pp. 135–141, 2021.
https://doi.org/10.1007/978-3-030-80103-8_14

In our opinion, it is necessary to firstly upgrade the process of burning to get the maximal amount of active MgO. The reason why the durability rises is in the fact that the active phase (MgO) of dolomite cement and caustic magnesite after the calcinations of raw material is different in quality. It is obvious that these differences are formed during the process of calcination and continue during the hydration and hardening. The establishment of such differences and their utilization will give the opportunity to utilize magnesite binding dolomites instead of magnesite more soundly. The replacement of magnesite by dolomite will allow to lower the cost of binder by reducing the cost of raw material and to use local natural resources.

Research aims. Considering the finite reserves of limestone and magnesite, the raw materials base for the production of binding substances is rapidly reducing. The problem may be solved by use of dolomites to get the magnesia cement. The latter, firstly code-named as Sorel's cement or caustic magnesite, caustic dolomite, has been used in the course of building to one extent or another for 150 years. The name «dolomite cement» was also stuck for the caustic dolomite. It focuses on assessing the calcination regime of dolomite for binder strength. The object of research is dolomite cement; the subject of research is calcination processes and internal structure of binder.

2 Materials and Methods

To establish the qualitative difference of active phase (MgO) in magnesite binders depending on the parameters of calcination the authors carried out the analysis of the products of calcination by using different types of magnesite and dolomite without admixtures. The chemical structure of magnesite and dolomite samples are shown in Table 1. To avoid the influence of secondary factors on calcination, the samples were selected with approximately identical composition of admixtures.

Table 1. The chemical composition of magnesite and dolomite.

Raw material	MgO	CaO	CO_2	SiO_2	Al_2O_3	Fe_2O_3	Loss in calcination
Dolomite	21.08	29.53	46.39	1.10	0.55	0.65	47.09
Magnesite	46.17	–	50.83	1.12	0.46	0.61	51.64

Calcination was held under the same specific power consumption for dissociation of 1 unit of raw material taking into account the value of endothermic effects of magnesite and dolomite dissociation. The temperature of calcination was 650–750 °C relatively. The materials of fractions 5–7 mm were calcinated within time interval of 0.5–3.5 h at constant ventilation of space under binding.

To establish physical and mechanical indexes of dolomite cement and caustic magnesite tests on durability as well as methods of qualitative chemical and X-ray phase analysis were used. The research took place immediately after the calcinations. When fully expanded, 260 MPa versus 160 MPa, respectively.

3 Results and Discussion

The kinetics of the process of magnesite and dolomite dissociation where the presence of magnesium oxide and of the product under calcination at the moment of taking information is shown in Tables 2–3 and Fig. 1–2.

The analysis of the results demonstrated shows that within 3 h practically complete dissociation of magnesite and dolomite takes place. It is established both by X-ray phase and chemical analyses. However, at the primary stage the dissociation of magnesite takes place with much higher speed than that of dolomite.

After 2 h of dissociation (Table 2, Fig. 1) practically all magnesite dissociates and later can't be displayed by X-rays. At maximum durability of caustic magnesite, the dissociation of magnesite is 100%. The quantity of the active phase of MgO formed as a result of dissociation of magnesite and defined by X-ray analysis, practically coincides with the quantity of active phase defined by chemical analysis (curves 1 and 2, Fig. 1).

Table 2. Kinetics of the process of dissociation of magnesite.

Duration of calcination, hour	Quantity of magnesite, %	Quantity of MgO, %	
		By chemical analysis	By X-ray phase analysis
0.5	69.4	31.0	24.0
1.0	55.5	48.0	42.4
1.5	35.1	68.6	64.5
2.0	10.0	92.0	90.6
2.5	1.2	97.1	95.5
3.0	0	100	100

Table 3. Kinetics of the process of dissociation of dolomite.

Duration of calcination, hour	Quantity of dolomite, %	Quantity of MgO, %	
		By chemical analysis	By Xray-phase analysis
0.5	73.5	24.5	7.3
1.0	58.0	42.4	24.5
1.5	40.8	61.2	48.5
2.0	21.2	81.6	73.5
2.5	9.0	94.9	91.4
3.0	0	100	100

Moreover, the amount of magnesite that did not decompose during the burning process (curve 3, Fig. 1), practically corresponds to the amount of formed MgO, determined by the same analysis (amount 100%). According to graph (Fig. 2) and data from Table 3, dolomite decomposition is slower during the initial period. By the end of the second hour of RFA burning, the presence of up to 20% of undecomposed dolomite is still noticeable (to simplify – the amount of calcite formed during the burning process was not taken into account).

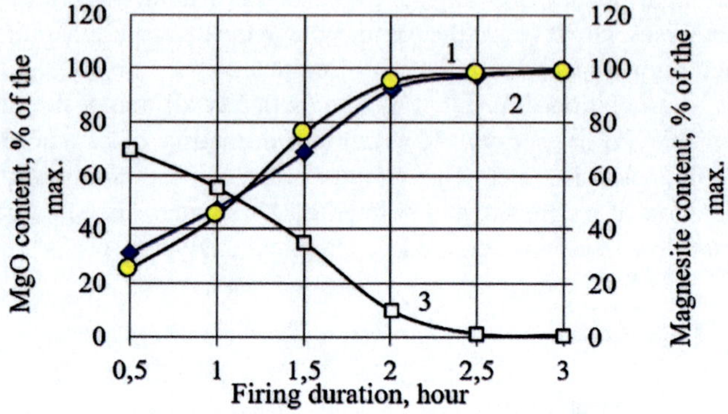

Fig. 1. Kinetics of magnesite decomposition: 1 – MgO content by chemical analysis; 2 – MgO content by X-ray phase analysis; 3 – magnesite content by X-ray phase analysis.

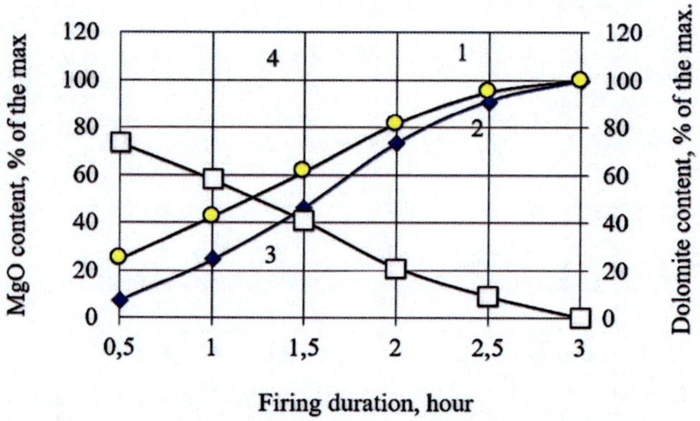

Fig. 2. Dolomite decomposition kinetics: 1 – MgO content by chemical analysis; 2 – MgO content by X-ray phase analysis; 3 – dolomite content by X-ray phase analysis; 4 – area of X-ray amorphousness of MgO.

When comparing the dolomite composition burned products content at different stages of burning (curves 1 and 2, Fig. 2), based on chemical and X-ray phase analysis, we can see, that during the initial period of burning, chemical analysis shows more magnesium oxide than X-ray diffraction analysis (for example, 42% and 24%, respectively, with a burning duration of 1 h). Moreover, the amount of undecomposed dolomite during burning (curve 3, Fig. 2), according to XRF, does not correspond to the amount of formed MgO, determined by the same analysis (the amount is less than a 100%). And, at the same time, it corresponds to the amount of MgO, determined by chemical analysis (100% sum).

At first, this fact was taken for an error in chemical analysis, but the results were similar after the repeated test analysis.

Tests on durability of dolomite cement at various stages of calcination were held and the factors of durability and quantity of MgO were compared. The comparison was made with graphical correlations (taken earlier and are not demonstrated in the present document) of durability of dolomite cement on the quantity of MgO in a wide range of factors. The result shows that durability indexes of dolomite cement correspond to quantity of MgO, defined by chemical analysis.

When the results were analyzed, the authors came to a conclusion that unlike chemical analysis which shows both the quantity of crystal and amorphous MgO in dolomite cement, X-ray-phase analysis on defractional maxima of MgO fixes only the quantity of crystal MgO. The other amorphous part of MgO, during the qualitative X-ray phase analysis is not taken into account. That's why according to X-ray phase analysis the substance of residual dolomite and newly-formed crystal MgO in calcinated sample is equal to less than 100%. The area of the largest divergence of curves 1 and 2 in the graph in Fig. 2 is defined as X-ray-amorphous area of MgO.

Thus, the results of the analyses show the qualitative and quantitative differences between active phases MgO of caustic magnesite and dolomite cement. Since the conditions of calcination are equal thus the reasons of high durability of dolomite cement in comparison with the durability of caustic magnesite should be found in the peculiarities of active phase of MgO of both binders. The actuality of this supposition is proved in the works dedicated to the study of reactionary ability of hard substances.

Additional Xray graphic research and the analysis of the rate of crystallinity of MgO (Table 4) proved that what was said above is right. In fact, MgO which is formed at thermal dissociation of dolomite has lower rate of crystallinity, than MgO, formed at thermal dissociation of magnesite.

Table 4. Average size of crystals of active phase of caustic magnesite and caustic dolomite.

№	Time of calcination, hour	MgO, %		Size of crystals, nm	
		In caustic magnesite	In dolomite cement	MgO of caustic magnesite	MgO of dolomite cement
1	0.5	32	25	22	15
2	1.0	48	42	26	20
3	1.5	70	60	30	25
4	2.0	90	80	38	31
5	2.5	95–100	95	45	36
6	3.0	100	95–100	47	39
7	3.5	100	100	50	42

The results of comparison of the sizes of crystals of magnesium oxide prove the abovementioned (Table 4). The average size of crystals of active phase of MgO of caustic dolomite, measured on half the wideness of its defractional maxima is 15 to 42 nm depending on the time of calcinations since the same index for caustic magnesite is 22 to 50 nm.

Consequently, at similar duration of calcination, the forming phase of MgO in dolomite cement is more active during the hardening, if compared with the phase of caustic magnesite; the smaller size of its crystal, is the proof to it. We can exclude the fact that one of the reasons of this effect is the presence of non-dissociated carbonate phase in the form of $CaCO_3$ in dolomite cement, which can prevent the formation and the grow of crystals of MgO owing to epitaxional and endotaxial phenomena.

4 Conclusions

Thus, the active phase of dolomite cement presented as magnesium oxide, owing to above mentioned reasons is more dispersed, has highly abnormal surface power and, relatively, highly abnormal reactional ability, and thus magnesium oxide of dolomite cement at other equal conditions of calcination of raw material is more active than magnesium oxide of caustic magnesite. And as a binder substance dolomite cement is more effective than caustic magnesite.

References

1. Shelikhov, N., Rakhimov, R., Sagdiev, R.: Inorganic Mater. **52**, 1172–1178 (2016). https://doi.org/10.1134/S0020168516110169
2. Shelikhov, N., Sagdiev, R.: IOP Conf. Series: Mater. Sci. Eng. **890** (2020). https://doi.org/10.1088/1757-899X/890/1/012097
3. Rakhimova, N.R., Rakhimov, R.Z., Morozov, V.P., Eskin, A.A.: Periodica Polytechnica Civil Eng. **65**(1), 204–214 (2021). https://doi.org/10.3311/PPci.14084
4. Ermilova, E., Kamalova, Z., Ravil, R.: IOP Conf. Series: Mater. Sci. Eng. **890**(1) (2020). https://doi.org/10.1088/1757-899X/890/1/012087
5. Sychugov, S., Tokarev, Y., Plekhanova, T., Mikhailova, O., Pudov, I., Faizullin, R., Gaifullin, A., Sagdiev, R.: Proc. Eng. 982–990 (2017). https://doi.org/10.1016/j.proeng.2017.02.124
6. Rakhimov, R.Z., Rakhimova, N.R., Gaifullin, A.R., Bikmukhametov, A.R., Morozov, V.P.: News KSUAE **3**(49), 172–180 (2019)
7. Shelikhov, N., Sagdiev, R., Timirgaliev, M.: News KSUAE **4**(50), 394–400 (2019)
8. Hailong, Y.: Cement Concr. Res. **139** (2021). https://doi.org/10.1016/j.cemconres.2020.106267
9. Xu, J., Lu, D., Zhang, S., Xu, Z., Hooton, R.D.: Constr. Build. Mater. **270** (2021). https://doi.org/10.1016/j.conbuildmat.2020.121375
10. Shaaban, M.: Heliyon **7**(2021). https://doi.org/10.1016/j.heliyon.2021.e06311
11. Rakhimova, N.R.: Geosyst. Eng. **23**(5), 287–298 (2020). https://doi.org/10.1080/12269328.2020.1713911
12. Baghrichea, M., Achourb, S., Baghrichec, O.: Case Stud. Constr. Mater. **13** (2020). https://doi.org/10.1016/j.cscm.2020.e00386
13. Chena, L., Wangab, L., Tsanga, D.C.W., Mechtcherineb, V., Poona, C.S.: J. Clean. Prod. **258** (2020). https://doi.org/10.1016/j.jclepro.2020.120997

14. Hamdy Abdel-Gawwad, A., Soltan Hassan, H., El-Kattan, I.M.: J. Clean. Prod. **28** (2019). https://doi.org/10.1016/j.jclepro.2019.119875
15. Yu, J., Qian, J., Wang, F., Li, Z., Jia, X.: Cement Concr. Res. **138** (2020). https://doi.org/10.1016/j.cemconres.2020.106235
16. Yu, J., Qian, J., Wang, F., Qina, J., Daia, X., Youb, C., Jia, X.: Constr. Build. Mater. **253** (2020). https://doi.org/10.1016/j.conbuildmat.2020.119147
17. Peng, L., Chen, B.: Constr. Build. Mater. **281** (2021). shttps://doi.org/10.1016/j.conbuildmat.2021.122609
18. Liu, Z., Wang, S., Huang, J., Wei, Z., Guan, B., Fang, J.: Constr. Build. Mater. **85**, 247–255 (2015). https://doi.org/10.1016/j.conbuildmat.2015.01.056
19. Wu, C., Chen, C., Zhang, H., Tan, Y., Yu, H.: Constr. Build. Mater. **172**, 597–607 (2018). https://doi.org/10.1016/j.conbuildmat.2018.04.005
20. Guan, Y., Hu, Z., Zhang, Z., Chang, J., Bi, W., Cheeseman, C.R., Zhang, T.: Cement Concr. Res. **143** (2021). https://doi.org/10.1016/j.cemconres.2021.106387

Improving the Dynamic Properties of Pedestrian Overpasses by Including a Reinforced Concrete Slab in Structural Behavior

Valery Eremeev[2], Gennady Shmelev[1], Pavel Eremeev[1](✉), and Daniil Eremeev[1]

[1] Kazan State University of Architecture and Engineering, 1 Zelenaya Street, 420043 Kazan, Russian Federation

[2] JSC Spetsremproekt, 5 Zaslonova Street, 420097 Kazan, Russian Federation

Abstract. Aboveground pedestrian overpasses are generally vulnerable to dynamic impact, due to their high slenderness ratio. The situation is aggravated by the possibility of matching of the natural oscillation frequency and the frequency of walking of people, which can lead to the appearance of resonant phenomena. The article considers one of the possible ways to reduce the period of natural oscillations of an aboveground pedestrian overpass by including a reinforced concrete slab structure in structural behavior. There is a well-known design of the pedestrian overpass consisting of the main longitudinal and transverse beams on which the profiled flooring filled with reinforced concrete is laid. Dynamic tests of the structure were carried out in order to determine the period of natural oscillations. Two structural designs have been developed. In the first one, the reinforced concrete slab is not included in structural behavior and is set by nodal masses. In the second structural design, the reinforced concrete slab is modeled with volumetric finite elements and acts in collaboration with steel plate finite elements. Based on the results of comparing the theoretical periods of natural oscillation of the two structural designs, the influence of the reinforced concrete slab on the period of natural oscillation of the structure is determined.

Keywords: Pedestrian overpasses · Dynamics · Oscillation · Dynamic loading · Construction · Structures

1 Introduction

Ensuring the dynamic comfort of pedestrians during the operation of aboveground overpasses is an important factor that must be taken into account when designing such structures. The task is complicated by the lack of a fundamental theory that fully describes the interaction of a person and a pedestrian overpass [1–3]. It is also necessary to take into account the possibility of the phenomenon of unconscious synchronization [4–6], as a result of which the oscillations of the structure can affect the frequency of the pedestrian's step [7–10].

N. Vatin (Ed.): STCCE 2021, LNCE 169, pp. 142–149, 2021.
https://doi.org/10.1007/978-3-030-80103-8_15

Usually, to provide dynamic comfort is to limit the amplitudes and frequencies of natural oscillations of the structure, by changing the bending stiffness, span length, material, and the introduction of dampers.

There is a well-known design of the pedestrian overpass consisting of the main longitudinal and transverse beams on which the profiled flooring filled with reinforced concrete is laid. Two different variants of such design are considered. In the first version, the reinforced concrete slab acts together with the steel part of the structure. In the second version the reinforced concrete slab acts separately from the steel part. It is known that concrete has better dissipative qualities compared to steel [11, 12].

The article considers a two-span continuous pedestrian overpass with spans of 20 m. Dynamic tests of the structure were carried out in order to determine the period of natural oscillations.

Two structural designs were developed. In the first one, the reinforced concrete slab was not included in structural behavior and was set by nodal masses. In the second structural design, the reinforced concrete slab was modeled with volumetric finite elements and acted in collaboration with steel plate finite elements.

The regulatory documents provide two criteria for assessing the dynamic state of pedestrian overpasses – the period of natural oscillations of structures, the limitation of which allows preventing resonant phenomena, and the acceleration of oscillations of the structure, which ensures the comfort of the pedestrian [13–15].

According to Russian regulatory standards, the periods of natural oscillations of a pedestrian bridge in the vertical plane for the two lowest modes should not fall within the range of 0.45–0.6 s.

According to the international standards ISO 10137:2007 (E) [16], the limit value of the oscillation acceleration is determined by the base curve with the multiplication of the value obtained from this graph by a multiplier equal to 60 for pedestrian overpasses.

There are many known studies about the dynamic properties of steel-reinforced concrete structures [17–26]. In this study, the structural design was created without taking into account the shear stops.

2 Materials and Methods

2.1 Experimental Determination of the Dynamic Characteristics of a Pedestrian Overpass

The purpose of the dynamic tests was to measure the oscillations with vibration sensors installed in the middle of the span of the structure. The simulation of oscillations was carried out by walking people.

The oscillations were recorded using the vertical and horizontal vibration sensors "VEG" and the NI SCXI-1000 docking station with the NI 1317 module (Figs. 1 and 2).

2.2 Theoretical Determination of the Dynamic Characteristics of a Pedestrian Overpass

The determination of the theoretical frequency of natural oscillations of a pedestrian overpass before and after the inclusion of a concrete slab in the structural design was carried out on the basis of a modal analysis of two structural designs.

Fig. 1. General view of vertical and horizontal vibration sensors

Fig. 2. General view of the NI SCXI-1000 docking station with the NI 1317 module

In the structural design №1 (Fig. 3), the steel elements were modeled by plate finite elements, and the reinforced concrete slab was set by nodal masses.

In the structural design №2 (Fig. 4), steel transverse beams were excluded; the reinforced concrete slab was modeled with volumetric finite elements.

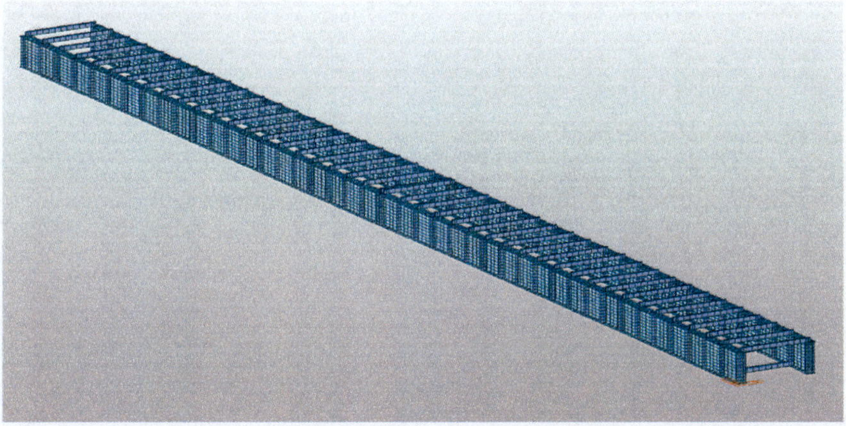

Fig. 3. General view of the structural design before the inclusion of the reinforced concrete slab in structural behavior

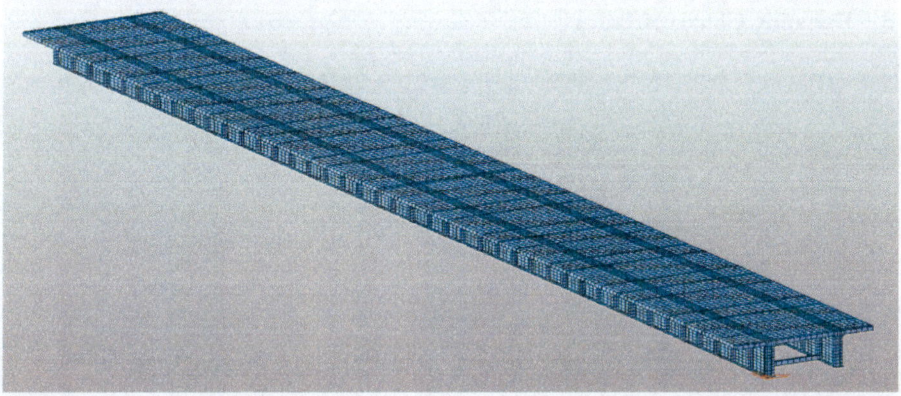

Fig. 4. General view of the structural design after the inclusion of the reinforced concrete slab in structural behavior

3 Results and Discussion

3.1 Dynamic Test Results

Based on the results of dynamic tests, an oscillation graph was obtained on the basis of which the natural oscillation frequencies were determined. The actual period of natural oscillations was 0.178 s, which corresponds to the requirements of the regulatory standards (Figs. 5, 6, 7, 8 and 9).

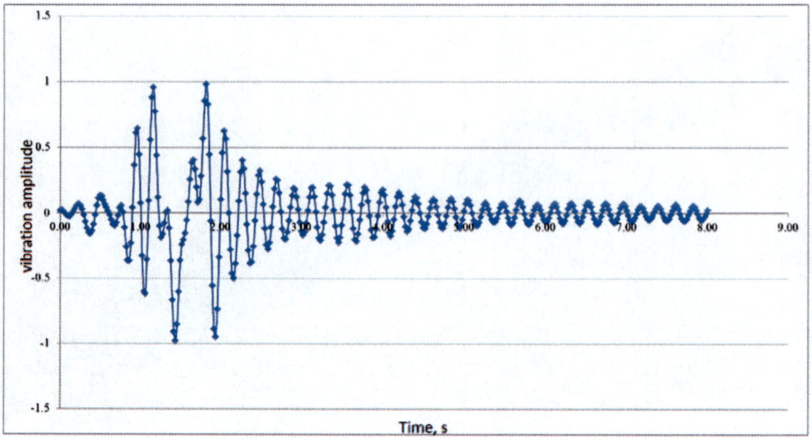

Fig. 5. Diagram of actual oscillations in the span of a pedestrian bridge, with a span of 20 m

3.2 Results of Numerical Calculations in Order to Determine the Theoretical Dynamic Characteristics

Based on the results of the modal analysis, the modes and frequencies of natural oscillations were obtained (Table 1).

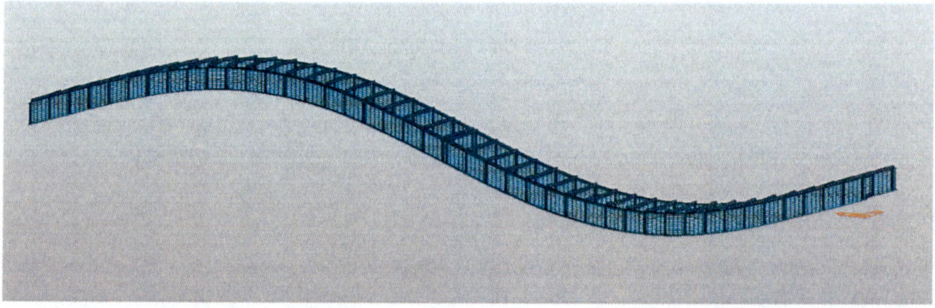

Fig. 6. The first mode of oscillations of the structural design № 1

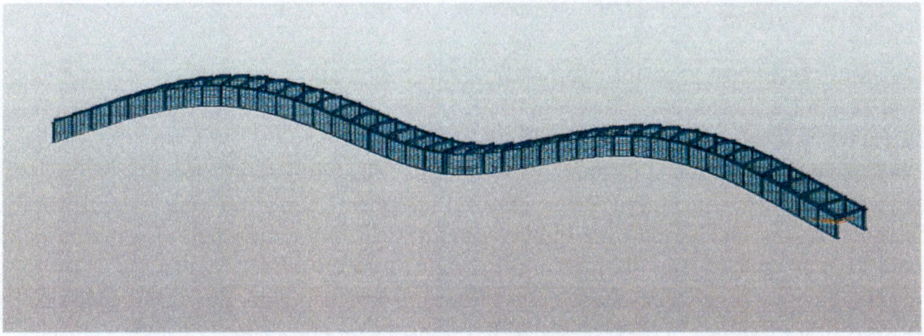

Fig. 7. The second mode of oscillations of the structural design № 1

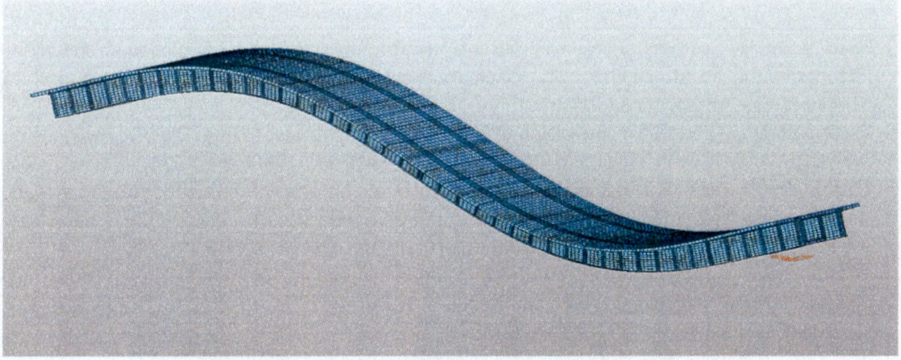

Fig. 8. The first mode of oscillations of the structural design № 2

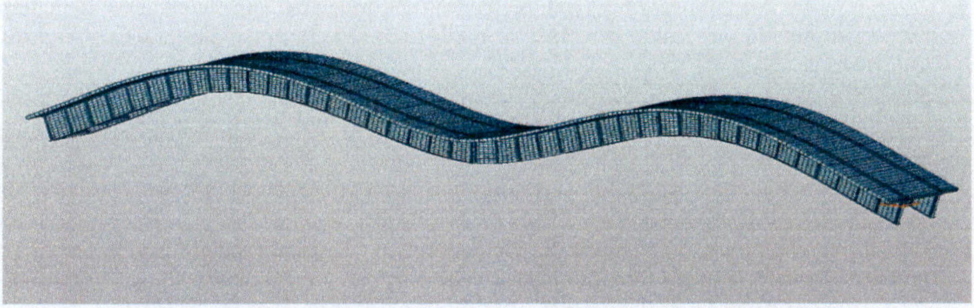

Fig. 9. The second mode of oscillations of the structural design № 2

Table 1. Results of determining the period of natural oscillations

Structural design	The period of oscillations in the first mode, s	The period of oscillations in the second mode, s
1	0.227	0.149
2	0.179	0.118

4 Conclusions

According to the results of the comparison of the calculation according to the two structural designs, a decrease in the period of natural oscillations by 21% was discovered for the two lowest vertical modes of oscillations. If while designing pedestrian overpasses, it is known that the period of natural oscillations falls within an unacceptable range or dangerously close to it, there is an option to include a reinforced concrete slab in the design to reduce the period of natural oscillations. Such solution will also increase the load-bearing capacity and dissipative properties of the structure.

The theoretical periods of natural oscillations obtained from the calculation results showed similar values in comparison with the experimental ones.

References

1. Bachmann, H., Ammann, W.: Vibrations in structures induced by man and machines. Structural Engineering Document 3e, International Association for Bridge and Structural Engineering, Zürich, 2(A) (1987)
2. Newland, D.E.: Pedestrian excitation of bridges. Proc. Inst. Mech. Eng. Part C J. Mech. Eng. Sci. **218**(5), 477–492 (2004). https://doi.org/10.1243/095440604323052274
3. Fujino, Y., Pacheco, B.M., Nakamura, S.-I., Warnitchai, P.: Synchronization of human walking observed during lateral vibration of a congested pedestrian bridge. Earthq. Eng. Struct. Dynam. **22**, 741–758 (1993). https://doi.org/10.1002/eqe.4290220902
4. Bocian, M., Macdonald, J.H.G., Burn, J.F.: Biomechanically inspired modeling of pedestrian-induced vertical self-excited forces. J. Bridge Eng. **18**(12), 1336–1346 (2013). https://doi.org/10.1061/(asce)be.1943-5592.0000490
5. Carroll, S.P., Owen, J.S., Hussein, M.F.M.: Modelling crowd-bridge dynamic interaction with a discretely defined crowd. J. Sound Vib. **331**(11), 2685–2709 (2012). https://doi.org/10.1016/j.jsv.2012.01.025
6. Qin, J.W., Law, S.S., Yang, Q.S., Yang, N.: Pedestrian-bridge dynamic interaction, including human participation. J. Sound Vib. **332**(4), 1107–1124 (2013). https://doi.org/10.1016/j.jsv.2012.09.021
7. Hu, J., Bian, X., Jiang, J.: Critical velocity of high-speed train running on soft soil and induced dynamic soil response. Proc. Eng. **143**, 1034–1042 (2016). https://doi.org/10.1016/j.proeng.2016.06.102
8. Ntotsios, E.: Modelling of train induced vibration. In: Stephenson Conference Research for Railways 2015, April 2015, pp. 153–162 (2015)
9. Ahmadi, E., Caprani, C., Živanović, S., Heidarpour, A.: Vertical ground reaction forces on rigid and vibrating surfaces for vibration serviceability assessment of structures. Eng. Struct. **172**, 723–738 (2018). https://doi.org/10.1016/j.engstruct.2018.06.059
10. Van Nimmen, K., Maes, K., Živanović, S., Lombaert, G., De. Roeck, G., Van den Broeck, P.: Identification and modelling of vertical human-structure interaction. In: Caicedo, J., Pakzad, S. (eds.) Dynamics of Civil Structures, Volume 2. CPSEMS, pp. 319–330. Springer, Cham (2015). https://doi.org/10.1007/978-3-319-15248-6_34
11. Shardakov, I., Shestakov, A., Glot, I.: Elastic and dissipative properties of concrete under impact load. Proc. Struct. Integr. **9**, 199–206 (2018). https://doi.org/10.1016/j.prostr.2018.06.031
12. Zheng, L., Sharon Huo, X., Yuan, Y.: Experimental investigation on dynamic properties of rubberized concrete. Constr. Build. Mater. **22**(5), 939–947 (2008). https://doi.org/10.1016/j.conbuildmat.2007.03.005

13. Qin, S., Zhou, Y.-L., Kang, J.: Footbridge serviceability analysis: from system identification to tuned mass damper implementation. KSCE J. Civ. Eng. **23**(2), 754–762 (2018). https://doi.org/10.1007/s12205-018-0985-7
14. Wang, D., Wu, C., Zhang, Y., Li, S.: Study on vertical vibration control of long-span steel footbridge with tuned mass dampers under pedestrian excitation. J. Constr. Steel Res. **154**, 84–98 (2019). https://doi.org/10.1016/j.jcsr.2018.11.021
15. Miguel, L.F.F., Fadel Miguel, L.F., Lopez, R.H.: A firefly algorithm for the design of force and placement of friction dampers for control of man-induced vibrations in footbridges. Optim. Eng. **16**(3), 633–661 (2014). https://doi.org/10.1007/s11081-014-9269-3
16. ISO 10137:2007(E) International standard, Bases for design of structures—Serviceability of buildings and walkways against vibrations, vol. 112 (2010)
17. Ranzi, G., Zona, A.: A steel-concrete composite beam model with partial interaction including the shear deformability of the steel component. Eng. Struct. **29**(11), 3026–3041 (2007). https://doi.org/10.1016/j.engstruct.2007.02.007
18. Ranzi, G., Zona, A.: Finite element models for nonlinear analysis of steelconcrete composite beams with partial interaction in combined bending and shear. Finite Elem. Anal. Des. **43**(2), 98–118 (2011). https://doi.org/10.1016/j.finel.2010.09.006
19. Chiorean, C.G., Buru, S.M.: Practical nonlinear inelastic analysis method of composite steel-concrete beams with partial composite action. Eng. Struct. **134**(1), 74–106 (2017). https://doi.org/10.1016/j.engstruct.2016.12.017
20. Misiurek, K., Śniady, P.: Vibrations of sandwich beam due to a moving force. Compos. Struct. **104**, 85–93 (2013). https://doi.org/10.1016/j.compstruct.2013.04.007
21. Hou, Z., Xia, H., Zhang, Y.L.: Dynamic analysis and shear connector damage identification of steel-concrete composite beams. Steel Compos. Struct. **13**(4), 327–341 (2012). https://doi.org/10.12989/scs.2012.13.4.327
22. Tan, Z.X., Thambiratnam, D.P., Chan, T.H.T., Gordan, M., Abdul Razak, H.: Damage detection in steel-concrete composite bridge using vibration characteristics and artificial neural network. Struct. Infrastruct. Eng. **16**(9), 1247–1261 (2020). https://doi.org/10.1080/15732479.2019.1696378
23. Kożuch, M., Lorenc, W.: Stress concentration factors of shear connection by composite dowels with MCL shape. Arch. Civil Mech. Eng. **19**(1), 32–46 (2018). https://doi.org/10.1016/j.acme.2018.08.006
24. Kozma, A., Odenbreit, C., Braun, M.V., Veljkovic, M., Nijgh, M.P.: Push-out tests on demountable shear connectors of steel-concrete composite structures. Structures **21**, 45–54 (2019). https://doi.org/10.1016/j.istruc.2019.05.011
25. Ivanov, A.N., Yashnov, A.N.: Improving the construction and calculation methods of span structures of bridges with load-bearing elements made of composite materials. Sci. Probl. Transp. Siberia Far East **1–2**, 142–146 (2014)
26. Bogdan, T., Chrzanowski, M., Odenbreit, C.: Mega columns with several reinforced steel profiles – Experimental and numerical investigations. Structures **21**, 3–21 (2019). https://doi.org/10.1016/j.istruc.2019.06.024

Check for updates

Method of Calculating the Strength of Clay Soils Under Triaxial Regime Loading

Ilizar Mirsayapov(✉)

Kazan State University of Architecture and Engineering, 420043 Kazan, Russia

Abstract. The main purpose of the research is to study the changes in the strength of clay soil in conditions of triaxial compression, under block mode cyclic loads. To date, there are no data on the results of studies of changes in the strength of clay soils under cyclic loading for the considered modes. Theoretical studies of the strength of clay soils under conditions of triaxial compression $\sigma_1 \neq \sigma_2 = \sigma_3$ under block regime cyclic loading are carried out. A distinctive feature of the conducted studies is that the change in strength is described from the point of view of the mechanics of destruction of elastic-plastic bodies and the main criterion is the formation and development of micro-and macro-cracks to the maximum values. The regularities of changes in the main strength characteristics of the soil during transitions to blocks with high and low stress levels are established. New data on the regularities of changes in the strength of clay soils under stationary and non-stationary modes of cyclic loading are obtained. In the study of changes in the fatigue resistance to the destruction of clay soils under triaxial regime block cyclic loading. It is established that the stress-strain state and fatigue strength of soils vary depending on the sequence of alternating blocks with different values of the maximum load of the cycle.

Keywords: Block-cyclic loading · Stationary and non-stationary loading · Clay soil · Micro-cracks · Macro-cracks · Triaxial loading

1 Introduction

Analytical equations for describing changes in the strength of the soil under regime block cyclic loading are developed on the basis of the spatial calculation model of the soil under triaxial compression, proposed by I.T. Mirsayapov and I.V. Koroleva. Theoretical studies conducted by I.T. Mirsayapov [1, 2] show that under regime cyclic loadings, changes in the angle of internal friction and specific adhesion between clay soil particles occur due to the formation and development of micro - and macro-cracks in the planes of extreme equilibrium, leading to a decrease in the strength of the soil (Fig. 1).

To determine the fatigue strength under non-stationary loading conditions, the concept of damage accumulation is usually introduced [3–7], while it is meant that in the process of repeatedly repeating cyclic stress changes in the ground, damage accumulates, which then leads to destruction. In the general case of damage, it depends on the level of the loading mode and the duration of the action (number of cycles) of the cyclic

N. Vatin (Ed.): STCCE 2021, LNCE 169, pp. 150–159, 2021.
https://doi.org/10.1007/978-3-030-80103-8_16

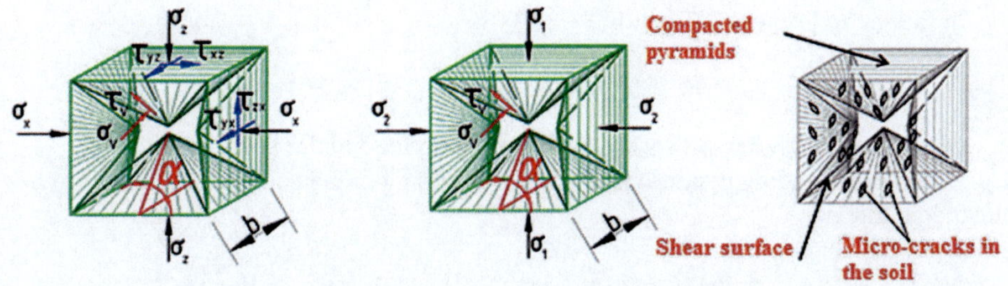

Fig. 1. a) Stress state in the elementary volume of the soil in the space X, Y, Z at an arbitrary time t in the pre-limit state; b) In the space of the main stresses at the stage of the limit equilibrium; Diagram of the development of cracks in the planes of the limit equilibrium within the elementary volume of the soil.

load [8–12]. For comparison with the limit value of damage, it must have a numerical or analytical description [13, 14].

Usually, the simplest hypothesis is accepted, that each unchanging cycle suppresses the same value to the damage, and the damage is summed from cycle to cycle, i.e., the linear hypothesis of summing fatigue damage. In reality, this process looks much more complicated [15–17]. Fatigue failure of the soil is a process that includes the origin, localization, and fusion of microcracks, the formation of microcracks, and its development to critical dimensions. At the same time, the resulting micro and macro cracks affect the formation of the following ones, and the subsequent ones affect the development of the previous ones [18, 21]. For this reason, the linear fatigue damage summation hypothesis gives a very large range of durability values, even for such homogeneous materials as steels and alloys. The application of the corrected linear hypothesis of fatigue damage summation also does not give the desired results.

2 Materials and Methods

We consider the equation of the objective strength (endurance) of the soil under non-stationary conditions, taking into account the above kinetics of fatigue failure of the soil. In case of non-stationary cyclic loading, the actual change in compressive loads leads to block loading. The entire loading process can be reduced to the sum of a series of blocks alternating in a certain order, depending on the characteristics of the operating mode (Fig. 2). We consider the following two basic loading sequences (Fig. 2) all other loads are obtained under a certain combination of these two main cases (Fig. 2). Let's define the following loading sequence.

γ_1 – for the number of cycles N_1; γ_2 – for the number of cycles N_2; γ_3 – for the number of cycles N_3; γ_{n-1} – for the number of cycles N_{n-1}; γ_n – for the number of cycles N_n; γ_{n+1} – for the number of cycles N_{n+1}.

In the first scheme (Fig. 2a):

$$\gamma_1 < \gamma_2 < \gamma_3 \dots \dots \dots \gamma_{n-1} < \gamma_n < \gamma_{n+1}$$

In the second scheme, (Fig. 2b)

$$\gamma_1 > \gamma_2 > \gamma_3 \cdots \cdots \cdots \gamma_{n-1} > \gamma_n > \gamma_{n+1}$$

where $\gamma = P_{max}/P_{failure}$, $P_{failure}$ is the breaking load under static loading.

Under such loading conditions, the stresses arising in the compressed soil also obey the inequalities:

$$\sigma_{\text{gr1}}^{\text{max}} < \sigma_{\text{gr2}}^{\text{max}} < \sigma_{\text{gr3}}^{\text{max}} \ldots\ldots \sigma_{\text{grn}-1}^{\text{max}} < \sigma_{\text{grn}}^{\text{max}} < \sigma_{\text{grn}+1}^{\text{max}}$$

and

$$\sigma_{\text{gr1}}^{\text{max}} > \sigma_{\text{gr2}}^{\text{max}} > \sigma_{\text{gr3}}^{\text{max}} \ldots\ldots \sigma_{\text{grn}-1}^{\text{max}} > \sigma_{\text{grn}}^{\text{max}} > \sigma_{\text{grn}+1}^{\text{max}}$$

In the general case:

$$N_1 \neq N_2 \neq N_3 \neq \cdots \ldots\ldots \ldots N_{n-1} \neq$$
$$N_n \neq N_{n+1};$$

$$\mathcal{P} = \frac{P_{min}}{P_{max}} \neq const$$

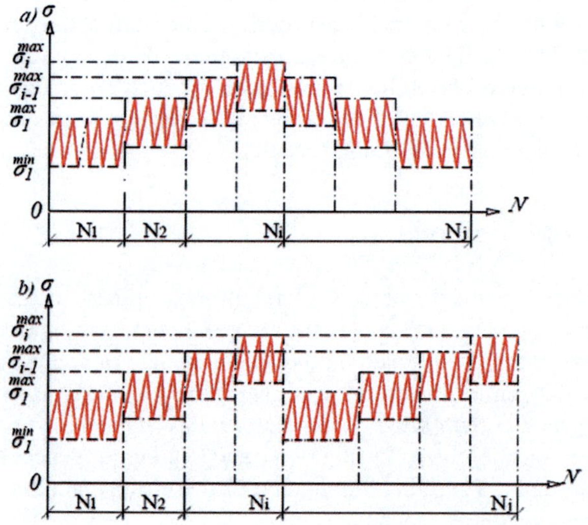

Fig. 2. Options of block configuration under non-stationary repeatedly repeated cyclic loading.

To determine the fatigue strength of the soil under these conditions, it is necessary to calculate the values of the creep measure, the elastic modulus, critical and current values of the stress intensity coefficient under variable cyclic load conditions.

It should be noted that for variable (non-stationary) cyclic load modes, changes in the creep measure $C_{(t,\tau)}$, the elastic modulus $E_{gr(t,\tau)}$, the stress intensity coefficient

$K_{cf(t)}$, and $K_{Ii}{}^{max}(t)$ the crack development significantly differs from changes in the same parameters for cyclic loads with constant characteristics, and therefore they are described by other dependencies, in this case, the accumulation of fatigue damage in the soil, leading to its destruction, occurs gradually and depends on the nature of the non-stationary loading mode.

In this case, the interaction of cycles with different amplitudes plays a significant role, if the cyclic loading process is first carried out at a lower value of σ_{gr1}^{max}, and then at a higher value of σ_{gr1}^{max}, then the overall durability increases compared to the one in which the loading is carried out only with a higher value of σ_{gr2}^{max}. This is due to an increase in the effective surface energy of soil destruction in the plastic pre-fracture zone at the crack tip due to repeated hydration of montmorillonite and the redistribution of internal intrinsic structural stresses. To determine the effect of the decreasing loading mode and the pattern of macro-crack development, it is necessary to be able to predict the number of loading cycles N_D and the crack length, Δl_D, which characterize the effect of a high level of loading on the development of macro-cracks with a subsequent lower load level.

The solution of this problem makes it necessary to establish the dependence of the change in the rate of crack development when it is applied in the zone of influence of the loading unit with a higher stress level $V^n(t) = f_i(l_i)$, i.e. it is necessary to establish the regularities of the influence of single and multiple overloads on the delay in the development of the fatigue crack N_D. Initially, cyclic loading is performed up to point A in the mode with the maximum coefficient of intensity, stress, K_{Ii-1}^{max}, minimum stress intensity, K_{Ii-1}^{min}, by span, $\Delta k_{i-1} = K_{Ii-1}^{max} - K_{Ii-1}^{min}$, then at point A the loading mode changes and a higher level load is applied with the maximum value of the stress intensity coefficient K_{In-1}^{max}, after loading in this mode N_n. cycles a lower loading mode is set.

As a result of the action of the overload loading unit, the rate of fatigue crack development decreases, and only after a certain number of loading cycles, N_D (the so-called development delay) becomes at point B equal to or close to the value of V_{i+1} in the absence of the overload mode. The increment of the macro-crack length α_D for the number of loading cycles N_D shows the size of the zone of influence of the overload loading unit on the further development of the crack at subsequent lower load levels.

3 Results

Experimental studies have established that the area of influence of overload is determined by the size of the end plastic zone, at the top of the microcrack formed during the overload mode. Then the size of the zone of influence of the overload mode α_D is calculated by the equation:

$$a_D = 2l_y = \frac{1}{\pi}\left[\frac{K_{In}^{max}}{R_{grt}}\right]^2, \tag{1}$$

where K_{In-1}^{max} is the maximum value of the stress intensity coefficient at the overload loading block (Fig. 3).

To determine the crack development rate functions $V_{(t)}{}^n = f_{(li)}$ in the zone of influence of the overload loading unit, we use the function. Then the function of the change

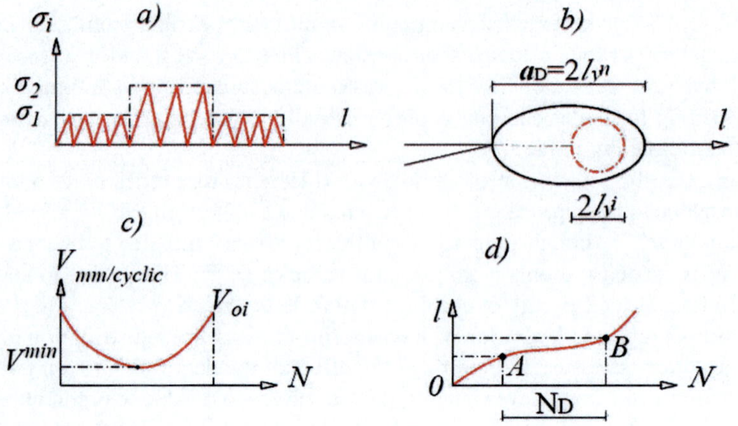

Fig. 3. Diagram for determining the rate of development of a fatigue crack in the area of influence of the overload loading block.

in the rate of development of a macro-crack in the area of influence of the overload block $V_j^n(l)$ at the i-th loading block is represented as (Fig. 1):

$$V_i^n(l) = V_{oi} - \widehat{V_i^n} f\left(\frac{l}{a_D}; \text{ʑ}; \gamma\right) \text{ при } 0 \leq \frac{l}{a_D}, \tag{2}$$

where V_{oi} is the rate of macro-crack development in the *i-th* loading block with K_{Ii-1}^{max} and P_i under the assumption that there is no overload loading block at $0 < l/a_D \leq 1 < \gamma < n$:

$$f\left(\frac{l}{a_D}; \text{ʑ}; \gamma\right) = \left\{\frac{\Gamma(\gamma + \eta)}{\Gamma(\gamma) \cdot \Gamma(\eta)} \cdot \left(\frac{l}{a_D}\right)^{n-1} \cdot \left(1 - \frac{l}{a_D}\right)^{\eta - 1}\right. . \tag{3}$$

Here Γ is the gamma function.

In this case, the condition is met:

$$\int_0^{a_D} \bar{V}^n f\left(\frac{1}{a_D}; \eta; \gamma\right) dl = S^{\Pi}, \tag{4}$$

where S^n is the area bounded the curve; (l) and $V_{oi}(l)$; $\overline{V_j^n} = S^n / a_D$ mm^2/cycle is the dimension coefficient.

For this reason, a compression region occurs at the crack tip, and as a result, the plastic zone at the crack tip is pre-compressed at the load level with $k_{In}{}^{max}$. Therefore, after the overload mode for some time, the crack develops extremely shallow or does not develop at all, since, depending on the difference in the levels of the overload and normal loading modes, the energy of the tensile stresses is balanced by the energy of the compressive stresses, ΔW_c (Fig. 5).

The compressive stress energy ΔW_c is initially fixed, and its felting decreases with increasing number of loading cycles and is calculated by the equation:

$$\Delta W_c = \int_0^{\varepsilon_y} \sigma_{11}(t) d\varepsilon - \int_0^{\varepsilon_y} \sigma_{11i}(t) d, \tag{5}$$

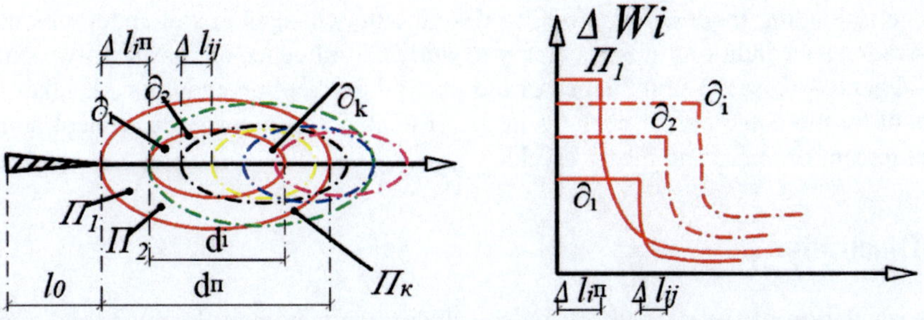

Fig. 4. Diagram of accumulation of the specific energy of inelastic deformation within the plastic zone at the top of the macro-crack in the area of the influence of overload, in Fig. 4, d^n is the size of the plastic zone in the overload mode; d_i is the size of the plastic zone in the considered mode; $\Delta l_i{}^n$ is the size of the trace elements in the plastic zone in the overload mode; Δl_{ij} is the size of the trace elements in the plastic zone in the considered mode; $\Pi_1 \ldots \Pi_k$ are the numbers of trace elements in the overload mode; $\partial_1 \ldots \partial_k$ are the numbers of trace elements in the considered mode.

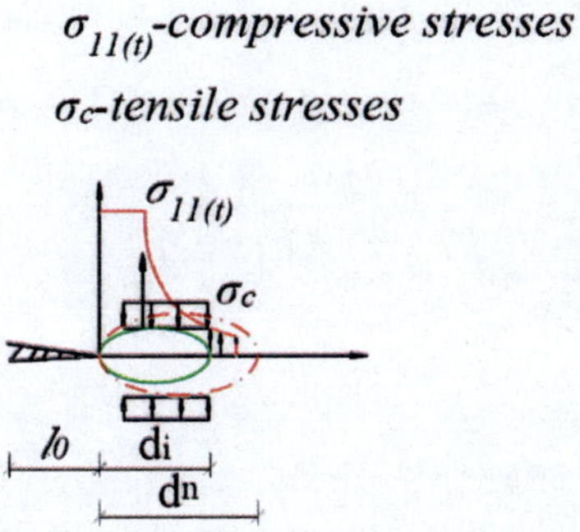

Fig. 5. Diagram of the calculated stress state at the crack tip in the area of overload influence: Σ_{11t} – compressive stresses, σ_c – tensile stresses.

where $\sigma_{11\Pi}(t)^0$ – stresses at the crack tip at the value of the stress intensity coefficient $k_{In}{}^{max}$ in the overload loading mode; $\sigma_{11i}(t)$ – the same, at the stress intensity coefficient $k_{In}{}^{max}$, corresponding to the *i*-th loading block $k_{In}{}^{max} < k_{In}{}^{max}$.

In the case when the loading is carried out in the reverse sequence, i.e., first at a higher value of σ_{gr}^{max} and then at a lower value of σ_{gr1}^{max}, the reduction in the fatigue strength of the soil at the second stage i.e., at, σ_{gr1}^{max} will be extremely slow. This is due to the fact that during the overload loading unit, a large zone of plastically deformed soil appears at the top of the micro-and macro-fatigue cracks. In the next loading block, when the maximum stress level decreases to σ_{gr1}^{max}, the zone of plastically deformed soil must fit into the surrounding elastic medium. After unloading, the elastic part of the soil tends to take its original position and «compresses» the plastically deformed soil at the top of the crack. Residual compressive stresses tend to close (clamp) the crack tip.

The subsequent process of cyclic loading can cause the development of micro-and macro-cracks only if the value of the residual stresses is exceeded so that the crack tip can open again. This explains the low rate of reduction of fatigue strength with such a change in loading modes. Therefore, to describe the changes in soil endurance under non-stationary conditions, it is necessary to study the influence of consistently high and low variable stresses. The influence of the loading gradient on endurance is taken into account by the equations of changes in the critical stress intensity coefficient and the development of micro and macro cracks.

4 Discussion

The calculation of the strength of the soil under non-stationary loading conditions is performed according to the equations:

$$R_{gr}(t,\tau)=\sqrt{\sqrt{\frac{2E_{grt}\left\langle \begin{array}{l} \int_0^{\varepsilon_R}\sigma_{ti}+\sum_1^3\left[m_{in}(\varepsilon_R-\varepsilon_{ti})^n\right]d\varepsilon- \\ -\left\{\sum\sum\left(\left(\begin{array}{l} C_\partial\prod_{k=1}^{k=g}k_k a\psi_{V11}\left(\sigma_{grt1}^{\max}\right)^2+\left(C_\partial\prod_{k=1}^{k=g}k_k\right)_i^{\prime}a\psi_{Vi1}\Delta\sigma_{grti}^2- \\ -C_\partial\prod_{k=1}^{k=g}k_k a\psi_{V1j}\left(\sigma_{grt1}^{\max}\right)^2(1-\rho_1)^2\left[1+\left(1-a\psi_{V1j}\right)^{N_i-1}\right]- \\ -\left(C_\partial\prod_{k=1}^{k=g}k_k\right)_i^{\prime}a\psi_{Vij}\left[1+\left(1-a\psi_{Vij}\right)^{N_i-1}\right]\Delta\sigma_{grti}^2 \end{array}\right)\right)+\right. \\ \quad \sum_1^N C_\partial\prod_{k=1}^{k=g}k_k a\psi_{V1j}\left(\sigma_{grt1}^{\max}\right)^2(1-\rho_1)^2\left[1+\left(1-a\psi_{Vij}\right)^{N-1}\right]+ \\ \quad +\sum_{i=1}^n\sum_1^{N_i}\left(C_\partial\prod_{k=1}^{k=g}k_k\right)_i^{\prime}a\psi_{Vij}\left[1+\left(1-a\psi_{Vij}\right)^{N_i-1}\right]\Delta\sigma_{grti}^2+ \\ \quad \left.+\sum_{i=1}^n\rho_i\left(\sigma_{grti}^{\max}\right)^2C_\infty(t,\tau)f_i(t,\tau)\right\}- \\ -\sum_1^{ki1}\Delta W_{npij(j-1)}-\Delta W_{npi(i-1)} \end{array}\right\rangle}{k_0^2R_{grt.sez}k_{\rho gr}ds}}}\;\frac{2}{\sqrt{\pi}\,aY(l)}\left(l_u+4l_{sh}\cdot tg\,\alpha\cdot\sin\alpha\right) \tag{6}$$

$$\begin{cases} l(t_0,\tau)+\sum_{i=1}^n l_{\text{мик}}+\sum_{j=1}^{ж}\Delta l+ \\ +\sum_1^n\sum_1^{ж}\dfrac{\dfrac{\left\langle\left[k_{Ii}(t)\varphi_{11}(\sigma_i)+k_{IIi}(t)\varphi_{21}(\sigma_i)\right]\left\langle\left[\dfrac{k_{Ii}(t)\varphi_1(\sigma_i)}{E_{grt}}+\dfrac{k_{IIi}(t)\varphi_2(\sigma_i)}{E_{grt}}\right]-\int_{t_0}^{t}\left[k_{Ii}(t)\varphi_{11}(\sigma_i)+k_{IIi}(t)\varphi_{21}(\sigma_i)\right]\dfrac{\partial}{\partial\tau}C(t,\right.}{2\pi\left(k_{\rho gr}R_{grt,t}\right)^4m_j^4(t,\tau)\left[\dfrac{1}{E_{grt}}+C_\partial\prod_{k=1}^{k=g}k_k a\psi_{Vi}\right]^2}}{\dfrac{\int_0^{\varepsilon_R}\left[\sigma_{ti}+\sum_1^n m_{in}(\varepsilon_R-\varepsilon_{ti})^n\right]d\varepsilon-\dfrac{1}{2}\sigma_{ti}\varepsilon_{ti}-\sum_1^n\sum_1^{ж}W_{npi(i-1)}-\sum_1^{ж}\Delta W_{np1j(j-1)}}{\left[k_{Ii}(t)\varphi_{11}(\sigma_i)+k_{IIi}(t)\varphi_{21}(\sigma_i)\right]\left\langle\left[\dfrac{k_{Ii}(t)\varphi_1(\sigma_i)}{E_{grt}}+\dfrac{k_{IIi}(t)\varphi_2(\sigma_i)}{E_{grt}}\right]-\int_{t_0}^{t}\left[k_{Ii}(t)\varphi_1(\sigma_i)+k_{IIi}(t)\varphi_2(\sigma_i)\right]\dfrac{\partial}{\partial\tau}C(t,\tau)\right.}} \end{cases}$$

where $A^*=\left(k_{\rho gr}R_{grt}\right)^2m_j^2(t,\tau)\left[\frac{1}{Egrt}+C\partial\prod_{k=1}^{k=g}k_k a\psi_{Vij}\right].$

The calculation of the strength of the soil under non-stationary modes of successively decreasing loading mode is performed by the equation:

$$R_{gr}(t,\tau)=\sqrt{\sqrt{\frac{2E_{grt}\left\langle \begin{array}{l}\int_0^{\varepsilon_R}\sigma_{ti}+\sum_1^3\left[m_{in}\left(\varepsilon_R-\varepsilon_{ti}\right)^n\right]d\varepsilon- \\ -\left\{\sum\sum\left(\left(\begin{array}{l}C_\partial\prod_{k=1}^{k=g}k_k a\psi_{V11}\left(\sigma_{grt1}^{\max}\right)^2+\varepsilon_{xi+1}^{nc}\sigma_{grti+1}^{\max}\left(1-\rho_{i+1}\right)- \\ -\left(C_\partial\prod_{k=1}^{k=g}k_k\right)_i^{\prime}a\psi_{Vij}\left(\sigma_{grti}^{\max}\right)^2\left(1-\rho_i\right)^2\left[1+\left(1-a\psi_{Vij}\right)^{N_i-1}\right]\end{array}\right)\right)+ \begin{array}{l} \\ +\sum_{i=1}^{n}\sum_1^{N_i}\left(C_\partial\prod_{k=1}^{k=g}k_k\right)_i^{\prime}a\psi_{Vij}\left(\sigma_{grti}^{\max}\right)^2\left(1-\rho_i\right)^2\left[1+\left(1-a\psi_{Vij}\right)^{N_i-1}\right]+ \\ +\sum_{i=1}^{n}\rho_i\left(\sigma_{grti}^{\max}\right)^2C_\infty(t,\tau)f_i(t,\tau)-\sum_{i=2}^{n}\varepsilon_{xi}^{nc}\sigma_{grti}^{\max}\left(1-\rho_i\right)\end{array}\right\}- \\ -\sum_1^{k_1-1}\Delta W_{npij(j-1)}-\Delta W_c\end{array}\right\rangle \mathrm{g}}{\mathrm{g}k_0^2\,\mathrm{g}R_{grt.ses}\,\mathrm{g}k_{\rho gr}\,\mathrm{g}ds\,\mathrm{g}}}} \tag{7}$$

$$\mathrm{g}\frac{2}{\sqrt{\pi}\,a_1Y(l)}\left(l_u+4l_{sh}\cdot tg\,\alpha\cdot\sin\alpha\right)\mathrm{g}$$

$$\mathrm{g}\left\{\begin{array}{l} l\left(t_{0,\tau}\right)+\sum_{i=1}^{n}l_{\text{мак}}+ \\ +\sum_1^{n}\sum_1^{\text{ж}}\dfrac{\dfrac{\left\langle\left[k_{Ii}(t)\varphi_{11}(\sigma_i)+k_{IIIi}(t)\varphi_{21}(\sigma_i)\right]\left\langle\left[\dfrac{k_{Ii}(t)\varphi_1(\sigma_i)}{E_{grt}}+\dfrac{k_{IIIi}(t)\varphi_2(\sigma_i)}{E_{grt}}\right]-\int_{t_0}^{t}\left[k_{Ii}(t)\varphi_{11}(\sigma_i)+k_{IIIi}(t)\varphi_{21}(\sigma_i)\right]\dfrac{\partial}{\partial\tau}C(t,\tau\right.\right.}{2\pi\left(k_{\rho gr}R_{grt,t}\right)^4m_j^4(t,\tau)\left[\dfrac{1}{E_{grt}}+C_\partial\prod_{k=1}^{k=g}k_k a\psi_{Vi}\right]^2}}{\dfrac{\int_0^{\varepsilon_R}\left[\sigma_{ti}+\sum_1^{n}m_{in}\left(\varepsilon_R-\varepsilon_{ti}\right)^n\right]d\varepsilon-\dfrac{1}{2}\sigma_{ti}\varepsilon_{ti}-\sum_1^{n}\sum_1^{\text{ж}}W_{npi(i-1)}+\sum_1^{\text{ж}}\Delta W_c}{\left[k_{Ii}(t)\varphi_{11}(\sigma_i)+k_{IIIi}(t)\varphi_{21}(\sigma_i)\right]\left\langle\left[\dfrac{k_{Ii}(t)\varphi_1(\sigma_i)}{E_{grt}}+\dfrac{k_{IIIi}(t)\varphi_2(\sigma_i)}{E_{grt}}\right]-\int_{t_0}^{t}\left[k_{Ii}(t)\varphi_1(\sigma_i)+k_{IIIi}(t)\varphi_2(\sigma_i)\right]\dfrac{\partial}{\partial\tau}C(t,\tau)d\right.}}\end{array}\right.$$

where $A^*=\left(k_{\rho gr}R_{grt,t}\right)^2m_j^2(t,\tau)\left[\frac{1}{Egrt}+C_\partial\prod_{k=1}^{k=g}k_k a\psi_{Vi}\right]$.

where σ_{grti}^{max} is the maximum tensile stresses of the cycle at the crack tip.

5 Conclusion

1. The conducted studies allowed us to establish the regularities of changes in the strength of the soil under three-axis block cyclic loading, according to which the destruction and nonlinear deformation of the soil are characterized by the formation and development of micro - and macro-cracks in the planes of extreme equilibrium, initiated by structural defects in the form of pores or voids and shrinkage microcracks.
2. Equations for the change in the strength of clay soil under three-axis regime cyclic loading for stationary, successively increasing and successively decreasing modes are developed, based on the theory of crack mechanics.
3. The resulting equations describe the changes in the strength of the soil under the considered conditions, taking into account the processes of hardening and softening observed in experiments and the effect of braking at different stages of cyclic triaxial

compression, and allow us to assess accurately the load-bearing capacity of the foundation bases under regime loads in order to obtain reliable and economic design solutions.

References

1. Andersen, K.H.: Bearing capacity under cyclic loading-offshore, along the coast, and on land. In: The 21st Bejerrum Lecture presented in Oslo, vol. 46 (2007). https://doi.org/10.1139/T09-003
2. Mirsayapov, I.T., Koroleva, I.V.: Bearing capacity of foundations under regime cyclic loading. In: 15th Asian Reg. Conference on Soil Mech. Geotech. Eng. ARC (2015). https://doi.org/10.3208/jgssp.KAZ-18
3. Far, A., Davie, J.: Tank settlement due to highly plastic clays. In: 6th International Conference on Case Histories in Geotechnical Engineering, August. Arlington (2008)
4. Ling, X., Li, P., Zhang, F., Zhao, Y., Li, Y.: Permanent deformation characteristics of coarse-grained subgrade soils under train-induced repeated load. Hindawi Adv. Mater. Sci. Eng. **15** (2017). https://doi.org/10.1155/2017/6241479
5. Mirsayapov, I.T., Koroleva, I.V., Ivanova, O.A.: Low-Cycle endurance and deformation of clay soils under triaxial cyclic loading. Housing construction (2012)
6. Kutergin, V.N., Pankov, K.V., Kalbergenov, R.G., Karpenko, F.S., Manukin, V.B.: Assessment of changes in soil strength under cyclic loads modeling the impact of storm waves on the structure. Eng. Geo. Hydrogeol. Geocryol. **5** (2015)
7. Hussein, H.K., Zeena, W.S., Adel, H.J.: Behaviour of soft clayey soil improved by fly ash and geogrid under cyclic loading. Civil Eng. J. IRAQ **2** (2020). https://doi.org/10.28991/cej-2020-03091466
8. Elia, G., Rouainia, M.: Investigating the cyclic behaviour of clays using a kinematic hardening soil model. School of Civil Eng and Geosc, vol. 88. Newcastle University, NE1 7RU Newcastle upon Tyne (2016). https://doi.org/10.1016/j.soildyn.2016.06.014
9. Gu, C., Wang, J., Cai, Y., Sun, L., Wang, P., Dong, Q: Deformation characteristics of overconsolidated clay sheared under constant and variable confining pressure, soils and foundations. Jpn Geotech. Soc. **56** (2016). https://doi.org/10.1016/j.sandf.2016.04.009
10. Hicher, P.Y.: Experimental study of viscoplastic mechanisms in clay under complex loading. Geotechnique **66** (2016). https://doi.org/10.1680/jgeot.15.P.203
11. Hu, C., Liul, H.: A new bounding-surface plasticity model for cyclic behaviors of saturated clay. Commun. Nonlinear Sci. Numer. Simul. **22** (2015). https://doi.org/10.1016/j.cnsns.2014.10.023
12. Wang, Y.: Cyclic response of natural soft marine clay under principal stress rotation as induced by wave loads. Ocean Eng. **129** (2017). https://doi.org/10.1016/j.oceaneng.2016.11.031
13. Ni, J., Indraratna, B., Geng, X.Y., Carter, J. P., Chen, Y.L.: Model of soft soils under cyclic loading. Int. J. Geotech. **15** (2015). https://doi.org/10.1061/(ASCE)GM.1943-5622.0000411
14. Lei, H., Li, B., Lu, H., Ren, Q.: Dynamic deformation behavior and cyclic degradation of ultra soft soil under cyclic loading. J. Mater. Civ. Eng. **28** (2016). https://doi.org/10.1061/(ASCE)MT.1943-5533.0001641
15. Ren, X.W., Xu, Q., Teng, J., Zhao, N., Lv, L.: A novel model for the cumulative plastic strain of soft marine clay under long-term low cyclic loads. Ocean Eng. **149** (2018). https://doi.org/10.1016/j.oceaneng.2017.12.028
16. Khan, I., Nakai, K., Noda, T.: Undrained cyclic shear behavior of clay under drastically changed loading rate. Int. J. Geomate. **66** (2020). https://doi.org/10.21660/2020.66.07893

17. Hirai, H.: Assessment of cyclic response to suction caisson in clay using a three-dimensional displacement approach. Mar. Georesources Geotechnol. **36** (2018). https://doi.org/10.1080/1064119X.2017.1386743
18. Liu, Y., Huang, M., Ma, S.: A simplified calculation method for axial cyclic degradation of offshore wind turbine foundations in clay. Mar. Georesources Geotechnol. **38** (2020). https://doi.org/10.1080/1064119X.2019.1566296
19. Feng, J., Wu, X.Y., Zhu, B.L., Yang, Q.X.: Analytical solution to one-dimensional consolidation in unsaturated soils under sinusoidal cyclic loading. Cent. South Univ. **2** (2015). https://doi.org/10.1007/s11771-015-2566-y
20. Zhao, M.-H., Heng, S., Zheng, Y.: Numerical simulation on behavior of pile foundations under cyclic axial loads. J. Central South Univ. **24**(12), 2906–2913 (2017). https://doi.org/10.1007/s11771-017-3704-5
21. Kayumov, R.A., Tazyukov, B.F., Mukhamedova, I.Z.: Identification of mechanical characteristics of a nonlinear-viscoelastic composite by results of tests on shells of revolution. Mech. Compos. Mater. **55**(2), 171–180 (2019). https://doi.org/10.1007/s11029-019-09802-3

Additional Settlement of the Raft-Pile Foundation, Taking into Account the Deformations of the Pile Under Cyclic Loading

Ilizar Mirsayapov(✉)

Kazan State University of Architecture and Engineering, Kazan 420043, Russia

Abstract. The aim of the work is to study the additional settlement of the raft-pile foundation under cyclic loading due to the deformation of the reinforced concrete pile material under the conditions of redistribution of the compression forces of the reinforced concrete pile, taking into account the redistribution of forces between the reinforcement and the concrete of the pile, as well as the pile and the surrounding soil. There are no research results on this problem in the existing literature. Theoretical studies are carried out and the redistribution of forces between the main elements of the raft–pile foundation is considered. The change in the stress-strain state of a compressed reinforced concrete element due to the manifestation of deformations of the vibro-creep of concrete and reinforcement under cohesive conditions is considered. Based on the research, the equation of changes in stresses and constrained deformations of concrete and pile reinforcement under cyclic loading is proposed for the development of a method for calculating the settlement of a raft-pile foundation. For the first time the proposed method allows us to estimate the settlement of the raft-pile foundation taking into account the deformation of the reinforced concrete pile and the compression of the pile under cyclic loading, which is a significant contribution to the theory of calculating pile foundations and provides concrete savings of up to 15% compared to the standard method.

Keywords: Raft-pile foundation · Reinforced concrete pile · Concrete · Reinforcement · Vibration creep · Constrained deformations · Settlement · Soil · Cyclic loading

1 Introduction

In the general case, the settlement of the raft-pile foundation is represented as the sum of: the settlement of a conditional foundation, the pressure and the deformed compression of the reinforced concrete pile [1–5]. The settlement due to the compression of the pile depends on the conditions of joint deformation of all elements of the system «pile cap – piles – soil of the inter-pile space – soil below the pile toe», on the strength and deformation properties of concrete and reinforcement of the pile, on the parameters of cyclic

N. Vatin (Ed.): STCCE 2021, LNCE 169, pp. 160–166, 2021.
https://doi.org/10.1007/978-3-030-80103-8_17

loading. In the case of cyclic loading, the regularities of the formation and development of the raft-pile foundation settlement due to the compression of the reinforced concrete pile have not been studied [6–10]. In this regard, the aim of the study is to develop a method for calculating the settlement of the raft-pile foundation due to the deformation of the pile under cyclic loading [11]. At the same time, we have to establish the regularity of the development of the raft-pile foundation settlement by the compression of the pile under cyclic loading; to develop the equation of joint deformation of concrete and reinforcement of the pile and the surrounding soil; to develop the calculation model of piles as a part of a raft-pile foundation under cyclic loading.

2 Materials and Methods

Under the action of cyclic loads, the deformation of the pile leads to the appearance and development of additional settlement of the raft-pile foundation caused by deformations of the concrete pile during the first loading cycle and deformations of the vibration creep of the concrete during subsequent loading cycles.

This process is considered in the tridimensional formulation, taking into account the joint deformation of all elements of the system «pile cap – piles – soil of the inter-pile space – soil below the pile toe» with a rigid connection of the piles and the pile cap. When determining the stresses in piles we take into account following aspects: the redistribution of forces between the elements of the system in the process of cyclic loading; the joint deformation of the pile cap; piles; the soil of the inter-pile space and the soil below the pile toe; as well as the manifestations of deformations of the vibro-creep of the soil and concrete piles in cramped conditions (Fig. 1).

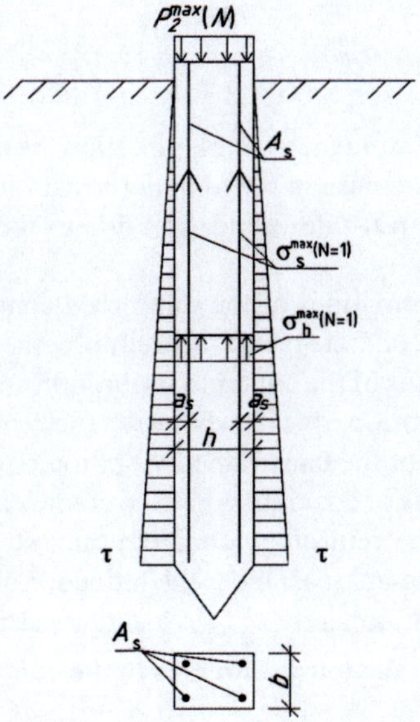

Fig. 1. Stress state of a reinforced concrete pile at the first loading.

In this case, two stages are considered. The first stage corresponds to the first loading cycle up to the maximum load of the cycle and the calculations are made as for static loading. The longitudinal deformations of the reinforcement and concrete piles in conditions close to the central compression due to the adhesion of the materials will be the same (1). Stresses in the reinforcement are calculated by (2).

$$\varepsilon_s^{max}(N=1) = \varepsilon_b^{max}(N=1) = \frac{\sigma_b^{max}(t, t_0)}{\mathrm{E_b}}, \tag{1}$$

$$\sigma_s^{max}(N=1) = \varepsilon_s^{max}(N=1) \cdot E_s = \sigma_b^{max}(N=1) \cdot \frac{\alpha}{V}, \tag{2}$$

where $\alpha = \frac{E_s}{E_b}$, E_S is the Young modulus of steel, E is the Young modulus of concrete, V is the elasticity coefficient of concrete.

The role of transverse bars is reduced to ensuring the stability of the longitudinal compressed bars and therefore does not affect the development of vertical deformations of the pile. The equilibrium equation of external loads and internal forces in concrete and reinforcement of the pile:

$$\begin{aligned} N_2^{max}(N=1) &= \sigma_b^{max}(N=1) \cdot A_p + \sigma_s^{max}(N=1) \cdot A_s \\ &= \sigma_b^{max}(N=1) \cdot A_p\left(1 + \frac{\mu\alpha}{V}\right) \end{aligned} \tag{3}$$

where $\mu = \frac{A_s}{A_p}$ is the percentage of longitudinal reinforcement of the pile, A_p is the cross-sectional area of the pile.

Hence the compressive stress in the concrete at the first loading up to the maximum load of the cycle is:

$$\sigma_{(N=1)}^{max} = \frac{P_2^{max}(N=1)}{A_p\left(1 + \frac{\mu\alpha}{V}\right)}, \tag{4}$$

At the second stage, the influence of the vibration creep deformations of the reinforced concrete pile on the increase in the total deformations of the pile and, as a result, additional settlement of the raft-pile foundation due to the compression of the pile is considered (Fig. 2).

The vibration creep deformations of the reinforced concrete pile are a consequence of the vibration creep of the concrete pile. Steel reinforcement becomes an internal bond that prevents free deformations of the concrete's vibration creep. In a reinforced concrete pile under load, the redistribution of forces between the reinforcement and the concrete of the pile is a consequence of the constrained vibration creep of the concrete.

Constrained deformations of concrete vibration creep lead to the appearance of internally balanced stresses in the reinforced concrete pile: stretching in the concrete and compressing in the reinforcement (5). Under the influence of the deformation difference of the free vibration creep of concrete $\varepsilon_{b\,pl}^{vib}(N)$ and constrained vibration creep of the reinforced element $\varepsilon_{pl}^{vib}(N)_{bs}$ the tensile stresses in the concrete arise (6).

$$\varepsilon_{b\,pl}^{res}(N) = \varepsilon_{b\,pl}^{vib}(N) - \varepsilon_{pl}^{vib}(N)_{bs}, \tag{5}$$

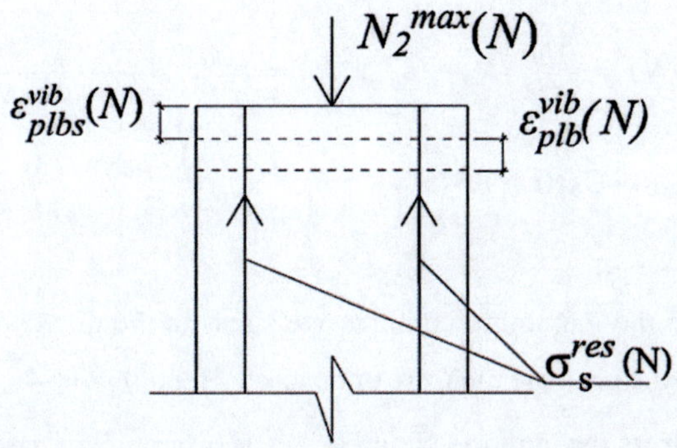

Fig. 2. Residual stress state in the pile cross section under cyclic loading.

$$\sigma_{bt}^{add}(N) = \varepsilon_{b\,pl}^{res}(N) \cdot E_b(N), \tag{6}$$

where $E_b(N)$ is the Young modulus of concrete under cyclic loading.

The highest stress values are located in the area of contact with the reinforcement. Deformations for rebar $\varepsilon_{pl}^{vib}(N)_{bs}$ are elastic and therefore compressive stresses occur in it:

$$\sigma_s^{res}(N) = \varepsilon_{pl}^{vib}(N)_{bs} \cdot E_S. \tag{7}$$

The equilibrium equations of internal forces of a concrete pile with two-sided symmetrical reinforcement have the form:

$$\sigma_s^{res}(N) * A_s = \sigma_{bt}^{add}(N) \cdot A_p. \tag{8}$$

Residual $\sigma_s^{res}(N)$ stresses in the rebar are determined from Eq. (9).

$$\sigma_s^{res}(N) = \sigma_{bt}^{add}(N)\frac{A_b}{A_s} = \frac{\sigma_{bt}^{add}(N)}{\mu}. \tag{9}$$

Substituting in (9) the deformations expressed in terms of the stress in (6)–(8):

$$\frac{\sigma_{bt}^{add}(N)}{E_b(N)} = \varepsilon_{b\,pl}^{vib}(N) - \frac{\sigma_{bt}^{add}(N)}{\mu E_s}. \tag{10}$$

After the transformations from (10), we determine the value of the additional tensile stresses in the concrete, Eq. (11). Based on Eq. (5), we determine the constrained deformations of the vibration creep of the concrete pile under cyclic loading, Eq. (12). Deformations of the free vibration creep of concrete piles are determined by the method proposed by I.T. Mirsayapov [14–16], Eq. (13).

$$\sigma_{bt}^{add}(N) = \frac{\varepsilon_{b\,pl}^{vib}(N) \cdot E_s}{\frac{1}{\mu} + \frac{\alpha}{V_p}}, \tag{11}$$

$$\varepsilon^{vib}_{pl\ bs}(N) = \varepsilon^{vib}_{b\ pl}(N) - \varepsilon^{res}_{b\ pl}(N) = \varepsilon^{vib}_{b\ pl}(N) - \frac{\sigma^{add}_{bt}(N)}{E_b(N)}, \tag{12}$$

$$\varepsilon^{vib}_{b\ pl}(N) = \sigma^{max}_b(t, t_0) \cdot C_\infty(t, \tau) \cdot \left\langle \begin{array}{c} \left[1 - e^{-\gamma(t-t_0)}\right] \cdot \rho_b + \\ +\left[1 - (1-a)^N\right] \cdot (1 - \rho_b) \end{array} \right\rangle \cdot S_k\left(\frac{\sigma^{max}_b}{R_b}\right) \cdot f(N), \tag{13}$$

where $\sigma^{max}_b(t, t_0)$ is the maximum compressive stress in the pile concrete under cyclic loading, $C_\infty(t,\tau)$ is the ultimate measure of concrete creep, $\rho_b = \frac{\sigma^{min}_b}{\sigma^{max}_b}$ is the asymmetry coefficient of the stress cycle in the concrete, a and γ are creep parameters, $S_k\left(\frac{\sigma^{max}_b}{R_b}\right)$ is the creep nonlinearity function of concrete deformations, m_n and η_n are non-linearity parameters, $f(N)$ is the growth function of the nonlinear part of the vibration creep deformations.

The modulus of concrete deformation under cyclic loading is determined in accordance with the method proposed by I.T. Mirsayapov according to the Eq. (14).

$$E_b(t, \tau) = \frac{\sigma^{max}_b(t, \tau)}{\varepsilon_{be} + \varepsilon^{vib}_{b\ pl}(N)}, \tag{14}$$

where $\sigma^{max}_b(t, \tau)$ is the maximum stress in the concrete pile at time $t = N$ (where N is the number of loading cycles), ε_{be} and $\varepsilon^{vib}_{b\ pl}(N)$ are elastic and plastic deformations of the pile at the same time, respectively.

Given the formula (13) and that $\varepsilon_{be} = \frac{\sigma^{max}_b(t,\tau)}{E_b(t_o)}$, the expression (14) after some transformations is reduced to the form (15).

$$E_b(t, \tau) =$$

$$E_b(t_0) \cdot \left\{1 + C_\infty(t, \tau) \cdot E_b(t_0) \cdot \left\langle\left[1 - e^{-\gamma(t-t_0)}\right]\rho_b + \left[1 - (1-a)^{N_i}\right](1 - \rho_b)\right\rangle \cdot S_k\left(\frac{\sigma^{max}_b}{R_b}\right) \cdot f(N)\right\}^{-1}, \tag{15}$$

where $E_b(t_0)$ is the initial Young modulus of concrete.

3 Results and Discussions

Taking into account the above information, we obtain the equation of additional precipitation due to compression of the pile trunk under cyclic loading $\Delta S_C(N)$. In this case, the additional sediment $\Delta S_C(N)$ is represented as the sum:

$$\Delta S_C(N) = \Delta S_{C1} + \Delta S_{C2}(N), \tag{16}$$

where ΔS_{C1} is the additional settlement of the pile at the first loading to the maximum load, $\Delta S_{C2}(N)$ is the additional settlement of the pile under cyclic loading due to the development and accumulation of constrained deformations of concrete vibration creep.

In general, the additional settlement $\Delta S_C(N)$ (the additional settlement at the maximum load of the cycle is considered) is represented as:

$$\Delta S_c(N) = \left[\varepsilon_b^{max}(N=1) + \varepsilon_{pl\ bs}^{max}(N)\right](l - a_c), \tag{17}$$

where l is the length of the pile, a_c is the size of the cross-section of the pile.

Then taking into account (1), (4), (11)–(13) the expression (17) is reduced to the form:

$$\Delta S_c(N) = \left\{ \frac{P_2^{max}}{A_{CB}\left(1+\frac{\mu\alpha}{V}\right)\cdot E_b(t_0)} + \frac{P_2^{max}(N)}{A_{CB}\left(1+\frac{\mu\alpha}{V}\right)} \cdot c_\infty(t,\tau)\cdot f(N)\cdot S_k\left(\frac{\sigma_b^{max}}{R_b}\right)\cdot \right.$$
$$\left. \left\langle \left[1 - e^{-\gamma(t-t_0)}\right]\cdot \rho_b + [1-(1-a)^N]\cdot(1-\rho_b) \right\rangle \cdot \left[1 - \frac{E_S}{\left(\frac{1}{\mu}+\frac{\alpha}{V}\right)\cdot E_b(N)}\right] \right\} \cdot (l - a_c), \tag{18}$$

where P_{2max} is the longitudinal force in the pile at the first loading up to the maximum load of the cycle, P_{2max} (N) is the longitudinal force in the pile under cyclic loading, at $N > 1$ cycle.

4 Conclusions

The theoretical studies of the stress-strain state of a reinforced concrete pile as part of a raft-pile foundation allowed us to determine the main regularities of the development of the pile shaft settlement under cyclic loading conditions, according to which the settlement due to the deformation of the pile shaft occurs because of the cyclic deformation of the compression of concrete and the reinforcement of the pile in cramped conditions. In this case, there is a redistribution of forces between the concrete and the reinforcement of the pile.

Equations for the development of raft-pile foundation settlement are developed, taking into account the compression of the pile shaft under cyclic loading. The resulting equation of the development of raft-pile foundation settlement describes the main characteristic features of the behavior of such foundations observed in laboratory and field experimental studies, and allows us to reliably estimate the settlement of the raft-pile foundation due to the compression of the pile shaft under cyclic loading.

References

1. Katzenbach, R., Leppla, S.: Environment-friendly and economically optimized foundation systems for sustainable high-rise buildings. In: 19th International Conference on Soil Mechanics and Geotechnical Engineering (2011)
2. Katzenbach, R., Leppla, S.: Optimised design of foundation systems for high-rise structures. In: Proceedings of the 6th International Conference on Structural Engineering, Mechanics and Computation, pp. 2042–2047 (2016). https://doi.org/10.1201/9781315641645-338

3. Bokov, I.A., Fedorovskii, V.G.: On the calculation of groups of piles using mutual influence coefficients in the elastic half-space model. Soil Mech. Found. Eng. **54**(6), 363–370 (2018). https://doi.org/10.1007/s11204-018-9482-8
4. Bokov, I.A., Fedorovskii, V.G.: On the applicability of the influence function obtained from single-pile calculations for the calculation of pile groups. Soil Mech. Found. Eng. **55**(6), 359–365 (2019). https://doi.org/10.1007/s11204-019-09549-y
5. Hirai, H.: Assessment of cyclic response to suction caisson in clay using a three-dimensional displacement approach. Mar. Georesources Geotechnol. **36**, 805–817 (2018). https://doi.org/10.1080/1064119X.2017.1386743
6. Safin, D.R.: Experimental studies of a weak clay base reinforced with sand piles. In: IOP Conference Series: Materials Science and Engineering, vol. 962, no. 3, p. 032020. IOP Publishing Ltd. https://doi.org/10.1088/1757-899X/962/3/032020.
7. Ren, X.-W., Xu, Q., Teng, J., Zhao, N., Lv, L.: A novel model for the cumulative plastic strain of soft marine clay under long-term low cyclic loads. Ocean Eng. **149**, 194–204 (2018). https://doi.org/10.1016/j.oceaneng.2017.12.028
8. Ni, J., Indraratna, B., Geng, X.Y., Carter, J.P., Chen, Y.L.: Model of soft soils under cyclic loading. Int. J. Geomech. **15**, 212 (2015). https://doi.org/10.1061/(ASCE)GM.1943-5622.0000411
9. Mirsayapov, Ilizar T., Shakirov, M.I.: Combined plate-pile foundations settlement calculation under cyclic loading. In: IOP Conference Series: Materials Science and Engineering, vol. 890, p. 012069 (2020). https://doi.org/10.1088/1757-899X/890/1/012069
10. Ter-Martirosyan, Z.G., Sidorov, V.V.: Interaction of a long barrette with a single-layer and double-layer base. Housing Construct. **1**, 36–39 (2010)
11. Mirsayapov, Ilizar T., Shakirov, M.I.: Bearing capacity and settlement of raft-pile foundations under cyclic loading. In: Proceedings of the 1st International Conference on Energy Geotechnics, pp. 423–428. Energy Geotechnics (2016)
12. Mirsayapov, Ilizar T., Koroleva, I.V., Ivanova, O.A.: Low-Cycle endurance and deformation of clay soils under three-axis cyclic loading. Housing Construct. Moscow **9**, 6–8 (2012). 18083643/0044–4472
13. Mirsayapov, Ilizar T., Koroleva, I.V.: Bearing capacity of foundations under regime cyclic loading. In: Proceedings of the 15th Asian Regional Conference on Soil Mechanics and Geotechnical Engineering, pp. 1214–1217. ARC New Innovations and Sustainability (2016)

Calculation of the Endurance of Reinforced Concrete Bending Elements by the Method of Limit Stresses

Ilizar Mirsayapov(✉)

Kazan State University of Architecture and Engineering, Kazan 420043, Russia

Abstract. In reinforced concrete bent structures under cyclic loading of the stationary regime, inelastic deformations of vibration creep in the concrete of the compressed zone form and develop under connected conditions. For this reason, the conditions for the deformation of concrete in the compressed zone are non-stationary, even when the external load is stationary. Experimental and theoretical studies of the behavior of the reinforced concrete bending element were carried out. The deformation mode of the concrete of the compressed zone of the bending element was established under the stationary mode of cyclic loading. To assess the endurance of concrete compressed zone under such deformation conditions, studies were carried out using methods of fracture mechanics of elastic-plastic materials and equations of endurance of compressed zone concrete for non-stationary deformation conditions were obtained. On the basis of the conducted research, the equation of the endurance of concrete of the compressed zone is developed for practical calculations of reinforced concrete bending elements under stationary conditions of repeated cyclic loading. The proposed method allows the most accurate assessment of the stress-strain state of concrete in the compressed zone and the processes of concrete change from the point of view of fracture mechanics, which is a significant contribution to the theory of fatigue strength and provides concrete savings of up to 25% compared to existing methods.

Keywords: Reinforced concrete · Compressed concrete zone · Endurance · Cyclic loading · Stationary loading · Mechanics of fracture · Vibration creep · Inelastic deformations

1 Introduction

Under the action of repeated cyclic loads of stationary mode in the concrete of the compressed zone of reinforced concrete bending elements, inelastic deformations of vibration creep are manifested and develop. Due to the fact that vibration creep deformations development under cohesive conditions, additional stresses appear in the concrete of the compressed zone and the longitudinal stretched reinforcement as the number of loading cycles increases. In this case, simultaneously with the change in the stresses in the concrete of the compressed zone of the bent element, the coefficient of asymmetry of the loading cycle also changes [1–6]. In the process of cyclic loading in the concrete of the

N. Vatin (Ed.): STCCE 2021, LNCE 169, pp. 167–174, 2021.
https://doi.org/10.1007/978-3-030-80103-8_18

compressed zone of the bent element, the stresses decrease and the load cycle asymmetry coefficients increase with an increase in the number of loading cycles [7–14]. In this regard, there is a need to develop a new method for calculating the fatigue strength of concrete in the compressed zone of bent reinforced concrete elements under stationary cyclic loading conditions [14–19, 20].

2 Materials and Methods

When the load is applied repeatedly, the change in the stress-strain state over time of the reinforced concrete bending bar can lead to the fact that the limit state will occur as a result of the exhaustion of the resource of concrete or reinforcement. Therefore, to assess the endurance of a reinforced concrete element, it is necessary to be able to assess the bearing capacity of concrete in a compressed zone.

The current stresses $\sigma_b^{max}(t, \tau)$ in the concrete of the compressed zone are represented as the sum of the initial $\sigma_b^{max}(t_0)$ and additional stresses $\sigma_b^{add}(t)$. Additional stresses in the concrete of the compressed zone due to the vibration creep of the concrete under cohesive conditions are based on (1) and (3). It follows that, with an increase in the number of loading cycles, the stresses in the concrete of the compressed zone decrease.

$$\sigma_b^{max}(t, \tau) = \sigma_b^{max}(t_0) + \sigma_b^{add}(t) \tag{1}$$

$$\sigma_b^{max}(t_0) = \frac{M_{max}}{\omega \cdot b \cdot x(h_0 - \gamma \cdot x)} = \frac{2M_{max}}{\xi\left[(1+\lambda) - 0.33\xi\left(\lambda^2 + \lambda + 1\right)\right]bh_0{}^2} \tag{2}$$

$$\sigma_b^{add}(t) = \left\{-\frac{h_0}{x} \cdot E_s \int\limits_{t_0}^{t} \sigma_b^{max}(\tau) \frac{\partial}{\partial \tau}\left[\frac{1}{E(t)} + C(t, \tau)\right]dt\right\} \cdot A_s\left[\frac{1}{A_{red}} - \frac{e_0(h - x_p)}{\mathcal{J}_{red}}\right] \tag{3}$$

$$\xi = \frac{-\mu\alpha \pm \sqrt{(\mu\alpha^2) + 2\mu\alpha(1 - \lambda^2)}}{1 - \lambda^2}$$

Where ξ relative height of the compressed zone, $\mu = \frac{A_s}{bh_0}$, $\alpha = \frac{E_s}{E_b}$.

At the initial loading stage, the stress cycle asymmetry coefficient in the concrete of the compressed zone $\mathcal{P}_{bt_0}$ is equal to cycle asymmetry coefficient of the external load $\mathcal{P}_M$. Under the action of cyclic loads due to the manifestation of vibration creep of concrete in the associated conditions, there is a continuous change in $\mathcal{P}_{bt}$. At the time (t), the stress cycle asymmetry coefficient of the concrete of the compressed zone can be represented as:

$$\mathcal{P}_b(t) = \frac{\sigma_b^{max}(t)\mathcal{P}_M - \frac{h_0}{h}E_sA_s\left[\frac{1}{A_{red}} - \frac{e_0(h - x_p)}{\mathcal{J}_{red}}\right] \cdot \int\limits_{t_0}^{t} \sigma_b^{max}(\tau)\frac{\partial}{\partial \tau}\left[\frac{1}{E_b(t)} + C(t, \tau)\right]dt}{\sigma_b^{max}(t_0) - \frac{h_0}{h}E_sA_s\left[\frac{1}{A_{red}} - \frac{e_0(h_0 - x_p)}{\mathcal{J}_{red}}\right] \cdot \int\limits_{t_0}^{t} \sigma_b^{max}(\tau)\frac{\partial}{\partial \tau}\left[\frac{1}{E_b(t)} + C(t, \tau)\right]dt} \tag{4}$$

It follows from (4) that as the number of loading cycles increases, the stress cycle asymmetry coefficient of concrete in the compressed zone decreases.

It is known that the objective strength of concrete under repeated cyclic loads is less than the short-term strength. In the calculations, this circumstance is taken into account by introducing a coefficient $k_{b\mathcal{P}}$ to the short-term strength corresponding to the moment of application of the load τ.

$$k_{b\mathcal{P}} = \frac{R\prime_b(t)}{R_b(\tau_1)} \tag{5}$$

Where $R\prime_b(t)$ is the endurance of concrete at the time t, $R_b(\tau_1)$ is the short-term strength of concrete at the time of applying a repeated load (τ_1).

In the elements of reinforced concrete structures, the process of reducing strength is manifested in a more complex form. This is due to the fact that in the concrete of the compressed zone of the bent elements, the maximum stress level of the cycle and the stress cycle asymmetry coefficient $\mathcal{P}_{bt}$ change. The endurance curves of concrete can be constructed experimentally. It is practically impossible to obtain the endurance curve of concrete of the compressed zone in reinforced concrete bending elements by experimental means. There are no proven methods for determining stresses in concrete by direct experiment. The construction of such a curve is associated with the need to use certain additional prerequisites. Data on the fatigue strength of concrete in the compressed zone of reinforced concrete bending elements and, in particular, in over-reinforced elements, can be obtained on the basis of the experimental fatigue resistance curve of the reinforced concrete element and the calculated stress values in the concrete. By comparing the stresses in concrete of samples destroyed by a short-term static load at the time τ_1 and after repeated loading, it is possible to obtain data on the effect of previous variable stresses on the short-term strength of concrete in the compressed zone.

Even under stationary loads under conditions of stress varying in maximum level and in amplitude, it is necessary to establish the concept of endurance under these conditions. There are two possible approaches. The endurance of the concrete of the compressed zone can be found in the following way:

- the value of the stress at the time t immediately preceding the destruction (hereinafter - the moment of destruction) - $\sigma_{b*}^{max}(t, \tau_1)$;
- the value of the stress at the time of application of the maximum load of the cycle τ_1, at which the fatigue failure will occur at the time t - $\sigma_b^{max}(t, \tau_1)$.

The stress change mode relates the values $\sigma_{b*}^{max}(t, \tau_1)$ and $\sigma_b^{max}(t, \tau_1)$. The set of values make up the curves $R_b^*(t, \tau_1)$ and $R_b(t, \tau_1)$ characterising the endurance of concrete under non-stationary stress conditions. It seems that the definition of endurance in terms of $R_b^*(t, \tau_1)$ is more correct from a physical point of view, because it is associated with stresses at the moment of failure. The value $R_b(t, \tau_1)$ is somewhat conditional, since it is associated with the stresses that acted earlier, i.e. at the time of applying the maximum load of the first cycle, leading to destruction. However, when performing practical endurance calculations, the use of the $R_b(t, \tau_1)$ curve is more preferable, because it does not require determining the stresses formed at the time of failure. For $t \approx N = 2.10^6$

cycles the values of $R_b^*(t, \tau_1)$ and $R_b(t, \tau_1)$ characterize the endurance of the concrete of the compressed zone of the reinforced concrete structure. Coefficients $\eta_\sigma^*(t, \tau_1)$ and $n_\sigma(t, \tau_1)$ characterize the relative endurance of concrete in the compressed zone.

$$\eta_{\sigma_b}^*(t, \tau_1) = \frac{R_b^*(t, \tau_1)}{R_b(\tau_1)} \tag{6}$$

$$\eta_{\sigma_b}(t, \tau_1) = \frac{R_b(t, \tau_1)}{R_b(\tau_1)} \tag{7}$$

Due to the fact that the coefficient $\eta_\sigma^*(t, \tau_1)$ is associated with the stresses formed at the moment of destruction, the decrease in its values under the conditions of cyclic load action is associated with three factors: the accumulation of damages, the decrease of stresses and decrease of stress cycle asymmetry coefficient in the concrete of the compressed zone. For using the coefficient $\eta_\sigma^*(t, \tau_1)$, it is necessary to determine the stresses at the moment of failure, which is associated with certain computational difficulties. For calculations, it is more convenient to use the coefficient $n_\sigma(t, \tau_1)$ because it is related to the stresses at the time of load application. In addition, the decrease in the values of $n_\sigma(t, \tau_1)$ over time is due only to a decrease in the fatigue strength of concrete in the compressed zone under variable stress conditions. From the Eq. (3), it can be seen that $n_\sigma(t, \tau_1)$ shows what proportion of the short-term strength should be the stress in the concrete of the compressed zone by the end of loading. The construction provided that fatigue failure occurs at time $t \approx N = 2.10^6$ cycles. Using the Eqs. (8) and (9) we can obtain an Eq. (10) for determining the relative endurance limit of concrete in a compressed zone.

$$\sigma_b^*(t, \tau_1) = \sigma_b(t, \tau_0) \cdot H_\sigma(t, \tau) \tag{8}$$

$$\mathcal{P}_b(t, \tau) = \mathcal{P}_b(t, \phi_0) \cdot H_{\mathcal{P}b}(t, \tau)$$

$$\sigma_b^*(t, \tau_1) = R_b^*(t, \tau) \tag{9}$$

$$\eta_\sigma(t, \tau) = \frac{R_b^*(t, \tau) \cdot H_{\mathcal{P}b}(t, \tau)}{R(\tau) \cdot H_\sigma(t, \tau)} \tag{10}$$

3 Results

The relative endurance limit of compressed zone concrete in reinforced concrete elements does not coincide with the relative endurance limit of concrete prisms at a constant level of maximum cycle stresses, the stress cycle asymmetry coefficient.

For the analytical analysis of the fatigue strength of concrete in the compressed zone, it is necessary to determine the compressive strength of the loaf at variable values of the maximum stresses and the stress cycle asymmetry coefficient.

To determine the fatigue strength of concrete in the compressed zone of a bent reinforced concrete element under the conditions described above, we use the equation of objective strength under non-stationary conditions. To simplify the calculation process, the endurance evaluation is performed at the final stage of the structure operation, immediately preceding the exhaustion of the bearing capacity.

First we calculate the values $\sigma_b^{max}(t_0)$, $\sigma_b^{max}(t)$, $\mathcal{P}_{bt}$, $H_{\sigma_b}(t, \tau)$. Then, based on the fact that the decrease in $\mathcal{P}_{bt}$ and $\sigma_b^{max}(t)$ depends on the vibration creep of the concrete of the compressed zone, the change is divided into *n* number of stages (steps). In general, the breakdown into stages can be completely arbitrary.

In this case, we assume that the stresses during the transition from one block to another change by the same value $\Delta\sigma_{bi}(t)$. To determine $\Delta\sigma_b(t)$, we divide the value $H_{\sigma_b}(t, \tau)$ by *n* (where *n* is the number of loading stages in the block), and then multiply the initial stresses by the same value.

$$\Delta\sigma_b(t) = \frac{H_{\sigma_b}(t, \tau)}{n} \cdot \sigma_b^{max}(t_0) \tag{11}$$

Consider the following sequence of stress changes in the concrete of the compressed zone (Fig. 1).

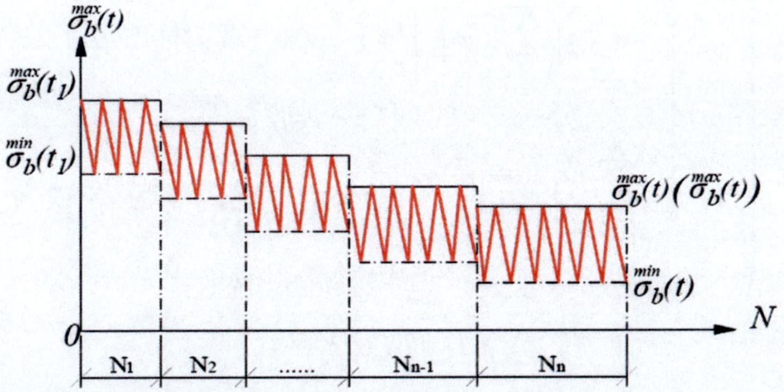

Fig. 1. Diagram of stress changes in compressed zone concrete under repeated loading with M^{max} and $\mathcal{P}_M$ are constant.

In Fig. 1, $\sigma_{b1}^{max} = \sigma_b^{max}(t_0) - \Delta\sigma_b$ – for the number of loading cycles N_1 and the stress cycle asymmetry coefficient P_{b1};

$\sigma_{b2}^{max} = \sigma_b^{max}(t_0) - 2\Delta\sigma_b$ – for the number of loading cycles N_2 and the stress cycle asymmetry coefficient $\mathcal{P}_{b2}$;

$\sigma_{b3}^{max} = \sigma_b^{max}(t_0) - 3\Delta\sigma_b$ – for the number of loading cycles $N3$ and the stress cycle asymmetry coefficient $\mathcal{P}_{b3}$;

$\sigma_{b_{n-1}}^{max} = \sigma_b^{max}(t_0) - (n-1)\Delta\sigma_b$ – for the number of loading cycles $N_{(n\text{-}1)}$ and the stress cycle asymmetry coefficient $\mathcal{P}_{b(n-1)}$;

$\sigma_{bn}^{max} = \sigma_b^{max}(t) - n\Delta\sigma_b$ – for the number of loading cycles N_n and the stress cycle asymmetry coefficient $\mathcal{P}_{bn}$.

With this stress distribution, the number of loading cycles in each block will obey the inequality, $N_1 < N_2 < N_3 < \cdots < N_n$. The stress cycle skewness decrease from block to block, i.e. $\mathcal{P}_{bt_0} > \mathcal{P}_{b_1} > \mathcal{P}_{b_2} > \cdots > \mathcal{P}_{b_n}$.

The objective strength of the concrete of the compressed zone by the time of the end of loading with the above scheme of changes in cyclic stresses is calculated by the formula:

$$R_b(t,\tau) = \frac{2}{b} \cdot \frac{k_{cf}(t)}{\sqrt{\pi \cdot l(t,\tau) \cdot \gamma(l)}} (l_a + 4l_{sh} \cdot tg\alpha \cdot \sin\alpha) =$$

$$= \sqrt{\sqrt{2E_{bt} \left(\int_0^{\varepsilon_R} \left[\sigma_{ti} + \sum_1^3 min(\varepsilon_R - \varepsilon_{ti})^n \right] d\varepsilon - \left\{ \begin{array}{l} \sum_1^n \sum_1^{Ni} \left(\left(\begin{array}{c} C\partial \prod_{k=1}^{k=g} \begin{array}{l} k_k \alpha \psi_{v1i} \sigma_{bt1}^2 + \\ + \mathring{a}_{xi+1}^{nc} \cdot \sigma_{bt1}^{max}(1 - \mathcal{P}_{i+1}) - \end{array} \\ - \left(C\partial \prod_{k=1}^{k=g} k_k \right)_i^1 \cdot \alpha\psi_{vij}(\sigma_{bti}^{max})^2 (1 - \mathcal{P}_i)^2 * \\ * \left[1 + (1 - \alpha\psi_{vij})^{Ni-1} \right] \end{array} \right) \right) + \\ + \sum_{i=1}^n \sum_1^{Ni} \left(C\partial \prod_{k=1}^{k=g} k_k \right)_i^1 \alpha\psi_{vij} (\acute{o}_{bti}^{max})^2 (1 - \mathcal{P}_i)^2 * \\ * \left[1 + (1 - \alpha\psi_{vij})^{Ni-1} \right] + \\ + \sum_{i=1}^n \mathcal{P}_i^2 (\sigma_{bti}^{max})^2 C_\infty(t,\tau) \cdot f_i(t,\tau) - \sum_{i=1}^n \varepsilon_{xi}^{nc} \cdot \sigma_{bti}^{max}(1 - \mathcal{P}_i) \end{array} \right\} - \sum_1^{k=1} \Delta W_{npij(j-1)} + \Delta W_c \right) - * \\ * k_0^2 \cdot R_{bt,su} \cdot k_{fs} \cdot ds}}$$

$$* \frac{2}{\sqrt{\pi} \cdot d_i\, \gamma(l)} \cdot (l_{cl} + 4l_{sh} \cdot tg\alpha \cdot \sin\alpha) *$$

$$* \left\{ \begin{array}{c} l(t,\tau) + \sum_{i=1}^h l_{MHK} + \\ + \sum_1^n \sum_i^{\text{ж}} \dfrac{\dfrac{\langle\langle [k_1\varphi_{11}(\sigma_i) + k_1\varphi_{12}(\acute{o}_i)] \rangle \langle \left[\frac{k_{1i}\varphi_1(\sigma_i)}{E_{bt}} + \frac{k_{1i}\varphi_2(\sigma_i)}{E_{bt}} \right] - \int_{t_0}^t [k_{1i}\varphi_{11}(\sigma_i) + k_1\varphi_{21}(\sigma_i)] \frac{\partial}{\partial\tau} C(t,\tau) d\tau \rangle\rangle^2}{2\pi (k_{\mathcal{P}b} R_{bt,\tau})^4 m_j^4(t,\tau) \left[\frac{1}{E_{bt}} C\partial \prod_{k=1}^{k=g} k_k \alpha \psi_{0ij} \right]^2}}{\dfrac{\int_0^{\varepsilon_R} \left[\sigma_{ti} + \sum_1^3 min(\varepsilon_R - \varepsilon_{ti})^2 d\varepsilon - \frac{1}{2}\sigma_{ti} \cdot \varepsilon_{ti} - \sum_1^n \sum_1^{\text{ж}} \Delta W_{npij(j-1)} + \sum_2^n \Delta W_c \right]}{[k_{zi}(t)\varphi_{11}(\sigma_i) + k_{zi}(t)\ddot{o}_{21}(\sigma_i)] \left\{ \left[\frac{k_{1i}\varphi_1(\sigma_i)}{E_{bt}} + \frac{k_{1i}\varphi_2(\sigma_i)}{E_{bt}} \right] - \int_{t_0}^t [k_{1i}\varphi_1(\sigma_i) + k_1\varphi_2(\sigma_i)] \frac{\partial}{\partial\tau} C(t,\tau) d\tau - A^* \right\}}} \cdot \Delta N_i \end{array} \right\}^{-95} \quad (12)$$

The endurance of structures on the concrete of the compressed zone is evaluated based on the condition:

$$\sigma_b^{max}(t, t_0) \leq R_b(t, \tau) \quad (13)$$

4 Discussions

The studies allowed us to establish that the mode of deformation of the concrete of the compressed zone in the composition of the reinforced concrete bending structure is non-stationary even in the stationary mode of a cyclic load.

The equation of the mechanical state of concrete in the compressed zone of a reinforced concrete bending structure under stationary repeated cyclic loads is developed on the basis of the theory of concrete vibration creep and the mechanics of destruction of elastic-plastic materials.

The resulting equation is adequate and sufficiently accurate, from the point of view of the requirements of practical calculations. It allows us to evaluate the endurance of concrete in a compressed zone under stationary cyclic loading conditions and to obtain reliable and at the same time economical solutions.

References

1. Kim, G., Loreto, G., Kim, J.-Y., Kurtis, K.E., Wall, J.J., Jacobs, L.J.: In situ nonlinear ultrasonic technique for monitoring microcracking in concrete subjected to creep and cyclic loading. Ultrasonics **88**, 64–71 (2018). https://doi.org/10.1016/j.ultras.2018.03.006
2. Li, Q., Liu, M., Lu, Z., Deng, X.: Creep model of high-strength high-performance concrete under cyclic loading. J. Wuhan Univ. Technol.-Mater. Sci. Ed. **34**(3), 622–629 (2019). https://doi.org/10.1007/s11595-019-2096-9
3. Chen, P., Zhou, X., Zheng, W., Wang, Y., Bao, B.: Influence of high sustained loads and longitudinal reinforcement on long-term deformation of reinforced concrete beams. J. Build. Eng. **30** (2020). https://doi.org/10.1016/j.jobe.2020.101241
4. Bouziadi, F., Boulekbache, B., Haddi, A., Hamrat, M., Djelal, C.: Finite element modeling of creep behavior of FRP-externally strengthened reinforced concrete beams. Eng. Struct. **204**, 109908 (2020).https://doi.org/10.1016/j.engstruct.2019.109908
5. Mirsayapov, I.T.: Detection of stress concentration regions in cyclic loading by the heat monitoring method. Mech. Solids. **1**(45), 133–139 (2010)
6. Song, L., Fan, Z., Hou, J.: Experimental and analytical investigation of the fatigue flexural behavior of corroded reinforced concrete beams. Int. J. Concrete Struct. Mater. **13**(1), 1–14 (2019). https://doi.org/10.1186/s40069-019-0340-5
7. Zamaliev, F.S., Zakirov, M.A.: Stress-strain state of a steel-reinforced concrete slab under long-term. Mag. Civil Eng. 12–23 (2018)
8. Tang, H., Chen, Z., Avinesh, O., Guo, H., Meng, Z., Engler-Pinto, C., Kang, H.: Notch insensitivity in fatigue failure of chopped carbon fiber chip-reinforced composites using experimental and computational analysis. Compos. Struct. **10**(16), 112280 (2020)
9. Choe, G., Shinohara, Y., Kim, G., Lee, S., Lee, E., Nam, J.: Concrete corrosion cracking and transverse bar strain behavior in a reinforced concrete column under simulated marine conditions. Appl. Sci. **5**(10) (2020)
10. Gambarelli, S., Ožbolt, J.: Interaction between damage and time-dependent deformation of mortar in concrete: 3D FE study at meso-scale. IOP Conf. Ser. Mater. Sci. Eng. **615**, 012013 (2019)
11. Augeard, E., Ferrier, E., Michel, L.: Mechanical behavior of timber-concrete composite members under cyclic loading and creep. Eng. Struct. **210**, 110289 (2020)
12. Trekin, N.N. Kodysh, E.N., Mamin, A.N., Trekin, D.N., Onana, J.: Improving Methods of Evaluating the Crack Resistance of Concrete Structures, vol. 326, pp. 93.1–93.6. American Concrete Institute, ACI Special Publication (2018)

13. Liang, J., Nie, X., Masud, M., Li, J., Mo, Y.L.: A study on the simulation method for fatigue damage behavior of reinforced concrete structures. Eng. Struct. **150**, 25–38 (2017)
14. Zhang, G., Zhang, Y., Zhou, Y.: Fatigue tests of concrete slabs reinforced with stainless steel bars. Adv. Mater. Sci. Eng. 1–5 (2018)
15. Barcley, L., Kowalsky, M.: Critical bending strain of reinforcing steel and the buckled bar tension test. ACI Mater. J. **3**(116), 53–61 (2019)
16. Luo, X., Tan, Z., Chen, Y.F., Wang, Y.: Comparative study on fatigue behavior between unbonded prestressed and ordinary reinforced reactive powder concrete beams. Mater. TEST **4**(61), 323–328 (2019)
17. Tang, S.W., Yao, Y., Andrade, C., Li, Z.: Recent durability studies on concrete structure. Cem. Concr. Res. **78**, 143–154 (2015)
18. Berrocal, C.G., Fernandez, I., Lundgren, K., Lofgren, I.: Corrosion-induced cracking and bond behaviour of corroded reinforcement bars in SFRC. Compos. B Eng. **113**, 123–137 (2017)
19. Chen, E., Berrocal, C.G., Löfgren, I., Lundgren, K.: Correlation between concrete cracks and corrosion characteristics of steel reinforcement in pre-cracked plain and fibre-reinforced concrete beams. Mater. Struct. **53**(2), 1–22 (2020). https://doi.org/10.1617/s11527-020-01466-z

Field Tests of Combined Pile Raft Foundation Under Cyclic Loading

Ilizar Mirsayapov, Marat Shakirov(✉), and Danil Sabirzyanov(✉)

Kazan State University of Architecture and Engineering, 420043 Kazan, Russia

Abstract. The purpose of the field tests is to study the effect of repetitive cyclic loads on the load-bearing capacity and the settlement development of the pile raft foundation. Experimental studies of this type of foundations were mainly carried out under short-term and stepped static loads. As a result of the research, data were obtained on stress-strain state of the soil mass in the inter-pile space and under the lower end of the piles, as well as information on the change in the forces in the piles during cyclic loading. Determining the influence of cyclic loading on the behavior of the pile raft foundations elements in the future opens up the possibility of creating a method for calculating the settlement and load-bearing capacity that takes into account the influence of cyclic loading on pile raft foundations.

Keywords: Pile raft foundation · Cyclic loadings · Loose soil · Field tests · Foundation settlement · Deformations · Forces · Inter-pile space

1 Introduction

Currently, the increasing density of urban development creates a tendency to an increase the number of buildings and structures storeys, and also leads to the active underground space development for the most efficient use of the urban areas allocated for building construction. In this regard, the loads on the foundation and the soil foundations underneath increase many times, while construction sites that were previously unsuitable for construction are being developed [1]. Therefore, combined pile raft foundations (CPRF) are widely used in construction as one of the ways to solve the problem of increasing the load-bearing capacity and reducing the deformations of the soil foundation. At the same time, the influence of cyclic loading on this type of foundations is poorly studied, since Russian and foreign researchers [2–5] have mainly considered the influence of short-term static loadings.

2 Materials and Methods

2.1 Experimental Research

Field tests for the study of the stress-strain state of the elements of the combined pile raft foundation under cyclic loading were carried out at the test site of the «Department of

N. Vatin (Ed.): STCCE 2021, LNCE 169, pp. 175–182, 2021.
https://doi.org/10.1007/978-3-030-80103-8_19

Fig. 1. General view of the pile raft foundation testing.

Foundations, Basements, Structural Dynamics and Engineering Geology» of KSUAE (Fig. 1).

Before the construction of the pile raft foundation, a pit with dimensions of 3 × 3 m and a depth of 4 m was first excavated, then soil strain sensors were installed at certain depths to determine the deformations and stresses of the soil foundation (Fig. 2).

Plastic sandy loam was used as a soil foundation. The soil was compacted in equal portions. The compaction mode was the same for all the studied samples. After compaction, the soil sample was kept under its own weight for 1 month before testing. To determine the initial physical and mechanical characteristics of the soil foundation, wells were drilled and soil samples were taken using the cutting ring method [6–10]. In the obtained values, the deviation from the specified values was within the range of 0.2% in terms of moisture content and 0.2 kN/m^3 in terms of specific gravity, which confirms the uniformity of the density distribution and moisture content of the prepared sample of clay soil $w = 12\%$, $\rho = 1.65$ g/cm^3, $E = 12.8$ MPa. The strength parameters of the soil sample were also determined: the internal friction angle $\varphi = 20°$, the specific cohesion of soil particles $c = 5$ kPa [11–16].

The piles of the pile raft foundation were made as follows: first, a well with a length of 2000 mm and diameter of 70 mm was drilled, then the pile was reinforced with A400 class reinforcing steel with a diameter of 10 mm. To measure the forces in the piles, strain sensors were glued to the reinforcing bars. After that, the well was filled with fine-grained concrete of class B15, W4. During the entire test, the vertical deformations and forces occurring in the piles were measured. As a raft, a monolithic reinforced concrete raft with dimensions of 2.0 × 2.0 m and a height of 0.2 m was arranged, made of heavy concrete of class B15, the reinforcement frame was made of A400 class reinforcing steel with a diameter of 10 mm and with a pitch of 150 × 150 mm [6, 17–19]. Diagrams of the location of soil pressure sensors and piles with strain gauges installed on them are shown in Fig. 2.

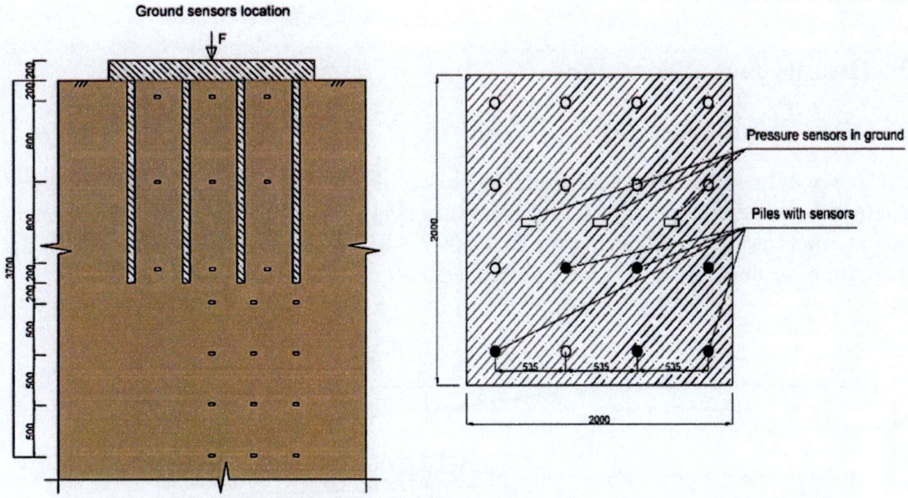

Fig. 2. Installation diagram of soil pressure sensors and piles with strain gauges.

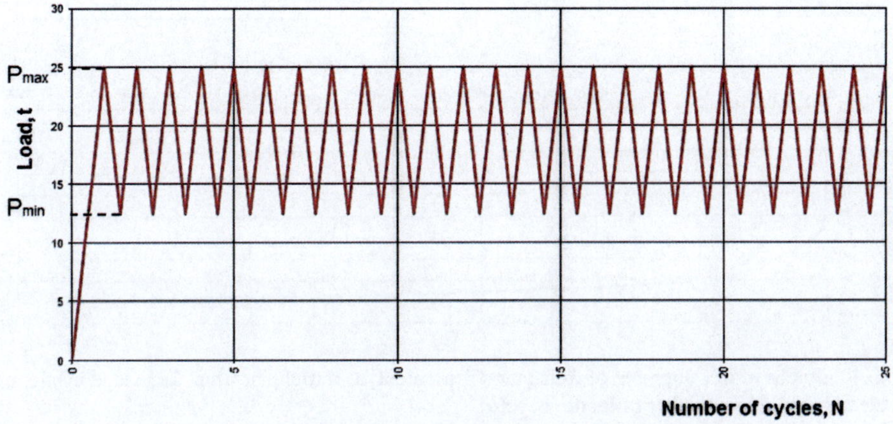

Fig. 3. Cyclic loading mode when testing a pile raft foundation.

The pile raft foundation settlement was also recorded using 6 PAO deflection indicators with a graduation mark of 0.01 mm [20–24]. The vertical loading of the pile raft foundation was carried out using a hydraulic jack. Soil tests under cyclic loading were carried out according to the following method (Fig. 3): first, a vertical static load was applied to the pile raft foundation by steps with interval of 30 min, then, when the settlement $s = 30$ mm, corresponding to the vertical loading $P_{max} = 25$ tons was reached, the unloading of the pile raft foundation to $P_{min} = 12.5$ tons began. The cycles were performed with a purity of 5×10^{-4} Hz (Fig. 3).

The value of 80 mm was taken as the maximum permissible settlement of the pile raft foundation [25–29].

3 Results and Discussions

3.1 Field Test Results

An increase in foundation settlement under cyclic loading was observed throughout the test. In the graphs S from F and S from N (Fig. 4 a, b), an intensive development of settlement is observed during the first 500 cycles of cyclic loading, after which the increase in settlement significantly decreased.

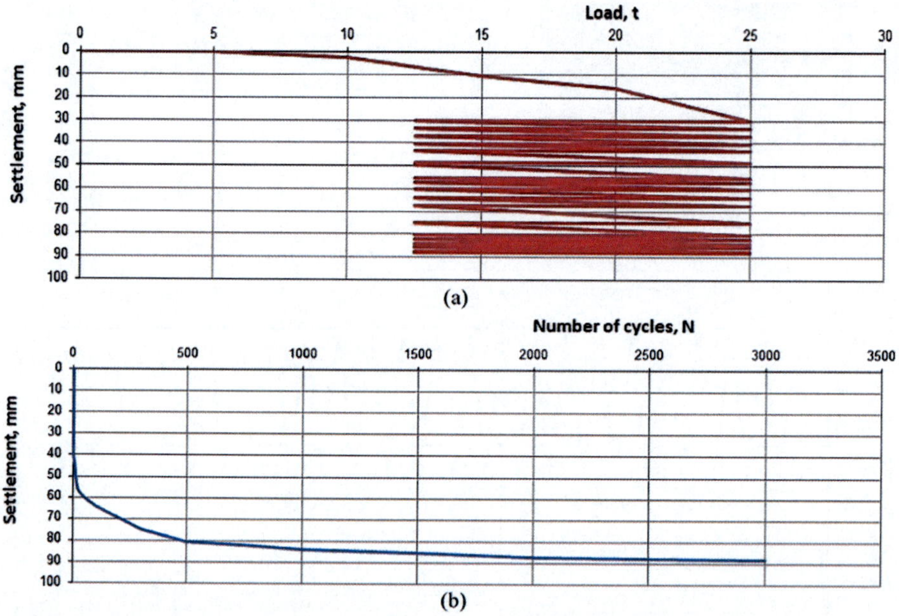

Fig. 4. Graphs of development of foundation settlement: a) settlement from the load at cyclic; b) settlement from the number of loading cycles.

Analysis of the change in the foundation settlement after a different number of cycles shows that an increase in settlement occurs mainly due to an increase in their residual part during cyclic loading (Fig. 4a). At the same time, the amount of settlement changes slightly during one cycle. During the tests, changes in these («elastic») settlements were recorded as the number of loading cycles increased. These settlements decrease slightly during the first 20 cycles. The decrease in the value of «elastic» settlements can be explained by compaction due to a decrease in the volume of soil pores, which is in the terms of development rate outstrips the decrease in the shear modulus of shear deformations between the piles and the soil of the inter-pile space [30, 31]. Since the greatest soil compaction occurs in the initial period of repeated loading, and the change in the adhesion between the soil and the piles is a longer process, the «elastic» settlements of the foundations begin to increase after the first 50 loading cycles. If the limiting state of the foundation is not reached, relative stabilization occurs by the time of 1000 loading cycles, i.e., the F – sup dependence becomes close to linear (Fig. 4a).

As can be seen from the graph S from N (Fig. 4 b), there is a transition of a curvilinear diagram to a straight one, which indicates the transition of soil deformation to a linear stage.

In this case, the foundation settlements change in the process of cyclic loading similar to the soil deformations of the inter-pile space. At the same time, the increase in the foundation settlements is up to 30% compared to the first loading cycle.

Figure 5 shows the changes in the forces in the piles. As the number of cycles increases, the forces in the piles increase due to the redistribution of deformations and stresses from the soil of the inter-pile space to the piles. It should be noted that the greatest forces occur in row and corner piles, and the least in central ones. This is explained by the fact that in the middle zone of the conditional foundation, the central piles are compressed by the most compacted soil, and the corner and row piles interact with areas of less compacted soil outside the grillage raft.

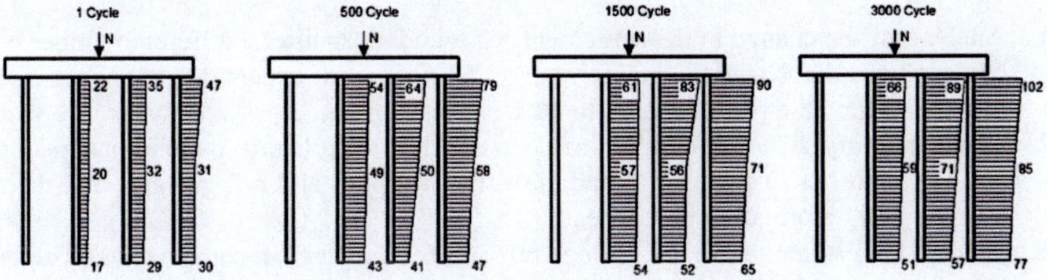

Fig. 5. Diagrams of forces in piles after a different number of cycles, N ($P_{min} = 12.5t$, $P_{max} = 25t$).

Under cyclic loading of a pile raft foundation, the deformations in the soil of the inter-pile space decrease throughout the test (Fig. 6). Thus, a significant decrease in overall deformations is observed under the raft part of the pile raft foundation (up to 2 times).

This is due to the fact that the growth of the general deformation of the soil foundation is limited to piles, in turn, piles deformation is restrained by the soil foundation compacted in the process of cyclic loading of the pile raft foundation.

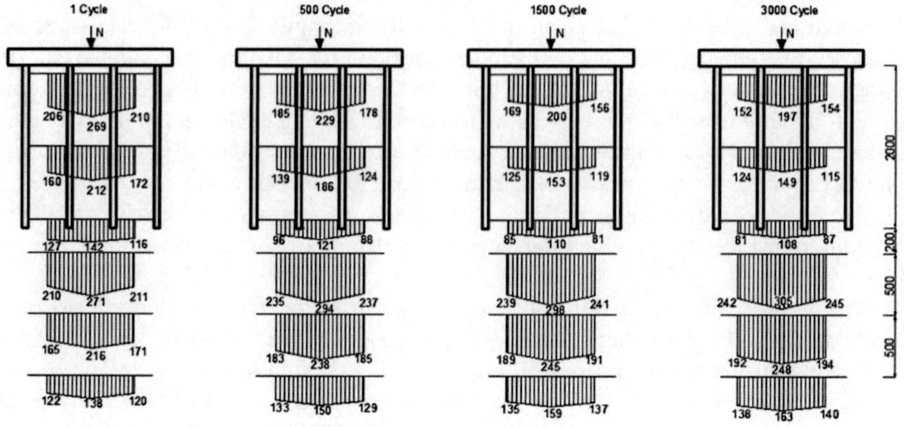

Fig. 6. Diagrams of soil deformations of the inter-pile space for different numbers of cycles, kPa ($P_{min} = 12.5t$, $P_{max} = 25t$).

4 Conclusions

Based on the results of field tests, the following conclusions can be drawn:

1. Analysis of the change in the settlements of foundations after a different number of cycles shows that basically the increase of the settlement occurs due to an increase in their irreversible part and is up to 30% of the original.
2. The maximum value of the cyclic load perceived by the pile raft foundation depends on the conditions of joint deformation of the soil, piles and raft grillage, and their strength and deformation properties.
3. When a load is applied to the grillage raft at a certain stage, a compacted soil zone (core) begins to form under the base of the grillage raft. Cyclic loading leads to an increase in the size of the compacted core and the density of the soil in this zone.
4. The conducted field tests show that during cyclic loading the soil of the inter-pile space is compacted up to 20%, which leads to an increase in the load-bearing capacity of the piles during cyclic loading on the lateral surface.
5. The phenomenon of the compacted zones formation is explained by the fact that the outer rows of piles, as well as piles located in the corners of the grillage, are significantly overloaded, in contrast to the central piles, due to the fact that they do not fall into the compacted zone.
6. The obtained results show that the formation of a compacted core leads to an increase in the load-bearing capacity of the pile in the central zone up to 20%.
7. The conducted experimental studies have shown that under cyclic loading there is a decrease of deformations in the soil of the inter-pile space with an increase in the number of loading cycles, but at the same time deformations and forces in piles increase. At the same time, the forces in the piles, depending on the location in the plan and the level of cyclic load, increase up to 60%.

References

1. Mirsayapov, I.T., Shakirov, M.I.: Combined plate-pile foundations settlement calculation under cyclic loading. IOP Conf. Ser. Mater. Sci. Eng. **890**, 1–8 (2020). https://doi.org/10.1088/1757-899X/890/1/012069
2. Katzenbach, R., Leppla, S., Ramm, H., Seip, M., Kuttig, H.: Design and construction of deep foundation systems and retaining structures in urban areas in difficult soil and groundwater conditions. Procedia Eng. **57**, 540–548 (2013)
3. Katzenbach, R., et al.: Soil-structure-interaction of tunnels and superstructures during construction and service time. Procedia Eng. **57**, 35–44 (2013)
4. Bhaduri, A., Choudhury, D.: Serviceability-based finite-element approach on analyzing combined pile-raft foundation. Int. J. Geomech. **2**(20), 43–51 (2020)
5. Boudaa, S., Khalfallah, S., Bilotta, E.: Static interaction analysis between beam and layered soil using a two-parameter elastic foundation. Int. J. Adv. Struct. Eng. **11**(1), 21–30 (2019). https://doi.org/10.1007/s40091-019-0213-9
6. Mirsayapov, I.T., Koroleva, I.V.: Long-term settlements assessment of high-rise building groundbase based on analytical ground deformation diagram. Procedia Eng. **165**, 519–527 (2016)
7. Mirsayapov, I.T., Koroleva, I.V.: The strength and deformability of clay soils under the regime spatial stress state in view of cracking. Grounds, foundations and soil mechanics 1, pp. 16–23 (2016)
8. Mirsayapov, I.T., Shakirov, M.I.: 1 st International conference on energy geotechnics. In: CRC Press. Conference 2016, ICEGT 724, pp. 423–428 (2016)
9. Mirsayapov, I.T., Koroleva, I.V.: Changes in physical and mechanical characteristics of soil under triaxial loading. In: CRC Press. Conference 2019, GFAC 466, pp. 193–196 (2019)
10. Khuziakhmetov, R., Nurieva, D.: Determination of the Reasons for the Fallof Pile Driving Machine Main Technical Near the Slope of the Foundation Pit. IOP Conf. Ser. Mater. Sci. Eng. **890**, 162304 (2020). https://doi.org/10.1088/1757-899X/890/1/012136
11. Mirsayapov, I.T., Aysin, N.N.: Influence of a deep construction pit on a technical condition of surrounding buildings. In: CRC Press. Conference 2019, GFAC 466, pp. 197–201 (2019)
12. Mirsayapov, I., Koroleva, I.: Calculation models of bearing capacity and deformation of soil foundations with vertical elements reinforced under regime cyclic loading. In: Ferrari, A., Laloui, L. (eds.) SEG 2018. SSGG, pp. 502–507. Springer, Cham (2018). https://doi.org/10.1007/978-3-319-99670-7_62
13. Mirsayapov, I., Sabirzyanov, D.: Bearing capacity of foundations base under combined alternating long-term static and cyclic loading. In: CRC Press. Conference 2018, IOP 700, pp. 43–51 (2018)
14. Mirsayapov, I.T., Koroleva, I.V.: Settlements assessment of high-rise building groundbase using transformed ground deformation diagram. In: IACMAG. Conference 2017, IACMAG 914, pp. 784–792 (2017)
15. Mirsayapov, I.T., Koroleva, I.V.: Calculation models of bearing capacity and deformation of soil foundations with vertical elements reinforced under cyclic loading. In: ICSMGE. Conference 2017, ICSMGE 3254, pp. 2599–2602. Seoul, (2017)
16. Mirsayapov, I.T., Koroleva, I.V.: Strength and deformability of clay soil under different triaxial load regimes that consider crack. Soil Mech. Found. Eng. **53**(1), 5–11 (2016)
17. Mirsayapov, I.T., Koroleva, I.V., Sabirzyanov, D.D.: Calculation model of foundation base settlement at the static and cyclic regime loading. In: ICEGT. Conference 2016, ICEGT 756, pp. 429–434 (2016)
18. Mirsayapov, I.T., Koroleva, I.V.: Computational model of the carrying capacity of a reinforced foundation with cyclic loading. Soil Mech. Found. Eng. **52**(4), 198–205 (2015). https://doi.org/10.1007/s11204-015-9328-6

19. Mirsayapov, I.T., Koroleva, I.V.: Settlements foundation bases under long-term regime loading. In: ARC. Conference 2015, ARC 1596, pp. 1398–1401 (2015)
20. Mirsayapov, I.T., Koroleva, I.V.: Bearing capacity of foundations under regime cyclic loading. In: ARC. Conference 2015, ARC 1596, pp. 1214–1217 (2015)
21. Mirsayapov, I.T., Koroleva, I.V.: Clayey soils rheological model under triaxial regime loading. In: ECSMGE. Conference 2015. ECSMGE 3976, pp. 3249–3254 (2015)
22. Mirsayapov, I.T., Koroleva, I.V.: Fourteenth international symposium on soil rheology prospective trends in theoretical and practical development in rheology and soil mechanics. Soil Mech. Found. Eng. **51**(6), 315–316 (2015)
23. Mirsayapov, I.T., Koroleva, I.V.: Experimental and theoretical studies of bearing capacity and deformation of reinforced soil foundations under cyclic loading. In: IACMAG. Conference 2014, IACMAG 1525, pp. 737–742 (2015)
24. Mirsayapov, I.T., Shakirov, M.I.: Bearing capacity and settlement of raft-pile foundations under cyclic loading. In: ICEGT. Conference 2016, ICEGT 756, pp. 423–428 (2016)
25. Mirsayapov, I.T., Koroleva, I.V., Mirsayapova, I.I.: Evaluation of seismic stability of layered soil bases in areas that are composed of clays and water-saturated sandstones. In: ARC. Conference 2015, ARC 1596, pp. 719–722 (2015)
26. Mirsayapov, I.T., Koroleva, I.V.: Bearing capacity and deformation of the base of deep foundations' ground bases. In: ISSMGE. Conference 2014, ISSMGE 1962, pp. 401–404 (2014)
27. Mirsayapov, I.T., Koroleva, I.V.: Prediction of deformations of foundation beds with a consideration of long-term nonlinear soil deformation. Soil Mech. Found. Eng. **48**(4), 148–157 (2011)
28. Mirsayapov, I.T.: A study of stress concentration zones under cyclic loading by thermal imaging method. Strength Mater. **41**(3), 339–344 (2009)
29. Shakirov, I.F., Gayfullina, V.A.: Researches of the effect from additives on the cement mortars injection properties used in strengthening soil pressure cementation : dig. of art. In: Technical Sciences – from Theory to Practice – XXI International Scientific-Practical Conference, pp. 37–41 (2017)
30. Mirsayapov, I.T., Shakirov, M.I.: Bearing capacity and settlement of raft-pile foundations under cyclic loading: dig. of art. In: 1 st International Conference on Energy Geotechnics, pp. 423–428 (2016)
31. Mirsayapov, I.T., Shakirov M.I.: Behaviour of pile–plate foundations under cyclic loading: dig. of art. In: Baltic Piling Days – Proceedings of the baltic piling days conference/Estonian Geotechnical Society ISSMGE, pp. 31–34 (2012)

Influence of Inelastic Deformations of Reinforcement on the Stress-Strain State of Reinforced Concrete Bending Elements Under Cyclic Loading

Ilizar Mirsayapov(✉)

Kazan State University of Architecture and Engineering, Kazan 420043, Russia

Abstract. The purpose of the study is to identify in the process of exploitation of reinforced concrete structures, buildings that experience various types of cyclic loads, while the behaviour of these structures under low-cycle loading, especially when the working extended reinforcement is deformed beyond the elastic limit, is poorly studied. Based on the results of the study, the development of inelastic deformations in the longitudinal extended reinforcement makes significant changes in the stress-strain state of normal sections of the bent reinforced concrete element under low-cycle loading in comparison with the case elastic deformation of the reinforcement. At the same time, there is an uneven development of reinforcement deformations due to the presence of cracks in the concrete of the extended zone, as well as secondary cracks that begin at the source of normal cracks and develop in directions parallel or slightly inclined to the compressed edge of the elements. Analytical studies of the influence of higher-specified factors on the stress-strain state of normal sections of a reinforced concrete element in the zone of compressed bending are performed. Based on the results study, the equations of the mechanical state concrete of the compressed zone and the extended longitudinal zone of the reinforcement under inelastic deformation of the reinforcement are developed. The results obtained are that the stress-strain state of normal sections in the zone of pure bending under cyclic loading, established on the basis of theoretical studies, allows us to assess most accurately the mechanical properties of the compressed zone and the longitudinal working reinforcement in the conditions of inelastic cyclic deformation of the reinforcement, which is a significant contribution to the theory of fatigue strength to provide savings of concrete and steel reinforcement up to 20% compared to similar calculation methods.

Keywords: Low-cycle endurance · Reinforced concrete · Reinforcement · Inelastic deformations yield strength · Normal sections · Cracks · Normal stresses · Compressed · Zone concrete · Cycle asymmetry coefficient

1 Introduction

In the process of exploitation, the reinforced concrete structures of industrial and energy buildings and structures are exposed to low-cycle loading, when the number of loading

N. Vatin (Ed.): STCCE 2021, LNCE 169, pp. 183–193, 2021.
https://doi.org/10.1007/978-3-030-80103-8_20

cycles does not exceed 50 -100 cycles [1–5]. Therefore, these structures should be calculated in the field of limited fatigue, which makes it possible to significantly increase the level of exploitative loads [6–11]. Currently, reinforced concrete structures experiencing low-cycle loading are calculated according to the calculation method developed for the case of multi-cycle loading, when the number of loading cycles is 2 × 106 [12–16]. For this reason, it is assumed that concrete and reinforcement are deformed in an elastic stage and therefore this approach doesn't correspond to the real nature of the inelastic work of such structures and doesn't allow properly to take into account the property of deforming reinforced concrete structures under low-cycle loading, and as a result prevents the reliable and economical design of such structures [17–21]. In this regard, there is a need to study the features of deforming of reinforced concrete structures under low-cycle loading in conditions of inelastic deforming reinforcement.

2 Materials and Methods

Deformed reinforcement at stresses above the yield strength σ_f leads to significant changes in the stress-strain state of the reinforced concrete bending element under low-cycle loading. Plastic or elastic-plastic deformation of the longitudinal working reinforcement of a reinforced concrete bending element under conditions of low-cycle loading leads to some features, one of which is unequal deformation in sections with cracks and sections between cracks along the length of the extended zone the bending element.

Based on the results of studies conducted under the leadership of A. A. Gvozdeva, uneven deformation of the reinforcement in conditions when the voltage is above the yield strength and steel is estimated by the coefficient: $\Psi_\varepsilon = \varepsilon_{sm}/\varepsilon_s$, where ε_{sm} is the relative deformations of the reinforcement in sections between cracks; ε_s is relative deformations of reinforcement in sections with cracks.

Inelastic deforming reinforcement occurs in the plastic and elastic-plastic stages. In those cases, when the stress in the extended reinforcement reaches a yield point in the sections with cracks, the intensive development of plastic deformations begins with an increase in the number of loading cycles, and in the sections between normal cracks, a reinforcement is deformed in the elastic stage and there is a less intensive development of deformations and therefore Ψ_ε decreases with an increase in the number of loading cycles.

This process will continue until the reinforcement is deformed in the plastic stage. In the elastic-plastic stage of reinforcement in sections with cracks, the stresses in these sections increase and as a result, the process of redistributing of efforts between sections with cracks and sections without cracks begins. Therefore, the development of deformations in sections with cracks is slowed down, and in sections between cracks increases, it leads to an increase in the coefficient unevenness of deformation along the length Ψ_ε.

In the case of inelastic deformation of the reinforcement under low-cycle loading, a self-balanced additional stress state occurs, leading to the formation and development of secondary cracks, which are formed at the source of normal cracks and develop in directions parallel and inclined with a small angle to the neutral axis of the bending element. The reason for the appearance of a self-balanced additional stress state in the form of transverse stresses σ_y and tangential stresses τ_{xy} (Fig. 1), and the appearance a system of secondary cracks is an uneven distribution stresses and deformations in sections with cracks and without cracks.

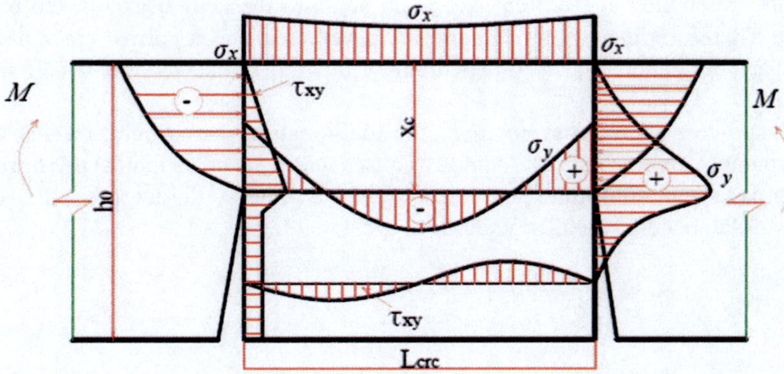

Fig. 1. Stress state of reinforced concrete beams in the block between normal cracks

Let us consider the regularities of the development additional stresses σ_y in the area between the tips of normal cracks. In the zone of the tip of a normal crack, tensile stresses σ_y occur, which, at some point in time, exceed the tensile strength of concrete R_{Bt}.

As the distance from the crack tip increases, the additional stresses σ_y become compressive in the area (*BO*) (Fig. 2) and in the future, this pattern is repeated. Plastic deformation of the reinforcement steel leads to an increase in the height and width of the opening normal cracks at a constant value of the maximum bending moment of the loading cycle.

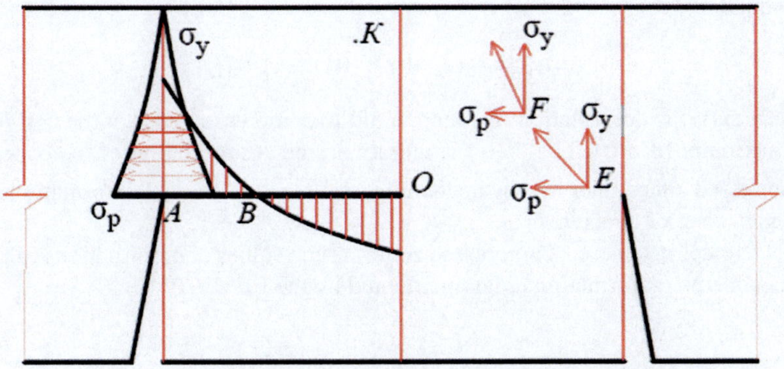

Fig. 2. Scheme of active stresses in various points of block

As a result, the normal crack develops to the boundary of the compressible zone, then its increment in height stops. With an increase in the height and crack opening, the difference between the laws of the distribution of normal stresses in the sections with the crack and between the cracks becomes much larger and consequently increases σ_y and τ_{xy}. In (Fig. 2). the picture of the stress state in the tip of a normal crack under inelastic deformation of the working extended reinforcement is shown. In this zone, there is a total tensile force, and an additional stress σ_y which have significant values and their resultants are applied at the point *(E),* at an angle to the neutral axis of the bending element. Therefore, there is a bend and the development of a normal crack along an inclined and horizontal trajectory, depending on the ratio of the values of tensile stresses σ_p and σ_y.

In the process of crack development, the additional stress σ_y is reduced (Fig. 2) and at some point *(K)* is equal to zero. And as a consequence, the development of a horizontal crack stops and, in the future, additional stresses σ_y and τ_x do not affect the bearing capacity of the bending element under low-cycle loading.

3 Results

Stresses in the Normal Sections of the Bending Element

Elastic-plastic deformation of the longitudinal extended reinforcement of the bending element under low-cycle loading conditions leads to a change in the stress-strain state of normal sections, the formation and development of inelastic deformations in the extended reinforcement leads to the occurrence and accumulation of plastic deformations, as a result, to an increase in the total deformations of the reinforcement under low-cycle loading, which also causes an intensive increment of deflections.

For this reason, there is an increase in total deformations in the concrete of compressed zone due to the development of vibration creep deformation of the concrete and due to the inelastic part deflection of the beam that occurs under the conditions of plastic deformation of reinforcement.

In this case, the total deformations of concrete of the compressed zone are determined by the equation:

$$\varepsilon_b(t_1) = \varepsilon_b(t_0) + \varepsilon_{pl}^{vbc}(t_1) + \varepsilon_{pl}^{adi}(t_1) \tag{1}$$

Where $\varepsilon_b(t_0)$ is deformation of concrete of the compressed zone at the first loading to the maximum load cycle; $\varepsilon_{pl}^{vbc}(t_1)$ is vibration creep deformation of the concrete of the compressed zone under low-cycle loading; $\varepsilon_{pl}^{adi}(t_1)$ is additional deformations of the compressed zone of concrete.

Total stresses in concrete compressed zone, at such values of deformations calculated from the concrete deformation diagram adopted by the FIP-ECB:

$$\sigma_b^{max}(t) = R_b(t)\frac{k\frac{\varepsilon_b(t)}{\varepsilon_{bR}} - \left(\frac{\varepsilon_b(t)}{\varepsilon_{bR}}\right)^2}{1 + (k-2)\left[\frac{\varepsilon_b(t)}{\varepsilon_{bR}}\right]} \tag{2}$$

It can be seen from Eq. (2) that with inelastic deformation of reinforcement under low-cycle loading of the stationary regime, the stresses in concrete compressed zone increase in proportion to the change in the total deformations.

The maximum stresses in the stretched reinforcement under conditions of plastic deformation are limited by the dynamic yield strength of the steel, which is determined by the rate of deformation under low-cycle loading, i.e.

$$\sigma_s^{max}(t_1, t_0) = \sigma_{sy}^{\partial} \tag{3}$$

Where σ_{sy}^{∂} is dynamic yield strength of reinforcement.

Elastic-plastic deformation of reinforcement changes the patterns of development of deformations and, as a result, stresses in the concrete of the compressed zone. In this case, the total deformations of the concrete of the compressed zone are provided in the form of:

$$\varepsilon_b(t) = \varepsilon_b(t_1) + \varepsilon_{pl}^{vbc}(t_2) + \varepsilon_{pl}^{adi}(t_2) \tag{4}$$

Where $\varepsilon_b(t_1)$ is deformations of concrete of the compressed zone corresponding to the deformations of reinforcement at the end a yield site; $\varepsilon_{pl}^{vbc}(t_2)$ is vibration creep deformations of concrete in the compressed zone; $\varepsilon_{pl}^{adc}(t_2)$ is additional deformations of concrete in the compressed zone.

At the end of plastic stage work, the bending element begins to fluctuate relative to the new position of neutral axis, corresponding to the value of the vertical displacements of element at the values of relative deformations at the boundary of the yield site.

At the elastic-plastic stage work of the extended reinforcement, the stresses in concrete zone increase similarly to the elastic stage. As a result of inelastic residual deformation under low-cycle loading, at the stages of load reduction, the compressed concrete fibers of normal section can't return to the first initial state that occurred during the deformations of reinforcement at the boundary yield site and prevent the return of the extended reinforcement to its original state. Due to the fact that the reinforcement is deformed in the elastic-plastic stage, the residual stresses from the elastic part of deformations tend to return it to its original state and create the effect of compression of normal section with extending of the upper drag.

Additional residual stresses in reinforcement in this case are represented as:

$$\sigma_s^{adi}(t_2) = \frac{h_0 - x}{X} \varepsilon_{pl}^{res}(t_2) \cdot E_s^{/}(t) \tag{5}$$

Where $\varepsilon_{pl}^{res}(t_2)$ is residual inelastic deformations of concrete corresponding to this stage. $E_s^{/}(t)$ is elastic-plastic modulus of reinforcing steel.

Based on the theory of an elastic creeping body, the development of residual deformations of vibration creep can be calculated from the equation:

$$\varepsilon_{pl}^{res}(t_2) = \sigma_b^{max}(t_1, t_0)C(t, \tau) + \int_{t_1}^{t_2} \frac{\partial \sigma_b(t_2, t_1)}{\partial \tau} \cdot \left[\frac{1}{E_b} + C(t, \tau)\right] dt \tag{6}$$

Where $\sigma_b^{max}(t_1, t_0)$ is the maximum cycle stress in concrete of the compressed zone, calculated from the deformations of reinforcement at the end of the yield site.

The residual stresses resulting from the accumulation of deformations in the working longitudinal reinforcement lead to the formation and development of tensile stresses in the concrete of the compressed zone, which are calculated by the equation:

$$\sigma_s^{res}(t_2) = A_s \frac{1-\xi}{\xi} E_s'(t) \left[\frac{1}{A_{red}} - \frac{\left(\frac{S_b}{A_b} - a_s\right)(h - x_p)}{J_{red}} \right] \cdot k \tag{7}$$

The stresses in the longitudinal stretched reinforcement are represented as:

$$\sigma_s^{max}(t_2, t_0) = \sigma_{sy}^{\partial} + \sigma_s^{\partial on}(t_2) + \Delta\,\sigma_s(t_2) \tag{8}$$

Where σ_{sy}^{∂} is dynamitic yield strength of the reinforcement; $\sigma_s^{adi}(t_2)$ is stress increment in the elastic– plastic stage.

To simplify the calculation process, the value of dynamic yield strength is determined by the equation:

$$\sigma_{sy}^{\partial} = \kappa_{\partial} \cdot \sigma_{sy} \tag{9}$$

The total stresses in concrete of the compressed zone bending element under low-cycle loading are represented as:

$$\sigma_b^{max}(t_2, t_0) = \sigma_b^{max}(t_1, t_0) - \sigma_b^{adi}(t_2) \tag{10}$$

Stress cycle asymmetry coefficient in section of the bending reinforced concrete element.

Inelastic deformation of the longitudinal stretched reinforcement in the normal section leads to a change in the values of the ratio between the minimum cycle stresses in the concrete of the compressed zone and the stretched reinforcement, and these changes occur according to different patterns in the plastic and elastic-plastic stages of deformation.

In the case of plastic deformation of the valve, the ratio of the minimum and maximum stresses of the valve cycle is constant, but differs from the ratio of the minimum and maximum stresses of the external load $p_m = Mmax/Mmin$. At the minimum values of external loads, the longitudinal stretched reinforcement is deformed elastically and therefore the stress depends on the load value.

In the same cycle, at the maximum load, the stresses in the valve above the yield strength and at this stage, the deformation are equal to the value of the dynamic yield strength, i.e.:

$$\sigma_s^{max}(t_1, t_0) = \sigma_{sy}^{\partial} \tag{11}$$

In this case the asymmetry coefficient of the stress cycle is calculated by the equation:

$$\rho_{st1} = \frac{\sigma_s^{min}(t_1, t_0)}{\sigma_{sy}^{\partial}} \tag{12}$$

Where $\sigma_s^{min}(t_1, t_0)$ is stresses in the reinforcement at the level of the minimum cycle load, it should be kept in mind that, $\rho_{st1} > \rho_M = \frac{M_{max}}{M_{min}}$.

During the elastic-plastic deformation of the reinforcement, due to the development of vibration creep deformations of concrete in compressed zone, additional residual stresses $\sigma_s^{ida}(t_2)$ appear and develop, and the stress cycle asymmetry coefficient is represented as follows:

$$\rho_{st2} = \frac{\sigma_s^{min}(t_2) + \sigma_s^{ida}(t_2)}{\sigma_s^{max}(t_2) + \sigma_s^{ida}(t_2)} \tag{13}$$

Where $\sigma_s^{max}(t_2) = \sigma_{sy}^{\partial} + \Delta\,\sigma_s(t_2)$. It should be added that, $\rho_{st1} > \rho_{st2} > \rho_M$, The ratio of minimum and maximum cycle stresses of concrete in the compressed zone decreases with increasing loading cycles and does not depend on the inelastic operation stage of the longitudinal tensile reinforcement.

In conditions of plastic deformation of longitudinal tensile reinforcement, the ratio of minimum and maximum stresses in the compressed zone of concrete depends on the value of plastic deformation of reinforcement and decreases when the amount of low-cycle loading increases.

In the plastic stage of reinforcement deformation, the stress cycle asymmetry coefficient in the concrete of the compressed zone is represented as follows:

$$\rho_b(t_1) = \frac{\sigma_b^{min}(t_1, t_0)}{\sigma_b^{max}(t_1, t_0)} \tag{14}$$

Where $\sigma_b^{min}(t_1, t_0)$ isstress in the concrete of the compressed zone at the level of the minimum cycle load; $\sigma_b^{max}(t_1, t_0)$ isstress in concrete of the compressed zone at the level of maximum cycle load.

Stresses in concrete of the compressed zone at maximum values of external load are calculated by Eq. (2) based on the total value of deformations:

$$\varepsilon_b^{max}(t_1, t_0) = \varepsilon_b^{max}(t_0) + \varepsilon_b^{\partial on}(t_1) \tag{15}$$

Where $\varepsilon_b^{ida}(t_1)$ isadditional deformations of concrete at the level of maximum cycle load due to manifestation of reinforcement vibration creep deformations.

Thus, at elastic plastic deformation of longitudinal tensile reinforcement the stress cycle asymmetry coefficient in concrete of the compressed zone is represented in the equation:

$$\rho_s(t_2) = \frac{\sigma_b^{min}(t_1, t_0) + \sigma_b^{adi}(t_2)}{\sigma_b^{max}(t_2) + \sigma_b^{adi}(t_2)} \tag{16}$$

Where $\sigma_b^{adi}(t_2)$ additional stresses in the concrete of the compressed zone due to the inelastic deformations of concrete in cohesive conditions.

Based on (7), let us represent expression (16) in the equation:

$$\rho_b(t_2) = \frac{\sigma_b^{min}(t_1, t_0) + \frac{1-\xi}{\xi} A_s E_s'(t) \cdot \left[\frac{1}{A_{red}} - \frac{\left(\frac{S_b}{A_b} - a_s\right)(h - x_p)}{J_{red}}\right]}{\sigma_b^{min}(t_1, t_0) + \frac{1-\xi}{\xi} A_s E_s'(t) \cdot \left[\frac{1}{A_{red}} - \frac{\left(\frac{S_b}{A_b} - a_s\right)(h - x_p)}{J_{red}}\right]} \cdot \frac{\left\{\sigma_b^{max}(t_1, t_0) C(t, \tau) + \int_{t_1}^{t_2} \frac{\partial \sigma_b(t, t_0)}{\partial \tau} \cdot \left[\frac{1}{E_b} + C(t, \tau)\right] dt\right\}}{\left\{\sigma_b^{max}(t_1, t_0) C(t, \tau) + \int_{t_1}^{t_2} \frac{\partial \sigma_b(t, t_0)}{\partial \tau} \cdot \left[\frac{1}{E_b} + C(t, \tau)\right] dt\right\}} \quad (17)$$

The obtained formulas (14 and 17) make it possible to conclude that with the increase of the number of loading cycles the asymmetry coefficient of the stress cycle in the concrete of the compressed zone decreases, i.e. $\rho_b(t_2) < \rho_b(t_1) < \rho_M = \frac{M_{max}}{M_{min}}$.

4 Discussion

During the deformation of the reinforcement in the plastic and elastic-plastic stages, the stress-strain state in the section with a crack and between the cracks is different inelastic and elastic, respectively, requires taking this pattern into account in practical calculations.

To estimate the endurance of normal sections of bendable reinforced concrete structures, under conditions of low-cycle loading under inelastic deformation of reinforcement, average sections corresponding to average deformations in concrete of the compressed zone and reinforcement are considered.

Average deformations in concrete of the compressed zone:

$$\varepsilon_{bm} = \varepsilon_b \cdot \Psi_b \quad (18)$$

Where ε_b isthe deformation of the section with the crack; Ψ_b is coefficient taking into account the uneven distribution of concrete deformations in the compressed zone in the area between the cracks; in practical calculations it is possible to take $\Psi_b = 0, 9$.

The averaged deformations of the reinforcement during its inelastic stage are represented as:

$$\varepsilon_{sm}(N) = \varepsilon_s(N) \cdot \Psi_\varepsilon \quad (19)$$

Where $\Psi\varepsilon$ is the coefficient taking into account the uneven distribution of reinforcement deformations under its working beyond the elastic limit.

In the elastic deformation stage, the average deformations of the reinforcement are represented as:

$$\varepsilon_{sm}(N) = \varepsilon_s(N) \cdot \Psi_s \quad (20)$$

Where Ψs is the coefficient taking into account the work of concrete of the tensile zone between cracks, calculated by Eq. (22), $\Psi\varepsilon 1$ is coefficient taking into account uneven distribution of reinforcement deformation during its plastic deformation.

The coefficient $\Psi\varepsilon 1$ for plastic deformation of the reinforcement is calculated by the equation:

$$\Psi_{s1} = \bar{\Psi}_s - \frac{\varepsilon_s(N) - \varepsilon_{sHy}}{\varepsilon_{sky} - \varepsilon_{sHy}}\left(\bar{\Psi}_s - \bar{\Psi}_{s1}\right) \tag{21}$$

Where $\bar{\Psi}_s$ is the coefficient taking into account the work of concrete of the tensile zone between cracks at elastic deformation of reinforcement at deformations $\varepsilon_{se,u}$; $\varepsilon_s(N)$ is deformations in the section with the crack at the considered moment of time; ε_{sHy} is deformation of the reinforcement at the beginning of the yield point; ε_{sky} isdeformation of the reinforcement at the end of the yield point; $\bar{\Psi}_{s1}$ is the limiting value of $\Psi_{\varepsilon 1}$ at $\varepsilon_s(N) = \varepsilon_{sky}$.

The coefficient $\bar{\Psi}_s$ is determined from the condition that the bending moment from the action of external load in the section with crack and between cracks is the same:

$$\bar{\Psi}_s = 1 - \frac{kN_{bt}(N)/E_sA_s}{\varepsilon_{sHy}} \cdot \beta \tag{22}$$

Where $N_{bt}(N)$ is force in the concrete tensile zone between cracks, taken equal to the force before the formation of cracks, calculated by (12) at values of concrete deformation $\varepsilon_{bt} = \varepsilon_{btu}$; in practical calculations, the values of β and k can be taken as 0.6 and 0.8, respectively. At elastic-plastic deformation of reinforcement coefficient $\Psi\varepsilon$ is calculated by Eq. (21) by replacement of ε_{sHy}, by ε_{sky}, and $\varepsilon_{sky} \approx 0.02$.

5 Conclusion

It has been determined that:

1. The deformation of longitudinal tensile reinforcement of a bendable reinforced concrete element under low-cycle loading is characterized by a number of features, one of which is the non-uniform development of deformations along the length of the bars due to the presence of cracks in the concrete of the extended zone;
2. Features of the stress-strain state of reinforced concrete bendable elements in a zone of bending moments action at deformation of reinforcement in a plastic and elastic-plastic stages which consist in the following are revealed:

   a) Intensive development of deformations along the yield point starts in the sections with crack, and in the sections between the cracks less intensive development of reinforcement deformations takes place, and as a result, non-uniformity of deformations along the length of reinforcement starts to increase;
   b) Secondary cracks appear, which begin in the truth of the normal cracks and develop in directions parallel or slightly inclined to the compressed face of the elements;
   c) Inelastic deformation of reinforcement and concrete of reinforced concrete bending element under low-cycle loading introduces significant changes in the character of stress state of concrete of compressed zone and tensile working reinforcement. Stress cycle asymmetry coefficients in concrete of compressed zone

decrease, and stresses and stress cycle asymmetry coefficients in working tensile reinforcement increase with increasing number of loading cycles at constant values of maximum cycle load and external load cycle asymmetry coefficient.

References

1. Atutis, E., Valivonis, J., Atutis, M.: Deflection determination method for bfrp pressurised concrete beams under fatigue loading. Compos. Struct. **226**, 111182 (2019). https://doi.org/10.1016/j.compstruct.2019.111182
2. Kim, G., Loreto, G., Kim, J.Y., Kurtis, K.E., Wall, J.J., Jacobs, L.J.: In situ nonlinear ultrasonic technique for monitoring microcracking in concrete subjected to creep and cyclic loading. Ultrasonic's **88**, 64–71 (2018). https://doi.org/10.1016/j.ultras.2018.03.006
3. Li, Q., Liu, M., Lu, Z., Deng, X.: Creep model of high-strength high-performance concrete under cyclic loading. J. Wuhan Univ. Technol. Sci. Ed. **34**(3), 622–629 (2019). https://doi.org/10.1016/j.ultras.2018.03.006.
4. Chen, P., Zhou, X., Zheng, W., Wang, Y., Bao, B.: Influence of high sustained loads and longitudinal reinforcement on long-term deformation of reinforced concrete beams. J. Build. Eng. **30**, 101241 (2020). https://doi.org/10.1016/j.jobe.2020.101241
5. Bouziadi, F., Boulekbache, B., Haddi, A., Hamrat, M., Djelal, C.: Finite element modelling of creep behavior of FRP-externally strengthened reinforced concrete beams. Eng. Struct. **204**, 109908 (2020). https://doi.org/10.1016/j.engstruct.2019.109908
6. Mirsayapov, I.T.: Detection of stress concentration regions in cyclic loading by the heat monitoring method. Mech. Solids **45**, 133–139 (2010). https://doi.org/10.3103/S0025654410010164
7. Song, L., Fan, Z., Hou, J.: Experimental and analytical investigation of the fatigue flexural behavior of corroded reinforced concrete beams. Int. J. Concr. Struct. Mater. **13**(1), 1–14 (2019). https://doi.org/10.1186/s40069-019-0340-5
8. Zamaliev, F.S., Zakirov, M.A.: Stress-strain state of a steel-reinforced concrete slab under long-term. Mag. Civil Eng. (2018). https://doi.org/10.18720/MCE.83.2
9. Tang, H., et al.: Insensitivity in fatigue failure of chopped carbon fiber chip-reinforced composites using experimental and computational analysis. Compos. Struct. **244**, 112288 (2020). https://doi.org/10.1016/j.compstruct.2020.112280
10. Choe, G., Shinohara, Y., Kim, G., Lee, S., Lee, E., Nam, J.: Concrete corrosion cracking and transverse bar strain behavior in a reinforced concrete column under simulated marine conditions. Appl. Sci. **20**(4), 1794 (2020). https://doi.org/10.3390/app10051794
11. Gambarelli, S., Ožbolt, J.: Interaction between damage and time-dependent deformation of mortar in concrete: 3D FE study at meso-scale. In: IOP Conference Series: Materials Science and Engineering, vol. 615 (2019). https://doi.org/10.1201/9781315182964-29
12. Augeard, E., Ferrier, E., Michel, L.: Mechanical behaviour of timber-concrete composite members under cyclic loading and creep. Eng. Struct. **210**, 110289 (2020). https://doi.org/10.1016/j.engstruct.2020.110289
13. Trekin, N., Kodysh, E.N., Mamin, A.N., Trekin, D.N., Onana, J.: Improving methods of evaluating the crack resistance of concrete structures. ACI Spec. Publ. **326**, 93 (2018)
14. Liang, J., Nie, X., Masud, M., Li, J., Mo, Y.L.: A study on the simulation method for fatigue damage behavior of reinforced concrete structures. Eng. Struct. **150**, 25–38 (2017). https://doi.org/10.1016/j.engstruct.2017.07.001
15. Zhang, G., Zhang, Y., Zhou, Y.: Fatigue tests of concrete slabs reinforced with stainless steel bars. In: Advances in Materials Science and Engineering (2018). https://doi.org/10.1155/2018/5451398

16. Zhang, G., Zhang, Y., Zhou, Y.: Fatigue tests of concrete slabs reinforced with stainless steel bars. In: Advances in Materials Science and Engineering, 1 (2018). https://doi.org/10.1155/2018/5451398
17. Barcley, L., Kowalsky, M.: Critical bending strain of reinforcing steel and the buckled bar tension test. ACI Mater. J. **116**(3), 63–61 (2019). https://doi.org/10.14359/51715583
18. Luo, X., Tan, Z., Chen, Y.F., Wang, Y.: Comparative study on fatigue behavior between unbonded prestressed and ordinary reinforced reactive powder concrete beams. Mater. Test. **61**(4), 323–328 (2019). https://doi.org/10.3139/120.111323
19. Tang, S.W, Yao, Y., Andrade, C., Li, Z.: Recent durability studies on concrete structure. Cem. Concr. Res. **78**, 143–154 (2015). https://doi.org/10.1016/j.cemconres.2015.05.021
20. Berrocal, C.G., Fernandez, I., Lundgren, K., Lofgren, I.: Corrosion-induced cracking and bond behavior of corroded reinforcement bars in SFRC. Compos. B Eng. **113**, 123–137 (2017). https://doi.org/10.1016/j.compositesb.2017.01.020
21. Chen, E., Berrocal, C.G., Löfgren, I., Lundgren, K.: Correlation between concrete cracks and corrosion characteristics of steel reinforcement in pre-cracked plain and fibre-reinforced concrete beams. Mater. Struct. **53**(2), 1–22 (2020). https://doi.org/10.1617/s11527-020-01466-z

Geometric Modeling of Coil Heat Exchanger Based on Spring-Twisted Channel

Yakov Zolotonosov[1], Iraida Krutova[1](✉), and Ekaterina Vachagina[2]

[1] Kazan State University of Architecture and Engineering, Kazan 420043, Russian Federation
[2] Institute of Energy and Advanced Technologies, Federal Research Center «Kazan Scientific Center of the Russian Academy of Sciences», Kazan 420043, Russian Federation

Abstract. Heat exchangers are widespread in industry, so it is important to have equipment that meets modern requirements. This raises questions on the improvement and modernization of obsolete and the development of new equipment. The authors proposed a heat exchange element made of tightly wound wire with subsequent welding of turns, called spring-twisted channel.

Also, based on the results of the analysis of scientific and technical literature and modern trends in the development of heat exchange equipment, a whole class of modern heat exchangers based on a spring-twisted channel has been proposed.

According to the authors, at present, conical heat exchangers such as «pipe in pipe» are of particular interest, so the object in this work is a high-speed heat exchanger in the form of a truncated cone on the basis of a spring-twisted channel. The purpose of this work is geometric modeling of the surface of the conical coil spring-twisted channel.

A general method of mathematical description of surfaces of spring-twisted channels, which refer to non-linear surfaces formed by the movement of a continuous closed curve along some curvilinear guide.

Equations for constructing the surface of the conical coil spring-twisted channel are given.

Keywords: Heat transfer · Heat exchanger · Conical coil · Spring-twisted channel

1 Introduction

Heat exchangers have become widespread in many industries. Designs of modern heat exchange equipment must ensure the design level of thermal efficiency, also, it must be technological and reliable during the entire service life, safe at its manufacture, installation and operation, provide for the possibility of cleaning heat exchangers, inspection and repair of the apparatus [1].

In a highly competitive market for heat exchange equipment, modern heat exchangers must be highly efficient, simple in design, have low hydraulic resistance and small weight and dimensions [1–3].

The main problem in the development of heat exchangers is the increase of the heat transfer coefficient k, W/ (m K), the most important characteristic of the device.

N. Vatin (Ed.): STCCE 2021, LNCE 169, pp. 194–202, 2021.
https://doi.org/10.1007/978-3-030-80103-8_21

This is possible to achieve this in several ways, which are divided into active, passive and mixed [3–5].

Active methods require direct energy costs from an external source. Such methods include: mixing of the coolant by mechanical means [5, 6], rotation of the heat transfer surface [5, 7], the use of ultrasound [8, 9], electric [10] or magnetic field [11, 12] and others.

Unlike active methods of heat transfer intensification, passive do not need any external force. They affect the flow form of the heat transfer surface. These methods include: the use of insert intensifiers (screw, local and plate flow curlers [5, 13, 14]), finned and other developed heat exchange surfaces on the side coolant with low heat transfer coefficient [15–18].

The analysis of the literature shows that the intensification of heat exchange occurs due to the twisting of the flow, which enhances convective heat transfer by introducing a vortex into the bulk flow and destroying the boundary layer on pipe surface.

According to the authors, the study and development of methods for intensifying heat exchange that are optimal from the point of view of technical and economic indicators is the central issue of modernizing heat exchangers, which becomes especially relevant in conditions of high physical and moral wear and tear of heat exchangers, which are widely used in production processes and housing and communal services.

Considering the fact that today the share of tubular heat exchangers is more than half of the entire fleet of heat exchange equipment, it is advisable to improve the shape of the tubular elements of the heating surfaces.

In this regard, practical interest is developed recently, the class of spring-twisted channels [19]. The straight spring-twisted channel shown in Fig. 1 is a spring of round section, the coils of which are rigidly connected by laser welding [19, 20]. With this technology of manufacturing a heat exchange element, the phenomenon of a sticker is excluded, leading to the embrittlement of the surface layer of metal that occurs in pipes with knurling.

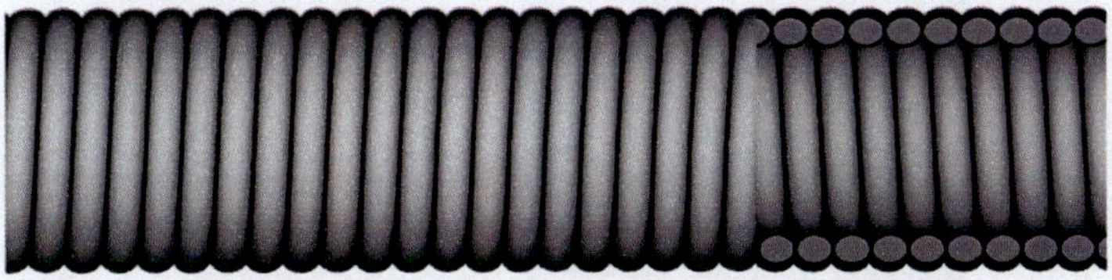

Fig. 1. Straight spring-twisted channel [20]

To intensify heat transfer due to turbulization of the flow of heat carriers, the authors proposed to install intensifiers between the coils of the spring or the flowing part of the spring-coiled pipe to be made in the form of projections 1 and 2 alternating with a given step [19, 21], Fig. 2. Also, the spring-twisted channel can be multi-inbound, Fig. 3 [22]. Wire for the manufacture of a spring-twisted channel may have a round, oval section [19–21] or ovoid [23].

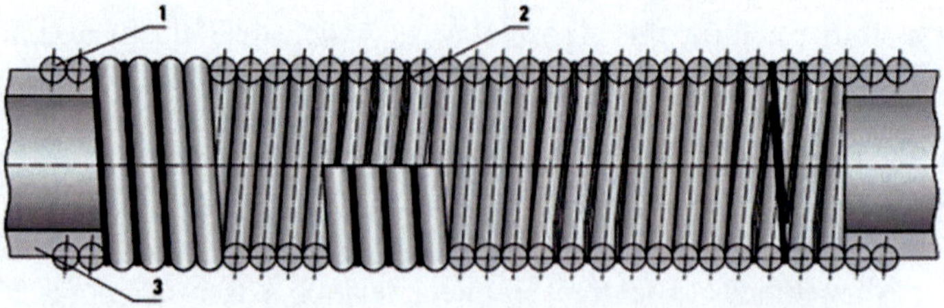

Fig. 2. Straight spring-twisted channel 1 with intensifier 2 [21]

Fig. 3. Heat exchange element in the form of a multi-start helical spiral [22]

On the basis of spring-twisted channels, a number of heat exchangers in the form of a coil, performed in the form of a round truncated cone [24–26], in the exponential curve [27], in the form of a ball [28, 29], oval configuration [30].

Figure 4 shows the appearance of a coil apparatus type «pipe in a pipe» made on a cone, with a heat exchange element in the form of a spring-total channel.

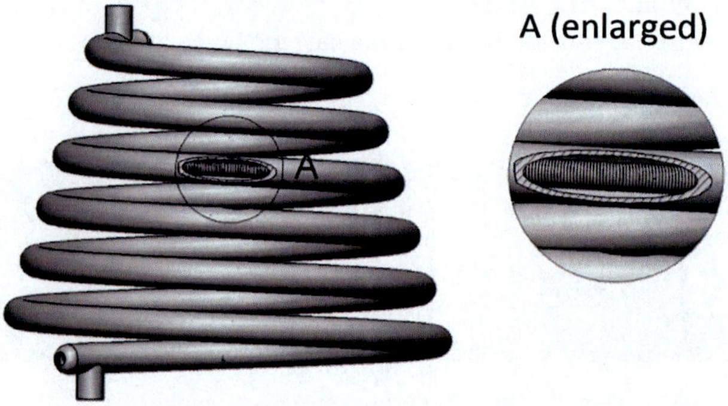

Fig. 4. Coil apparatus type «pipe in pipe» made on cone

According to the data [31–34], conical coil machines have increased efficiency compared to cylindrical heat exchangers such as «pipe-in-pipe».

A heat exchanger is an apparatus with coaxially installed pipes piled into a coil. Moreover, the inner pipe is made in the form of a spring-twisted channel of a round section, and the outer one is made of a smooth cylindrical pipe, Fig. 4.

The conical coil heat exchanger works as follows: with the counter flow of heat transfer, cold water through the fitting enters the inner pipe, simultaneously through the fitting, installed in an external steel coil, hot water is supplied to the annulus space.

With this pattern of motion, the heat agent (liquid) moves along a complex trajectory. First, on the turns of the flow part of the inner coil, where the twisted flow of liquid is realized along the cavities of the spring-twisted channel, and, secondly, along the screw line, determined by the turns of itself coil heat exchanger element.

The heating coolant fed into the intertube space, due to the external screw fining of the inner coil also makes a twisted current characterized by a complex three-dimensional vortex flow structure, which favorably affects heat exchange processes flowing in the annulus space.

When the fluid moves along the curvilinear trajectory of the coil, centrifugal forces appear, the values of which along the cross-section of the channel are different. On the axis of the pipe, where the velocity is maximum, these forces matter most. In the direction to the walls of the pipe, the coolant speed decreases and the effect of the centrifugal effect becomes less. This distribution of forces across the cross-section of the curved canal leads to the emergence of a transverse secondary circulation.

Due to the resulting transverse circulation, the actual fluid velocity in the curved pipe significantly exceeds the average axial flow rate, this results in a significant increase in energy exchange between the flow core and the laminar sublayer and, as a result, to a sharp increase in hydraulic resistance.

The main difficulty of the widespread introduction of this class of heat exchangers is the lack of reliable data on the efficiency of heat exchangers based on spring-twisted channels, research methods, their calculations and design.

The authors of the work [35] on the basis of integral calculus obtained equations for calculation of equivalent diameters of pipe and intertube space in a heat exchanger type “pipe in pipe” with spring-twisted channel. The results of this work can be used in the design and calculation of promising heat exchangers with intense heat transfer.

The purpose of this work is to develop a mathematical description of the geometric model of the conical coil spring-twisted channel with the purpose of further modeling of the heat exchange and hydrodynamic processes.

2 Methods

Surfaces and their geometric description play an important role in the development and design processes of new products.

Currently, in connection with the development and application of computer technology, the analytical method of surface specification has become widespread, allowing relatively easy analysis surface characteristics or quantitative surface - dependent physical characteristics, such as volume, surface area, moment of inertia, etc.

The simplest ways to get a surface are to rotate a two-dimensional object, such as a straight or plane curve around an axis in space, or to move an object, such as a line, polyline, or curve, along some line in space. Such surfaces are called surfaces of revolution and sweeping surfaces, respectively. Examples of surfaces of revolution include a sphere, cone, cylinder, ellipsoid, etc.

Spring-twisted channels refer to non-linear surfaces formed by the motion of a continuous closed curve, along some curvilinear guide. Consider the general method of mathematical description of surfaces of spring-twisted channels, given in the works [19, 21].

Let $\gamma : \vec{r} = \vec{r}(s)$ – guide curve, s is a natural parameter in the normal plane of the curve γ.

Let's represent the radius-vector of a surface point as a sum:

$$\vec{r}(s,\varphi) = \vec{r}(s) + \vec{\rho}(s,\varphi), \tag{1}$$

where φ is the polar angle in the normal plane of the curve γ, calculated from the main normal towards the binormal,

$\vec{\rho}(s,\varphi)$ - corresponding "polar radius", Fig. 5.

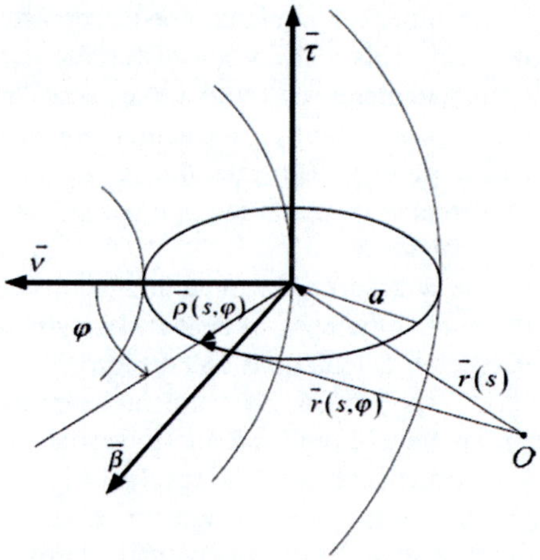

Fig. 5. Surface description scheme [19]

Then

$$\vec{\rho}(s,\varphi) = \rho(s,\varphi)\Big(\vec{\nu}(s)\cos(\varphi) + \vec{\beta}(s)\sin(\varphi)\Big), \tag{2}$$

where $\vec{\nu}(s)$ and $\vec{\beta}(s)$– unit vectors of the main normal and binormal at the point corresponding to the value of the s natural parameter,

$\rho(s,\varphi)$– variable, in general, in two parameters, the radius of the cross-sectional boundary of the channel.

The unit vectors of the tangent, normals, and binormals form a movable orthogonal basis moving along the curve, are computed by the following equations:

$$\vec{\tau}(s) = \vec{r}'(s),\ \vec{\nu}(s) = \frac{1}{k}\vec{r}''(s),\ \ \vec{\beta}(s) = \vec{\tau}(s) \times \vec{\nu}(s),\ k = \left|\vec{r}''(s)\right| \tag{3}$$

Substituting in the Eq. (1) for the radius vector of the $\vec{r}(s,\varphi)$ constraint Eq. (2,3), we get:

$$\vec{r}(s,\varphi) = \vec{r}(s) + \vec{\rho}(s,\varphi) = \vec{r}(s) + \rho(s,\varphi)\Big(\vec{\upsilon}(s)\cos(\varphi) + \vec{\beta}(s)\sin(\varphi)\Big)$$
$$= \vec{r}(s) + \rho(s,\varphi)\left(\frac{1}{k} \times \vec{r}''(s)\cos(\varphi) + \vec{\tau}(s) \times \vec{\upsilon}(s)\sin(\varphi)\right) \quad (4)$$

If the guide curve $\gamma : \vec{r} = \vec{r}(t)$ is a function of some parameter *t*, then unit vectors tangent, normals, and binormals are computed by the equations [19, 21]:

$$\vec{\tau} = \frac{d\vec{r}/dt}{|d\vec{r}/dt|},\ \vec{\upsilon} = \frac{d\vec{\tau}/dt}{|d\vec{\tau}/dt|},\ \vec{\beta} = \vec{\tau} \times \vec{\upsilon}. \quad (5)$$

In this case, the position of surface points can be determined by equality:

$$\vec{r}(t,\varphi) = \vec{r}(t) + \rho(t,\varphi)\Big(\vec{\upsilon}(t) \times cos(\varphi) + \vec{\beta}(t) \times sin(\varphi)\Big), \quad (6)$$

3 Results and Discussion

Geometric modeling of a cylindrical coil spring-twisted channel is described in detail in the book [21].

On the basis of the above model, the surface of the coil spring-twisted channel formed by a circle of constant radius *r* with a center on the helical line is made.

In this case, the guide curve is a bishelix located on the surface of a round cone with the lower base *R*.

Then we get the following system of equations:

$$\vec{r}(t) = \begin{pmatrix} ((R - bt \cdot \tan\psi) + r \cdot \cos\omega t) \cdot cost \\ ((R - bt \cdot \tan\psi) + r \cdot \cos\omega t) \cdot sint \\ bt + rsin\omega t \end{pmatrix},\ 0 \le t \le 2\pi n; \quad (7)$$

$$\vec{\rho}(t,\varphi) = r_0\Big(\cos\varphi\vec{\upsilon}(t) + \sin\varphi\vec{\beta}(t)\Big),\ 0 \le \varphi \le 2\pi \quad (8)$$

Where *R* –radius of the lower base of the coil,
r_0– spring coil wire radius,
n – number of turns of a double helical line,
ω –number of turns of the bishelix, falling on one turn of the central helical line,
ψ – cone angle.

Calculating the unit vectors of tangent, normals and binormals according to Eq. (5) and substituting the resulting expressions in Eq. (6), we get parametric equations of the conical surface coil spring-twisted channel:

$$\vec{r}(t,\varphi) = \begin{pmatrix} ((R - bt \cdot \tan\psi) + r \cdot \cos\omega t)cost \\ ((R - bt \cdot \tan\psi) + r \cdot \cos\omega t)sint \\ bt + rsin\omega t \end{pmatrix} + r_0\cos\varphi \cdot \vec{\upsilon}(t) + r_0\sin\varphi \cdot \vec{\beta}(t), \quad (9)$$

In order to check the conformity of Eq. (9) geometry of the coil spring-twisted channel under consideration, the surface is constructed in the MatLab system, Fig. 6.

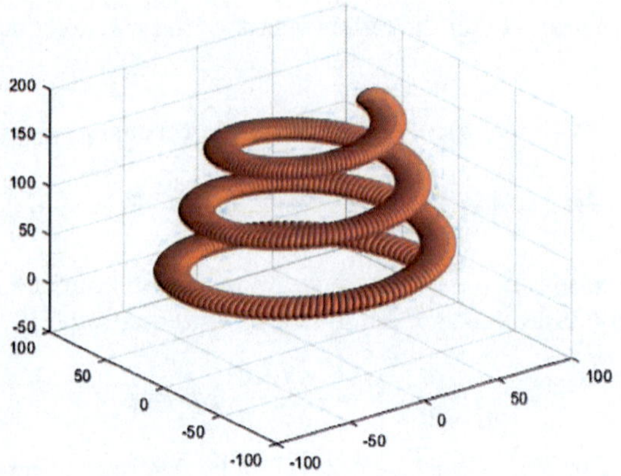

Fig. 6. Conical coil spring-twisted channel built at MatLab

4 Conclusions

Mathematical modeling is widely used in scientific disciplines, due to low costs compared to the rare nature experiment.

One of the stages of design development of innovative heat exchangers is the development of their geometric model.

In this paper, a mathematical description of the geometric model of the surface of the conical coil spring-twisted channel is given.

In the future, the results can be used for mathematical modeling of thermal engineering processes occurring in a conical coil heat exchanger based on a spring-twisted channel.

References

1. Li, Z., Shafee, A., Tlili, I., Jafaryar, M.: Nanofluid in Heat Exchangers for Mechanical Systems: Numerical Simulation. Elsevier (2020)
2. Roetzel, W., Luo, X., Chen, D.: Design and Operation of Heat Exchangers and their Networks. Academic Press (2019)
3. Sadik, K., Liu, H., Pramuanjaroenkij, A.: Heat exchangers: selection, rating, and thermal design. CRC Press (2020)
4. Pekar, L.: Advanced Analytic and Control Techniques for Thermal Systems with Heat Exchangers. Academic Press (2020)
5. Alam, T., Kim, M.H.: A comprehensive review on single phase heat transfer enhancement techniques in heat exchanger applications. Renew. Sustain. Energy Rev. **81**, 813–839 (2018). https://doi.org/10.1016/j.rser.2017.08.060
6. Li, Y., et al.: Analysis of the icing and melting process in a coil heat exchanger. Energy Procedia **136**, 450–455 (2017). https://doi.org/10.1016/j.egypro.2017.10.302
7. Ali, M.A., El-Maghlany, W.M., Eldrainy, Y.A., Attia, A.: Heat transfer enhancement of double pipe heat exchanger using rotating of variable eccentricity inner pipe. Alex. Eng. J. **57**(4), 3709–3725 (2018). https://doi.org/10.1016/j.aej.2018.03.003

8. Bulliard-Sauret, O., Ferrouillat, S., Vignal, A L.: Memponteil, andN. Gondrexon, Heat transfer enhancement using 2 MHz ultrasound. Ultrasonics sonochemistry, **39**, 262–271 (2017). https://doi.org/10.1016/j.ultsonch.2017.04.021
9. Asadi, A., et al.: Effect of sonication characteristics on stability, thermophysical properties, and heat transfer of nanofluids: a comprehensive review. Ultrason. Sonochem. **58**, 104701 (2019). https://doi.org/10.1016/j.ultsonch.2019.104701
10. Sheikholeslami, M., Bhatti, M.M.: Active method for nanofluid heat transfer enhancement by means of EHD. Int. J. Heat Mass Transf. **109**, 115–122 (2017). https://doi.org/10.1016/j.ijheatmasstransfer.2017.01.115
11. Hussain, T., Javed, M.T., Ansari, R.I.: A review on heat transfer enhancement using magnetic nanofluids. Nanosci. Nanotechnol. Asia. **10**(3), 266–278 (2020). https://doi.org/10.2174/2210681209666190412142721
12. Naphon, P., Wiriyasart, S., Arisariyawong, T., Nualboonrueng, T.: Magnetic field effect on the nanofluids convective heat transfer and pressure drop in the spirally coiled tubes. Int. J. Heat Mass Transf. **110**, 739–745 (2017). https://doi.org/10.1016/j.ijheatmasstransfer.2017.03.077
13. Sheikholeslami, M., Gorji-Bandpy, M., Ganji, D.D.: Review of heat transfer enhancement methods: focus on passive methods using swirl flow devices. Renew. Sustain. Energy Rev. **49**, 444–469 (2015).https://doi.org/10.1016/j.rser.2015.04.113
14. Mousa, M.H., Miljkovic, N., Nawaz, K.: Review of heat transfer enhancement techniques for single phase flows. Renew. Sustain. Energy Rev. **137**, 110566 (2021). https://doi.org/10.1016/j.rser.2020.110566
15. Gholamalizadeh, E., Hosseini, E., Jamnani, M.B., Amiri, A., Alimoradi, A.: Study of intensification of the heat transfer in helically coiled tube heat exchangers via coiled wire inserts. Int. J. Therm. Sci. **141**, 72–83 (2019). https://doi.org/10.1016/j.ijthermalsci.2019.03.029
16. Tuncer, A.D., Sözen, A., Khanlari, A., Gürbüz, E.Y., Variyenli, H.I.: Upgrading the performance of a new shell and helically coiled heat exchanger by using longitudinal fins. Appl. Thermal Eng. **191**, 116876 (2021)
17. Sepehr, M., Hashemi, S.S., Rahjoo, M., Farhangmehr, V., Alimoradi, A.: Prediction of heat transfer, pressure drop and entropy generation in shell and helically coiled finned tube heat exchangers. Chem. Eng. Res. Des. **134**, 277–291 (2018). https://doi.org/10.1016/j.cherd.2018.04.010
18. Kumar, E.P., Solanki, A.K., Kumar, M.M.J: Numerical investigation of heat transfer and pressure drop characteristics in the micro-fin helically coiled tubes. Appl. Thermal Eng. **182**, 116093 (2021). https://doi.org/10.1016/j.applthermaleng.2020.116093
19. Bagoutdinova, A.G., Zolotonosov, Ya. D.: Coil heat exchangers. Modeling, calculation, no. 245 (2016)
20. Zolotonosov, A. Ya., Zolotonosov, Ya. D., Konakhina, I.A.: Heat exchange element. Bulletin **34** (2007). Russian Federation patent 62694 Declared in 12 July 2006 Published in 27 April 2007
21. Zolotonosov, Y.D., Bagoutdinova, A.G., Zolotonosov, A.Y.: Tubular heat exchangers. Modeling. Calculation. Saint-Petersburg: Lan (2018)
22. Zolotonosov, A. Ya., Zolotonosov, Ya. D., Vachagina, E.K.: Heat exchange element. Bulletin **11** (2017). Russian Federation patent 170207 Declared in 17 August 2016 Published in 18 April 2017
23. Gorskaya, T., Zolotonosov, Y., Martynov, P., Khabibullina, A., Krutova, I.: Heat exchangers with spring-twisted heat-exchange elements made of wire with sections of various geometries. In: IOP Conference Series: Materials Science and Engineering, vol. 890, p. 012143 IOP Publishing (2020). https://doi.org/10.1088/1757-899X/890/1/012143
24. Zolotonosov, Ya. D., Tartygasheva, A.M.: Coil heat exchanger of the «pipe in pipe» type. Bulletin **19** (2019). Russian Federation patent 190475 Declared in 01 September 2019 Published in 07 February 2019

25. Zolotonosov, A.Ya., Zolotonosov, Ya. D., Knyazeva. I.A.: Coil heat exchanger. Bulletin **29** (2015). Russian Federation patent 155676 Declared in 02 December 2015 Published in 20 October 2015
26. Zolotonosov, A. Ya., Zolotonosov, Ya. D., Vachagina, E.K., Krutova, I.A.: Coil heat exchanger for heat transfer processes. Bulletin **16** (2017). Russian Federation patent 171543 Declared in 13 October 2016, Published in 06 June 2017
27. Zolotonosov. A. Ya., Zolotonosov. Ya. D., Vachagina, E.K.: Sectional coil heat exchanger. Bulletin **24** (2017). Russian Federation patent 173387 Published in 15 November 2017
28. Zolotonosov, A. Ya., Zolotonosov, Ya. D., Martynov, P.O., Krutova, I.A., Talynov, Sh. M., Shvetsov, M. V.: Coil heat exchange. Bulletin **29** (2017). Russian Federation patent 193127 Declared in 25 June 2019, Published in 06 June 2017
29. Zolotonosov, A. Ya., Zolotonosov, Ya. D., Martynov P.O.: Coil heat exchanger. Bulletin **29** (2020). Russian Federation patent 196872 Declared in 02 December 2019, Published in 18 March 2020
30. Ya. Zolotonosov, Ya. D. Zolotonosov, P.O.: Martynov Coil heat exchanger of the «pipe in pipe» type (2020). Russian Federation patent 201909 Declared in 23 July 2020
31. Krutova, I., Zolotonosov, Y.: Solution of conjugate problem in a conical coil heat exchanger. In: IOP Conf. Series: Materials Science and Engineering, vol. 890, p. 012156. IOP Publishing (2020). https://doi.org/10.1088/1757-899X/890/1/012156
32. Purandare, P., Lele, M., Gupta, R.: Experimental investigation on heat transfer and pressure drop of conical coil heat exchanger. Therm. Sci. **20** (6), 2087–2099 (2016). https://doi.org/10.2298/TSCI140802137P
33. Sheeba, A., Akhil, R., Prakash, J.: Heat Transfer and flow characteristics of a conical coil heat exchanger. Int. J. Refrig. **110**, 268–276 (2020). https://doi.org/10.1016/j.ijrefrig.2019.10.006
34. Radwan, M.A., Salem, M.R., Refaey, H.A., Moawed, M.A.: Experimental study on convective heat transfer and pressure drop of water flow inside conically coiled tube-in-tube heat exchanger. Eng. Res. J. (ERJ). **1**(39), 86–93 (2019)
35. Vorontsova, V.L., Bagoutdinova, A.G., Galemzianov, A.F.: Calculation of the geometrical parameters of the heat exchanger type "Pipe in Pipe" with a spiral-coiled channel. J. Comput. Theor. Nanosci. **16**(11), 4513–4518 (2019)

Deformation Features of Raft-Pile Foundation Models Under Cyclic Loading

Ilizar Mirsayapov and Marat Shakirov(✉)

Kazan State University of Architecture and Engineering, Tatarstan Republic, Zelenaya street, 1, Kazan, Russia

Abstract. At high load levels on the ground or adverse ground conditions, one way to increase the carrying capacity is the use of raft-pile foundations. Buildings and structures and their foundations, along with static exposed to various types of cyclic loads, which in many cases are the main determinants of the building elements security and integrity. The joint deformation of the «pile – raft grillage – ground between pile» with regard to redistribution of effort between the individual elements in the process of cyclic loading is practically unknown.

In this regard, authors conducted raft-pile foundation experimental research models under cyclic loading. Our studies have established the basic laws of raft-pile foundation soil deformation in between pile space.

Keywords: Raft-pile foundation · Soil · Cyclic loading · Tension · Stress · Between pile space · Trough test · Bearing capacity

1 Introduction

At high loading levels on groundbases or unfavorable soil conditions, one of the ways to increase the bearing capacity is to use raft-pile foundations.

A large number of works are devoted to study the raft pile foundations work. These studies are mainly limited to establish the main theoretical regularities and raft-pile foundation operation features under a short-term static load [1–5].

However, buildings and structures and their foundations, along with static ones, are exposed to various kinds of cyclic loads, which in some cases are the main factors determining the safety and serviceability of building elements [6–9].

The joint deformation of the system "piles - raft grillage - soil of the interpile space", taking into account the efforts redistribution between individual elements in the process of cyclic loading, has practically not been studied.

In connection with the above, it becomes necessary to study the behavior of raft-pile foundations under cyclic loading.

N. Vatin (Ed.): STCCE 2021, LNCE 169, pp. 203–212, 2021.
https://doi.org/10.1007/978-3-030-80103-8_22

2 Research Methods

Experimental studies were carried out in a volumetric laboratory tray with dimensions of 1.0 × 1.0 × 1.0 m (Fig. 1). A reinforced concrete raft with dimensions of 400 × 400 × 40mm, reinforced with wire reinforcement Ø3 Bp-I, was used as a grillage for the foundation model [5, 10–14].

Fig. 1. Test tray

The piles were modeled by hollow plastic tubes 7 mm in diameter, 400 mm long and 1 mm thick. Pile deformations were determined using strain gauges glued along the length.

The installation of the piles was carried out by layer-by-layer filling and compaction of the soil between the piles.

The base soil was semi-solid sandy loam (deformation modulus E = 4.1 MPa, internal friction angle = 15°, specific cohesion C = 3.3 kPa, density ρ = 14 kN/m2, humidity W = 11%). The pressure in the soil mass was determined using pressure sensors.

The measuring devices for determining the soil deformations are shown in Fig. 2.

Raft-pile foundation model and installation schemes of strain gauges in piles are shown in Fig. 2.

During the experimental studies, were recorded the movements of the foundation raft, vertical and horizontal movements of the piles, deformation in the soil base and in the piles.

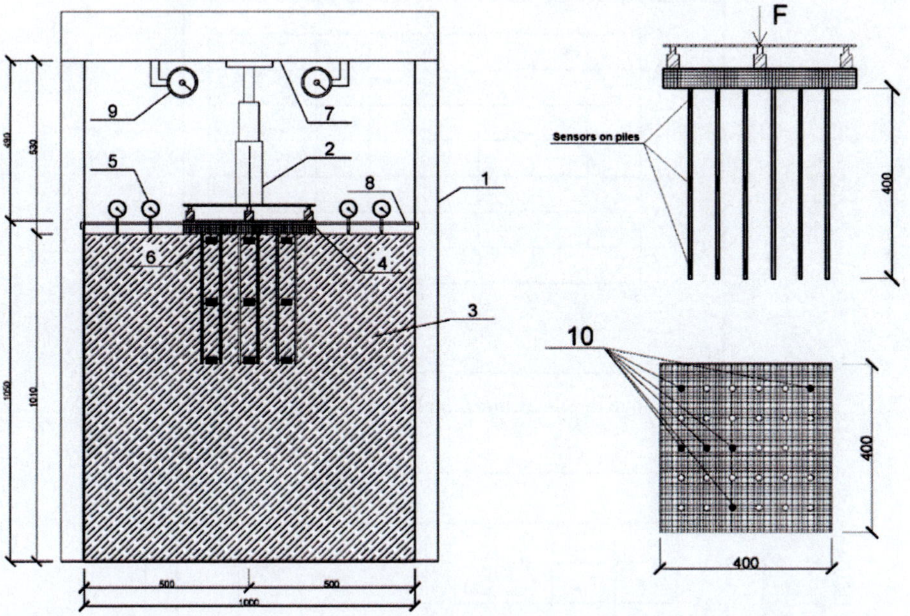

Fig. 2. Test scheme in laboratory conditions volumetric tray; 2. boot device; 3. foundation soil; 4. raft-pile foundation; 5. dial indicators; 6. pressure sensors in the ground; 7. construction of the loading device; 8. traverse of indicators; 9. deflection meters; 10. piles with sensors.

3 Results and Discussion

The studies carried out made it possible to establish the main regularities of the change in the raft-pile foundation base stress-strain state in the process of cyclic loading.

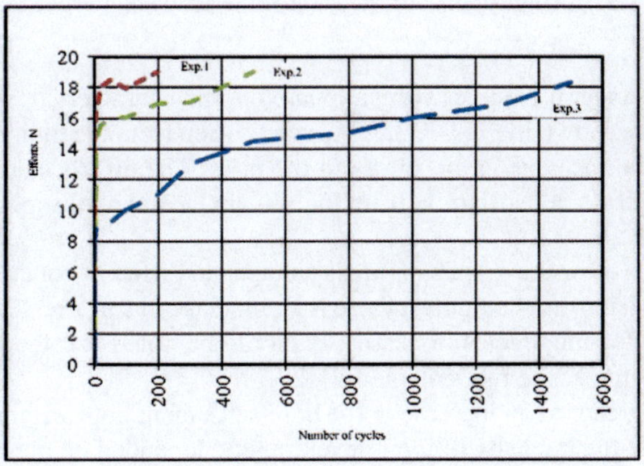

Fig. 3. Change in pile efforts 5 under the action of a cyclic load

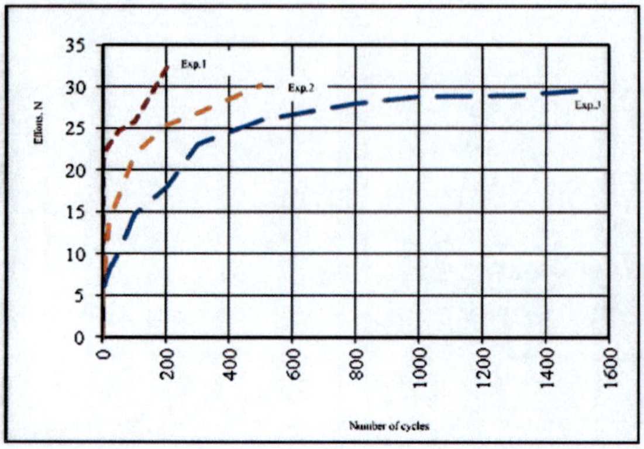

Fig. 4. Change in pile efforts 6 under the action of a cyclic load

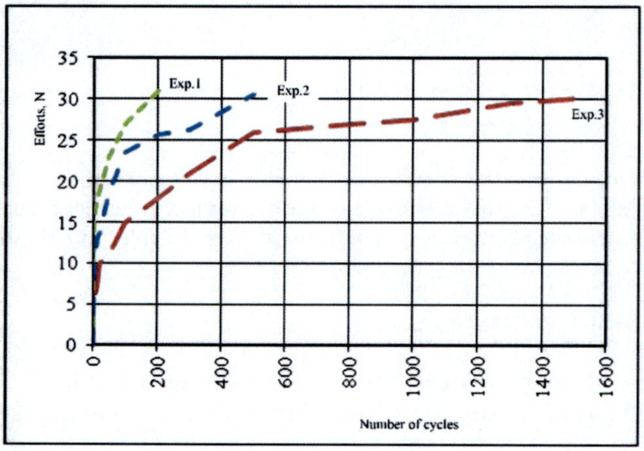

Fig. 5. Change in pile efforts 1 under the action of a cyclic load

The forces change in the piles, which located in the characteristic zones of the raft-pile foundation are shown in Figs. 3, 4, 5. As can be seen from the figures, cyclic loading does not lead to an increase in the forces in the piles. The efforts change nature in the piles shows the efforts redistribution from the soils in between pile space to the piles [6, 15–19].

Figures 6, 7, 8 show the graphs of stress changes in different zones soil between the piles. As can be seen from the figures, there is an increase in soil stresses in all soil zones as the number of loading cycles increases. It should be noted that the greatest increase in stresses occurs under the raft grillage [20–25].

Cyclic loading caused an increase in the base settlement both within the raft grillage and outside it, and the intensity of their development depended on the coordinate of the point under consideration. Figures 9, 10, 11, 12 show increase base settlement graphs depending on the number of loading cycles. As can be seen from the graphs above, the

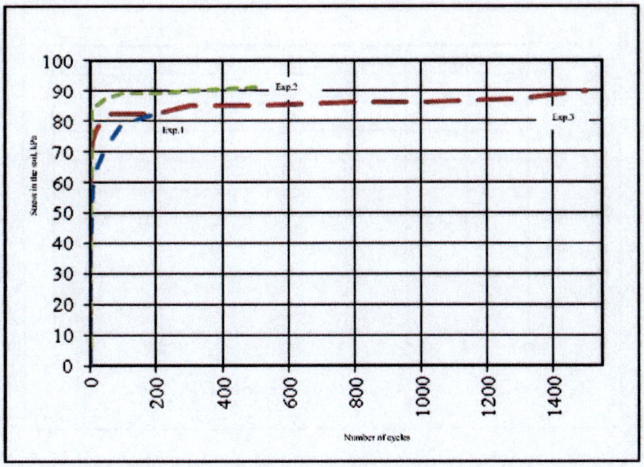

Fig. 6. Change in stress in the soil under the action of a cyclic load at the lower end of the piles

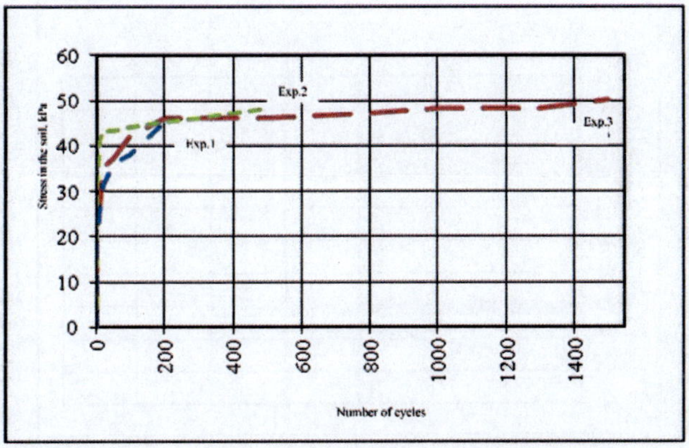

Fig. 7. Change in stress in the soil under the action of a cyclic load at the level 20 cm from the grillage

intensive settlements development, regardless of the considered base points coordinates, occurred during the first 200 cycles of repeated loading [26]. So, for example, for points 1, 2, 3, 4, 5, the total value of the base settlement for the first 200 cycles in relation to the initial ones in the first load increased by 417%, 395%, 377%, 381%, 411%, and subsequently for the entire period until the end of loading by 121% 117% 131% 119% 124%.

Hence, it can be seen that after 200 cycles of repeated loading, the settlement intensity growth decreased significantly, but their complete stabilization was not observed. This pattern of settlement development is explained by the fact that, in the general case,

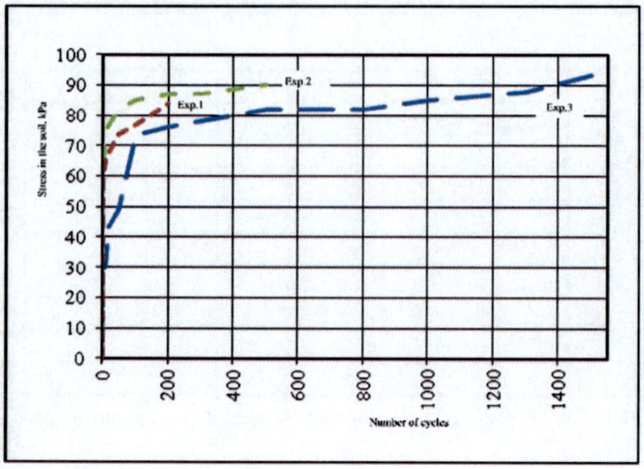

Fig. 8. Change in stress in the soil under the action of a cyclic load at the level 2 cm from the grillage

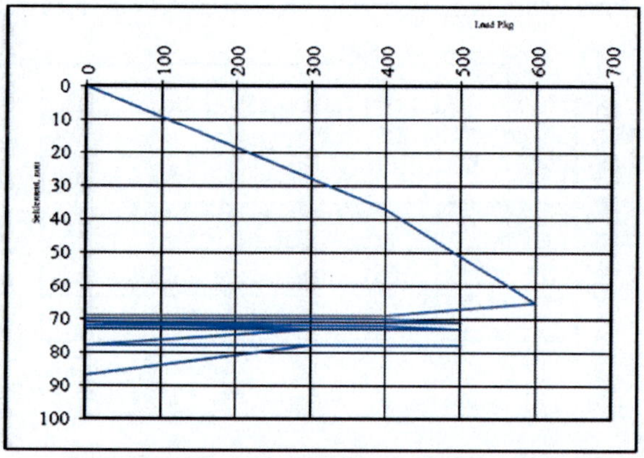

Fig. 9. Settlement of the foundation raft grillage after 200 cycles of step loading

the change in settlement depends on the development of soil deformations in the inter-pile space under the grillage, shear deformations between the soil and piles, and soil deformations in the zone under the lower end of the piles.

At the initial stage, compaction deformations are realized in almost all zones. The increase in compaction deformations after 200 loading cycles practically stopped, which caused a decrease in the intensity of settlement growth [17, 27–29].

Foundation settlements, measured in the course of stepwise static loading after different amounts of repeated loading, change similarly to the deformations of the soil in the interpile space. As an example, Figs. 9, 10, 11, 12 show the change in the base settlement within the raft grillage and along the edges of the raft grillage under stepped static loading.

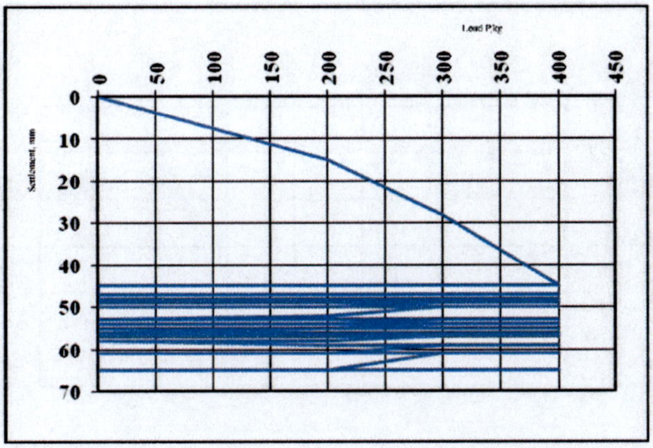

Fig. 10. Settlement of the raft grillage foundation base after 1500 cycles of step loading

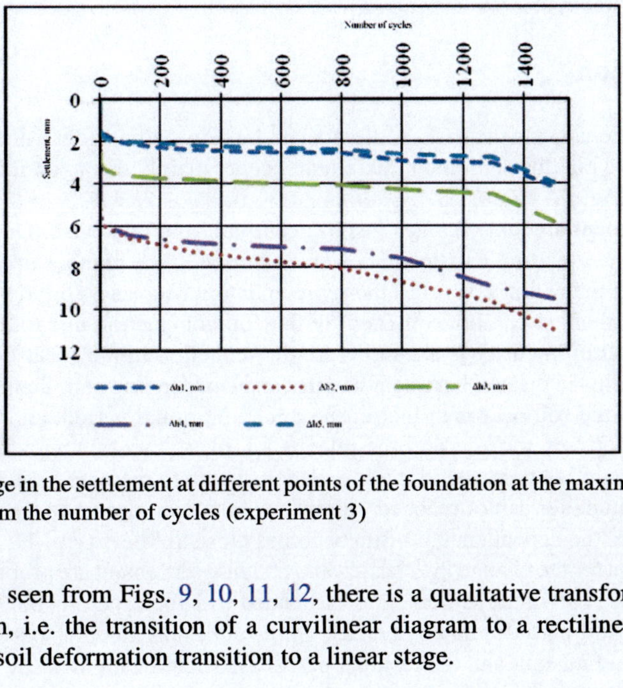

Fig. 11. Change in the settlement at different points of the foundation at the maximum load values in the cycle from the number of cycles (experiment 3)

As can be seen from Figs. 9, 10, 11, 12, there is a qualitative transformation of the P - S diagram, i.e. the transition of a curvilinear diagram to a rectilinear one, which indicates the soil deformation transition to a linear stage.

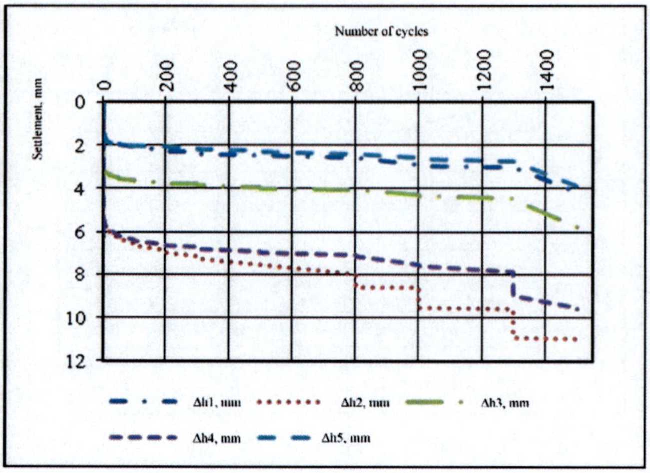

Fig. 12. Settlement at minimum cycle loads (experiment 3)

4 Conclusion

Analysis of the change in the bases settlement under step loading after a different number of cycles shows that the settlement increment occurs mainly due to an increase in their residual part (Fig. 11–12).

The settlement amount changes insignificantly during one cycle. During the tests, changes in these ("elastic") settlements were recorded as the number of loading cycles increased. During the first 20 cycles, these precipitation decreases slightly. The decrease in "elastic" settlement can be explained by the soil compaction due to a pore volume decrease, which in intensity is ahead before the reduction in the shear deformation of the shear modulus in the soil between the piles and surrounding soil. Since the intensive compaction of the soil occurs in the repeated loading initial period, and the change in the adhesion between the soil and the piles is a longer process, then after 20 loading cycles the "elastic" settlements of the foundations begin to increase [17]. If the limiting state of the foundation is not reached, relative stabilization occurs to the 1200 loading cycles time, i.e. the dependence F - Sup becomes close to linear (Fig. 11–12).

The reasons for the change in "elastic" and residual settlement are probably different.

The "elastic" part of the settlement is associated with the structural bonds destruction between solid soil particles and a decrease in the adhesion modulus between the piles and the surrounding soil, and the residual part is associated with inelastic deformations of the soil under the raft grillage due to vibration creep and the development of fatigue microcracks in plastically deformed local zones.

References

1. Mirsayapov, I.T., Shakirov, M.I.: Combined plate-pile foundations settlement calculation under cyclic loading. In: Socio-Technical Construction and Civil Engineering: IOP Conf. Series: Materials Science and Engineering, pp. 1–8 (2020). https://doi.org/10.1088/1757-899X/890/1/012069
2. Katzenbach, R., Leppla, S., Ramm, H., Seip, M., Kuttig, H.: Design and construction of deep foundation systems and retaining structures in urban areas in difficult soil and groundwater conditions. Procedia Eng. **57**, 540–548 (2013). https://doi.org/10.1016/j.proeng.2013.04.069
3. Katzenbach, R.: Soil-structure-interaction of tunnels and superstructures during construction and service time. Procedia Eng. **57**, 35–44 (2013). https://doi.org/10.1016/j.proeng.2013.04.007
4. Bhaduri, A., Choudhury, D.: Serviceability-based finite-element approach on analyzing combined pile-raft foundation. Int. J. Geomech. **2**(20), 43–51 (2020). https://doi.org/10.1061/(ASCE)GM.1943-5622.0001580
5. Boudaa, S., Khalfallah, S., Bilotta, E.: Static interaction analysis between beam and layered soil using a two-parameter elastic foundation. Int. J. Adv. Struct. Eng. **11**(1), 21–30 (2019). https://doi.org/10.1007/s40091-019-0213-9
6. Mirsayapov, I.T., Koroleva, I.V.: Long-term settlements assessment of high-rise building groundbase based on analytical ground deformation diagram. Procedia Eng. **165**, 519–527 (2016). https://doi.org/10.1016/j.proeng.2016.11.728
7. Siraziev, L.F.: Experimental studies of the stress-strain state of layered soil bases under the center of the stamp during short-term tests. Innov. Invest. **11**, 225–228 (2018). https://doi.org/10.1016/j.proeng.2018.04.007
8. Mirsayapov, I.T., Koroleva, I.V.: The strength and deformability of clay soils under the regime spatial stress state in view of cracking. Grounds Found. Soil Mech. **1**, 16–23 (2016). https://doi.org/10.1061/(ASCE)GM.1943-5622.0001580
9. Mirsayapov, I.T., Shakirov, M.I.: Combined plate-pile foundations settlement calculation under cyclic loading. In: 1 st International Conference on Energy Geotechnics. vol. 724, pp. 423–428 (2016). https://doi.org/10.1088/1757-899X/890/1/012069
10. Mirsayapov, I.T., Koroleva, I.V.: Changes in physical and mechanical characteristics of soil under triaxial loading. In: CRC Press. Conference, vol. 466, pp. 193–196 (2019).https://doi.org/10.1201/9780429058882-37
11. Khuziakhmetov, R., Nurieva, D.: Determination of the reasons for the fallof pile driving machine main technical near the slope of the foundation pit. In: IOP Conference Materials. Science Engeering, vol. 890, pp. 162–170 (2020). https://doi.org/10.1088/1757-899X/890/1/012136
12. Mirsayapov, I.T., Aysin, N.N.: Influence of a deep construction pit on a technical condition of surrounding buildings. In: CRC Press. Conference, vol. 466, 197–201 (2019). https://doi.org/10.1201/9780429058882-38
13. Mirsayapov I., Koroleva I.: Calculation models of bearing capacity and deformation of soil foundations with vertical elements reinforced under regime cyclic loading. In: Ferrari, A., Laloui, L. (eds) Energy Geotechnics. SEG 2018. Springer Series in Geomechanics and Geoengineering. Springer, Cham (2019). https://doi.org/10.1007/978-3-319-99670-7_62
14. Mirsayapov, I.T., Sabirzyanov, D.: Bearing capacity of foundations base under combined alternating long-term static and cyclic loading. In: CRC Press. Conference 2018, vol. 700, pp. 43–51 (2018).https://doi.org/10.1088/1757-899X/890/1/012069
15. Mirsayapov, I.T., Koroleva, I.V.: Strength and deformability of clay soil under different triaxial load regimes that consider crack formation. Soil Mech. Found. Eng. **53**(1), 5–11 (2016). https://doi.org/10.1007/s11204-016-9356-x

16. Mirsayapov, I.T., Koroleva, I.V., Sabirzyanov, D.D.: Calculation model of foundation base settlement at the static and cyclic regime loading, In: ICEGT. Conference 2016, ICEGT, vol. 756, pp. 429–434 (2016). https://doi.org/10.1201/b21938-68
17. Mirsayapov, I.T., Koroleva, I.V.: Computational model of the carrying capacity of a reinforced foundation with cyclic loading. Soil Mech. Found. Eng. **52**(4), 198–205 (2015). https://doi.org/10.1007/s11204-015-9328-6
18. Mirsayapov, I.T., Koroleva, I.V.: Settlements foundation bases under long-term regime loading. In: ARC. Conference 2015, ARC 1596, pp. 1398–1401 (2015). https://doi.org/10.1201/9780429058882-38
19. Mirsayapov, I.T., Koroleva, I.V.: Bearing capacity of foundations under regime cyclic loading. In: ARC. Conference 2015, ARC 1596, pp. 1214–1217 (2015). DOI: https://doi.org/10.3208/jgssp.KAZ-18
20. Mirsayapov, I.T., Koroleva, I.V.: Clayey soils rheological model under triaxial regime loading. In: ECSMGE. Conference 2015. ECSMGE 3976, pp. 3249–3254 (2015). https://doi.org/10.1016/j.proeng.2015.11.728
21. Mirsayapov, I.T., Koroleva, I.V.: Fourteenth international symposium on soil rheology prospective trends in theoretical and practical development in rheology and soil mechanics. Soil Mech. Found. Eng. **51**(6), 315–316 (2015). https://doi.org/10.1007/s11204-015-9296-x
22. Mirsayapov, I.T., Koroleva, I.V.: Experimental and theoretical studies of bearing capacity and deformation of reinforced soil foundations under cyclic loading. In: IACMAG. Conference 2014, IACMAG 1525, pp. 737–742 (2015). https://doi.org/10.1201/b17435-127
23. Mirsayapov, I.T, Shakirov, M.I.: Bearing capacity and settlement of raft-pile foundations under cyclic loading, In: ICEGT. Conference 2016, ICEGT 756, pp. 423–428 (2016). https://doi.org/10.1201/b21938-67
24. Mirsayapov, I.T., Koroleva, I.V., Mirsayapova, I.: Evaluation of seismic stability of layered soil bases in areas that are composed of clays and water-saturated sandstones, In: ARC. Conference 2015, ARC 1596, pp. 719–722 (2015). https://doi.org/10.3208/jgssp.OTH-29
25. Mirsayapov, I.T., Koroleva, I.V.: Bearing capacity and deformation of the base of deep foundations' ground base. In: ISSMGE. Conference 2014, ISSMGE 1962, pp. 401–404 (2014). https://doi.org/10.1201/b17240-74
26. Mirsayapov, I.T., Koroleva, I.V.: Prediction of deformations of foundation beds with a consideration of long-term nonlinear soil deformation. Soil Mech. Found. Eng. **48**, 148–157 (2011). https://doi.org/10.1007/s11204-011-9142-8
27. Mirsayapov, I.T.: A study of stress concentration zones under cyclic loading by thermal imaging method. Strength Mater. **41**, 339–344 (2009). https://doi.org/10.1007/s11223-009-9121-8
28. Shakirov, I.F., Gayfullina, V.A.: Researches of the effect from additives on the cement mortars injection properties used in strengthening soil pressure cementation: dig. of art. In: Technical Sciences – From Theory to Practice – XXI International Scientific-Practical Conference, pp. 37–41 (2017). https://doi.org/10.1201/9780429058882-38
29. Mirsayapov, I.T., Shakirov, M.I.: Bearing capacity and settlement of raft-pile foundations under cyclic loading: dig. of art. In: 1 st International Conference on Energy Geotechnics, pp. 423–428 (2016). https://doi.org/10.1201/b21938-67

Numerical Study of the Flow in a Symmetrical Ventilation Junction Tee with a Baffle Vane

Arslan Ziganshin[1] (✉), Svetlana Eremina[1], Guzel Safiullina[1], and Konstantin Logachev[2]

[1] Kazan State University of Architecture and Engineering, Zelenaya st., 1, 420043 Kazan, Russia

[2] Belgorod State Technological University Named After V.G. Shukhov, Kostyukov st., 46, 308012 Belgorod, Russia

Abstract. Ventilation systems are characterized by a large waste of energy due to air movement through the channels duct fitting. There is an effective way to reduce such losses – shaping along the outlines of vortex zones. For this, it is first necessary to determine such outlines and their dependence on the design and operating parameters. For the duct fittings in the form of symmetrical junction tees, there is a design option with a baffle vane, which has a reduced resistance along one of the side branches. The article presents the results of computer simulation of the flow in such a tee. The most optimal dimensions of the partition, leading to the least resistance, have been determined. It is shown that a decrease in resistance along a branch with a lower flow rate occurs due to the ejection effect and an increase in resistance along the other branch – with a higher flow rate. The outlines of vortex zones have been determined, the size of which does not depend on the baffle size, that is, it is possible to use them for shaping both for tees with and without baffles.

Keywords: Numerical study · Symmetrical tee · Baffle vane · Validation

1 Introduction

When designing air ducts for ventilation and air-conditioning systems, so-called the local pressure losses occur in places of sudden change of flow. This change leads to deformation of the flow and the formation of vortex zones. At the same time, unlike heating systems, for ventilation and air conditioning systems, the total pressure loss in the duct network, and, accordingly, energy, approximately 80% consists of local pressure losses. And therefore, the urgent task of reducing the resistance of ventilation systems is most effectively solved by improving the design of shaped elements and reducing their resistance. At the same time, reducing the total resistance of the ventilation system is important not only from the point of view of reducing the energy consumption of such systems. For exhaust ventilation systems of industrial buildings that remove polluted air, the reduced resistance of the ventilation network makes it possible to increase the efficiency of both local exhaust devices: exhaust hoods [1, 2], and devices for cleaning this exhaust air from pollution [3, 4]. To increase the efficiency of supply ventilation

N. Vatin (Ed.): STCCE 2021, LNCE 169, pp. 213–222, 2021.
https://doi.org/10.1007/978-3-030-80103-8_23

operating, a more detailed research and calculating of ventilation jets using computational fluid dynamics is possible [5]. At the nodes of exhaust ventilation systems, where two streams are connected with approximately the same air flow rate, for example, in front of the fan, a connection is made with the duct fitting – symmetrical tee at junction. In such a tee, according to the reference book [6], the aerodynamic resistance can be significantly reduced at one branch in comparison with the standard one (by about 4–5 times) by installing a dividing baffle vane. In [6], it is not indicated why such a significant decrease in resistance occurs, but this, apparently, occurs due to a change in the nature of the interaction of flows during impact. In a symmetrical tee, energy losses occur due to the impact of two streams, their mixing and rotation by 90°, and such losses can be called «functional», since they occur due to the main function of the tee – rotation and mixing of streams. In addition, due to the separation of the flow from the sharp edge in such a shaped element, there are also a «non-functional» energy losses associated with the presence of a vortex zone – for movement in the vortex zone, as well as losses when flowing around the separation zone – narrowing and expansion of the flow. Usually, reducing functional losses is a rather difficult task and it is necessary to pay attention to the possibility of reducing non-functional ones. But, in this case, the installing of the baffle vane, apparently separating the two processes – impact and mixing of flows, apparently leads to decreasing in functional losses.

There are not many studies of flows in symmetric tees; this is the already mentioned reference book [6], which contains the results of experiments carried out in the middle of the last century. There are similar, somewhat later experiments [7], which also show the test results, including symmetrical tees, but for round ducts.

In addition, there are attempts to analytically determine the local resistance coefficients of tees of various designs, for example, using the conformal mapping method [8], or analyzing the momentum equation [9]. However, due to the complexity of the flow, the dependences obtained analytically are further corrected using empirical coefficients to match the experimental data. Such studies are usually limited to the case of a symmetrical tee design without a baffle. Nevertheless, the use of various screens and guide vanes to reduce the resistance of asymmetric tees is studied both numerically and experimentally – in [10], by means of a numerical study, the optimal shape, length and angle of inclination of a guide vane installed in an asymmetric dividing tee were selected. At the same time, tees with guide vane options that show the most optimal results are also tested experimentally. The possibility of reducing the resistance by an amount from 4.3% to 263.9% depending on the flow rate of the air flowing through the branches is shown. Also, for the dividing tee, there is a variant of a different design of the guide vane [11], which leads to a more uniform distribution of the flow in the side channel, due to which the authors achieved a decrease from 5.2% to 38.4%. Also guide vanes in tees were used to decrease the resistance of the manifolds inlets [12]. A number of geometric parameters of guide vane were investigated numerically with experimental verification, and for the most optimal one, a decrease in resistance by up to 41% was obtained. There are examples of using guide vanes for asymmetric junction tees [13] used in ventilation shafts of high-rise buildings. A decrease in resistance is shown, but specific values and dependence of the LDC (local drag coefficient) on the flow rate ratio of flows along the branches of the tee are not indicated. In addition to installing guide vanes, to reduce aerodynamic drag, profiling of the tee walls along various outlines is also used, for example, in [14], it is shown the possibility of reducing the resistance of an asymmetrical junction tee by shaping its wall along the outlines of the river channel, at the point of joining the main stream of the tributary – to the passage by up to 248.2%, and on the lateral branch up to 817.9%. At the same time, the decrease of the local drag coefficient (LDC) significantly depends on the ratio of air flows, and in some cases of flow rates there is no reduction. In [15], in order to reduce the resistance of a dividing asymmetric tee, a rounding of the tee wall according to the shape of plant stems is already used, where there is an expansion in the places where the stem is divided.

The study was carried out numerically, with experimental verification of the results obtained. The opportunity to reduce the LDC by 36% and 21% in different directions was obtained. Also, one of the ways to reduce the aerodynamic resistance of duct fittings of various configurations is topological optimization. There are software modules, using special mathematical methods, search for the optimal shape of an element within predetermined limits and according to a predetermined parameter. In [16], the optimal shapes of the corresponding duct fittings were found using the Deformable Simplicial Complex (DSC) methodology applied to flows in channels with obstacles, elbows and dividing tees. For the tee, a reduction in resistance by 64.5% in one direction and 76.1% in the other is obtained. Another method of topological optimization was used for flows in channels with elbows and complex designs of crosses [17]. Unfortunately, the work does not indicate a comparative decrease in the resistance of the resulting shaped duct fittings. However, the duct fittings designed by topological optimization have one big and currently irreparable drawback – their shape is so complex that their production is possible only with use of additive technologies, which significantly increases their cost and makes them impossible to use in ventilation and air conditioning systems of buildings. In addition, the above methods of reducing the resistance of tees lead to an increase in their dimensions, which complicates the placement of air ducts in usually cramped technical spaces and therefore also limits their use. A method devoid of these drawbacks is the use of shaping inserts along simple outlines, installed inside the tee, or shaping the walls of the shaped part along these outlines without increasing their dimensions. In [18], a unit was numerically investigated, which consisted of a sequentially located elbow and an asymmetric diverging tee, into which a number of inserts of different shapes were installed – which were found in mechanical engineering, aircraft construction areas and ellipse outlines. It was found that all the profiles used showed a decrease in the resistance of the assembly, but only at a certain flow ratios. The best result was shown by the insert shaped along the ellipse outline – the maximum value of the reduction was about 40%. However, as the authors of the article themselves note, when choosing the outline of the shaping insert, it is necessary to take into account the peculiarities of the flow in the duct fitting element. The most rational for shaping is to use the outlines of the vortex zone, which occurs in the original design of the duct fitting. Using this method, positive results were obtained on the profiling of an asymmetric junction tee [19], where, by numerical simulation, a decrease in resistance was obtained by a factor of 2–3. For other shaped elements, there are also positive results – the reduction in the resistance of the exhaust openings by shaping along the inlet vortex zone reached 20% [20].

The information about use of a baffle to reduce a resistance of a symmetrical tee, from the available literature, is contained only in the reference book [6]. But there is no dependence for the LDC on the baffle size (c). The existing dependence, judging by the diagram, is given for the case $c = b$. At the same time, the resistance is reduced by about 3 times at the small values of a flow rate ratios. Therefore, it is of interest, based on the results of the study, to determine the size of the baffle vane that leads to the minimum resistance. In addition, it is clear that vortex zones (*VZ*) are formed in such a tee, that is, there is still some potential for reducing the resistance by shaping along the outlines of the *VZ*. At the same time, there are no studies to determine the outlines of the *VZ* and their dependence on the presence and size of the baffle. Thus, the task is to determine the dependence of the *LDC* and the outlines of the *VZ* on the dimensions of the baffle vane, in order to further reduce the resistance of such a duct fitting using its shaping. The most rational way is to solve this problem using the numerical modeling. Today, a large number of problems about flows in tees under complicated conditions are numerically solved, for example, the conditions of mixing in tees of cooling systems of nuclear power plants using Large Eddy Simulation [21]. Using the same method, the features of the flow in channels with different designs of tees located one after another are investigated [22].

2 Methods

The problem of air flow in a symmetrical tee with a baffle vane is solved numerically (Fig. 1). Numerical modeling is carried out using the software package [23] in a turbulent two-dimensional setting. The boundary condition (BC) «velocity inlet» with the velocity value $v_y = -10$ m/s (***GF***), at inlets – «pressure inlet» BC with variable gauge pressure (***AB, CD***), that will be used to modeling variable flow rate at the side branches. All other boundaries including baffle vane (*IJ*) are impermeable walls – «wall» BC. The channel width is $b = 0.1$ m, the length of channel before the tee is $l_b = 2$ m ($l_b/b = 20$), after - $l_c = 8$ m ($l_c/b = 80$). Unitless baffle vane sizes: $c/b = 0$; 0.2; 0.5; 1; 1.3.

To validate the computer model, a test study was carried out on a problem with a baffle size $c/b = 1$, using various versions of turbulence models – «standard» k-ε (*SKE*), Reynolds Stress Model (*RSM*) and «standard» k-ω (*SKO*) and methods of near-wall modeling – standard wall functions (*SWF*) and enhanced wall treatments (*EWT*) with the ratio of air flow rates flowing along the side branches (G_b) and the main channel (G_c) equal to $G_b/G_c = 0.5$. In this case, the values of the local drag coefficients (*LDC*) ζ and the outlines of vortex zones (*VZ*) were determined (Fig. 2). The method for determining the *LDC* on the results of the numerical solution is described in detail in [24–26]. From the distribution of the stream function at the inlet and outlet boundaries, the values of the stream function corresponding to the free streamline are found, and so on, the outlines of the vortex zones.

For the *RSM EWT* combination, no convergent solution was obtained, therefore, these results are absent here. It can be seen that, according to the *LDC* values, all models show close values during grid refinement (the maximum difference is about 13%), and stop changing at y^* (y+) < 8. Numerical modeling shows that two vortex zones are formed in such a tee – the first during breakdown flow from the sharp edge of the tee

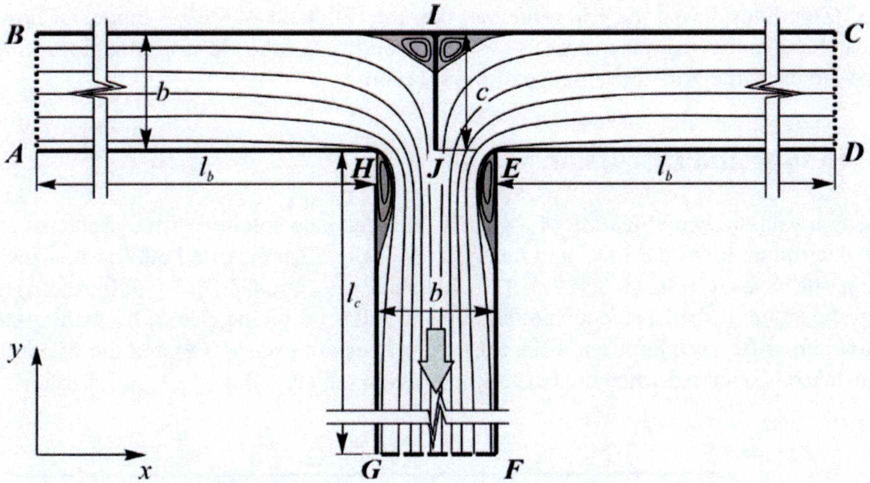

Fig. 1. Sizes of computational domain of the symmetrical tee with the baffle vane size $c/b = 1$.

(*VZ1*) and the second – in the corner formed by the baffle vane and the wall of the tee (*VZ2*) (Fig. 2b).The outlines of the vortex zones show that the *RSM SWF* and *SKO* models lead to a bigger vortex zones than the *SKE* models. At the same time it can be seen that the outline of *VZ* received by the *RSM EWT* models seems a nonphysical at a point of its closure at the wall of channel. The *SKE* models lead to the almost same outlines. So as a result of validation stage the *SKE EWT* model was accepted as appropriate for this study.

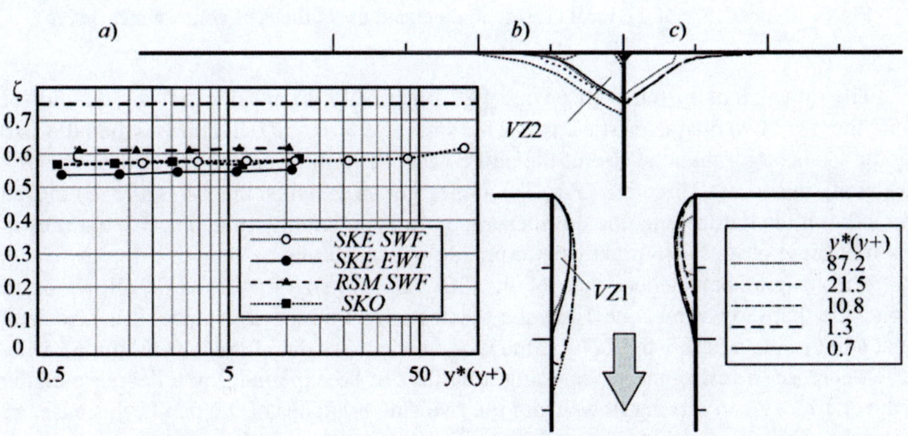

Fig. 2. Validation and verification of the problem with c/b = 1.

Using this model leads to a more accurate behavior during the grid refinement process. It should be noted that in the previous studies for the symmetry tee without baffle vane more appropriate was the model *SKE EWT* too [27]. On the Fig. 2c, the change in the *VZ* outlines during the grid refinement for the validated *SKE EWT* model is shown. It can be seen that from the value $y+ < 10$, the outlines of the *VZ* stop changing, which also indicates the grid convergence of the solution.

3 Results and Discussion

Using a validated combination of *SKE EWT* models, the solution of the problems and the determination of the *LDC* and the outlines of the *VZ* are carried out for all sizes of the baffle vane $c/b = 0.2, 0.5, 0.7, 1, 1.3$ with a flow rate ratio $G_b/G_c = 0.5$. It can be seen that the value of the LDC does not significantly depend on the size of the baffle – the maximum difference between different *LDC's* does not exceed 3%, and the minimum resistance is obtained when the baffle size is $c/b = 0.5$ (Fig. 3a).

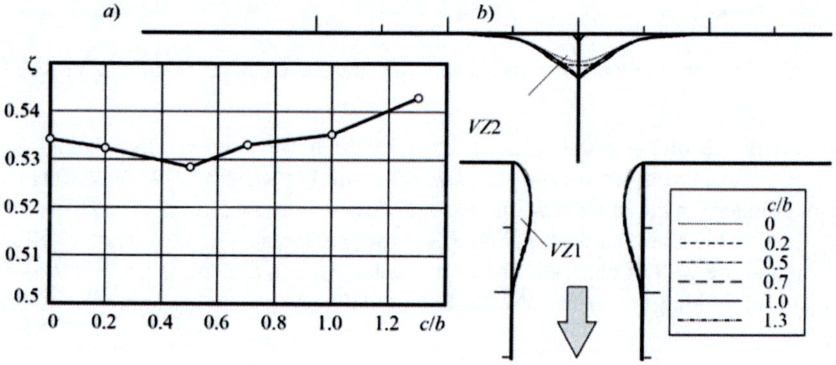

Fig. 3. Dependence of the local drag coefficient and the outlines of vortex zones on *c*/*b*.

The dimensions of the vortex zone *VZ1* practically do not depend on *c*/*b*, and for *VZ2* there are two characteristic cases – for small $c/b = 0$ and 0.2, that is, when the size of the *VZ* is larger than the size of the baffle vane, the outlines reach the central line of a tee along the normal direction (Fig. 3b). For larger sizes, when the *VZ* is already closed directly on the baffle vane, the outlines are closed tangentially to it. The *VZ* outlines in each of these groups also practically coincide with each other.

The insignificant dependence of the *LDC* on the size of the baffle allows us to conclude that the vortex zone 1 has the greatest effect on the resistance. It is also seen that when using a baffle, the *LDC* value is somewhat less than for a tee without a baffle. The decrease in resistance when using a baffle can be explained by a decrease of the impact effect when mixing flows from the two side branches of the tee. In this case, an increase and decrease in the size of the baffle relative to $c/b = 0.5$ leads to an increase of the *LDC*. Apparently, with an increase of the baffle length, the losses associated with the friction against the baffle increase. The smaller size of the baffle vane leads to an insufficient effect provided by the baffle.

Further, for the variant of the tee without a baffle ($c/b = 0$) and with a size of the baffle $c/b = 1$, the *LDC* was determined with a change in the flow rate ratio of 0.1 <

$G_b/G_c < 0.9$. As a result, the dependencies are obtained (Fig. 4). The figure also shows the data [7] (*Gardel*), *Ito & Imai* given in publication [9] for the tee without baffle vane, and the data of the reference [6] (*Idel'chik*) both for the case with the baffle and the case without it.

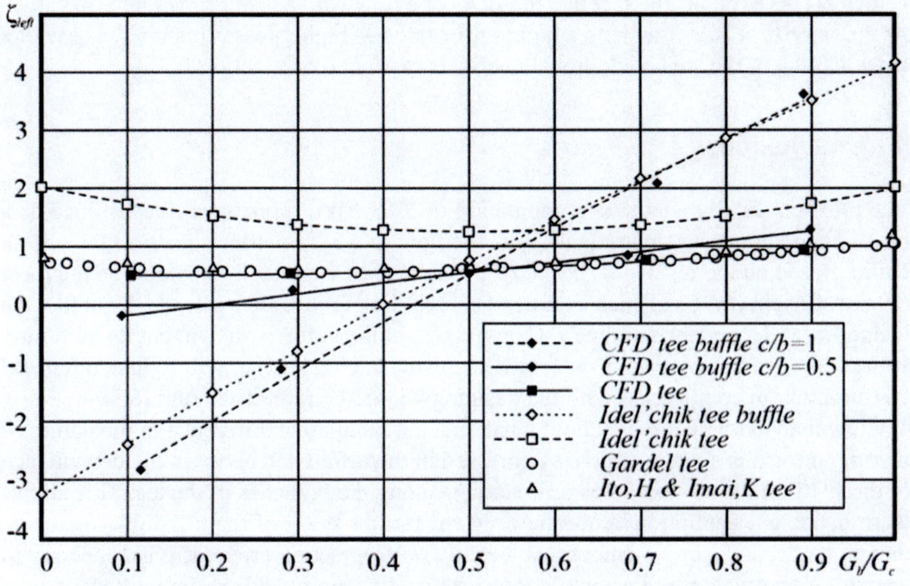

Fig. 4. Dependencies of local drag coefficient on the flow rates ratio G_b/G_c.

It can be seen that the *LDC* of a tee with a baffle where $c/b = 1$ coincides quite well in the nature of the dependence on the flow rates ratio with the data [6]. Quantitatively, the values differ by up to 30%. Differences were also observed in the study of a symmetrical junction tee without a baffle [27], where it was shown that the resistance of a symmetric tee without a baffle is significantly lower than in [6] (by 115.5%) and slightly lower than in experiment [27] (by 14.8%). This difference between the resistance obtained numerically and found experimentally can be explained by the fact that the computer simulation uses a two-dimensional formulation, as well as the solution of the Reynolds-averaged Navier-Stokes equations. That is, the impact that occurs when two flows meet from the side branches of a symmetric tee occurs as ideally as possible, in contrast to real conditions, where the flow is immediately mixed, with the formation of separate streams and vortices, which leads to a more intense exchange of momentum and energy losses of the flow. At the same time, such a significant difference from the reference book [6] can only be explained by the peculiarities of the experiment, which made a significant difference from the experimental conditions [27] and were not reflected in the reference book, and the original source, unfortunately, is not available at the time of writing this article.

The nature of the *LDC* change with a change of the flow rates ratio in tees in the presence and absence of baffle vane is significantly different. In a tee without baffles, as mentioned above, mixing and energy loss occur immediately during a head-on collision of two streams, so both streams lose energy. In the presence of a baffle, the flows initially lose energy per turn, and then move a certain distance along the baffle without mixing. Then there is a mixing of two parallel moving streams, while if the mixing streams have different velocities, then a stream with a higher velocity ejects a stream with a lower velocity. This explains the fact that the *LDC* of a flow with a lower flow rate is negative. At the same time, the flow with a higher flow rate has higher losses than in the case of a tee without a baffle, since additional energy is expended on ejection.

4 Conclusions

The most physically adequate combination of *SKE EWT* models was determined as a result of modeling and investigation of the flow in a symmetrical junction tee with a baffle. Based on the results of the study, dependencies were built to determine the local drag coefficient (*LDC*) of such a tee design, including for different sizes of the baffle. An insignificant dependence of the *LDC* on the size of the baffle is shown, but its minimum value was found at $c/b = 0.5$, which, when using such a design, will reduce the metal consumption in comparison with the design given in the reference book [6] with $c/b = 1$. A comparison with a tee without a baffle and a qualitative difference in the nature of mixing in them is shown, which is expressed in the difference between the dependence of the *LDC* on the air flow rates ratio flowing along the branches of the tee. This allows us to make an assumption about the need for further study of the possibility of introducing baffle vanes into symmetrical tees to use the ejection effect, if it is necessary to design symmetric tees in the ventilation system with significantly different flow rates of mixing flows. In addition, the presence of quite significant in size vortex zones shows the possibility of improving their design and reducing aerodynamic drag. Therefore, the further direction of the research development is seen in the construction of computer models of profiled tees, as well as experimental confirmation of the numerically found results. For better convergence with the experiment, it is also possible to research direction related to 3D-modeling of such duct fittings, as well as the use of turbulence models with fewer assumptions than RANS models – Detached Eddy Simulation (*DES*), Large Eddy Simulation (*LES*) and Direct Numerical Simulation (*DNS*), the development and application of which to real practical problems is still constrained by high computational costs.

Acknowledgments. This study would not have been possible without a grant support from the Russian Science Foundation (Project No. 18–79-10025).

References

1. Logachev, K.I., Ziganshin, A.M., Popov, E.N., Averkova, O.A., Kryukova, O.S., Gol'tsov, A.B.: Experiment determining pressure loss reduction using a shaped round exhaust hood. Build. Environ. **190**, 107572 (2021). https://doi.org/10.1016/j.buildenv.2020.107572

2. Logachev, K., Ziganshin, A., Kryukova, O., Averkova, O., Kryukov, I., Gol'tsov, A.: Improving dust capture efficiency with local exhaust hoods in manicure shops. Build. Environ. **181**, 107124 (2020). https://doi.org/10.1016/j.buildenv.2020.107124
3. Zamalieva, A.T., Ziganshin, M.G., Potapova, L.I.: Ob jeffektivnosti sushhestvujushhih metodov ciklonnoj fil'tracii pri osazhdenii melkodispersnyh chastic klassov PM10, PM2,5 [The analysis of efficiency of use of a cyclonic filtration for sedimentation of fine particles]. Izv. KazGASU [News KSUAE]. **42**, 415–423 (2017)
4. Belyaeva, G.I., Zamalieva, A.T., Ziganshin, M.G.: Numerical and experimental studies of gas cleaning in multi-cyclone elements with filter inserts. In: IOP Conference Series: Materials Science and Engineering (2019).https://doi.org/10.1088/1757-899X/481/1/012041
5. Kareeva, J., Bliznyakova, K., Ashadullina, D., Zakieva, R.: Influence of geometrical parameters of air inlet hole on the kinematic characteristics of jet. In: IOP Conference Series: Materials Science and Engineering (2020). https://doi.org/10.1088/1757-899X/890/1/012164
6. Idel'chik, I.E., Steinberg, M.O.: Handbook of Hydraulic Resistance. CRC Press, Boca Raton, FL (1994)
7. Gardel, A., Rechsteiner, G.F.: Les pertes de charge dans les branchements en Té des conduites de section circulaire. Bull. Tech. La Suisse Rom. **96**, 363–376 (1970)
8. Beilin, A.I., Messina, D.R., Richards, P.B.: Preliminary study of stability of flow from two ducts discharging into a common duct (1951)
9. Bassett, M.D., Winterbone, D.E., Pearson, R.J.: Calculation of steady flow pressure loss coefficients for pipe junctions. Proceedings of the Institution of Mechanical Engineers, Part C. Journal of Mechanical Engineering Science, vol. 215, pp. 861–881 (2001). https://doi.org/10.1177/095440620121500801
10. Gao, R., Liu, K., Li, A., Fang, Z., Yang, Z., Cong, B.: Study of the shape optimization of a tee guide vane in a ventilation and air-conditioning duct. Build. Environ. **132**, 345–356 (2018). https://doi.org/10.1016/j.buildenv.2018.02.006
11. Gao, R., Li, H., Li, A., Liu, K., Yu, S., Deng, B.: Applicability study of a deflector in ventilation and air conditioning duct tees based on an analysis of energy dissipation. J. Wind Eng. Ind. Aerodyn. **184**, 256–264 (2019). https://doi.org/10.1016/j.jweia.2018.11.025
12. Zhang, W., Li, A.: Resistance reduction via guide vane in dividing manifold systems with parallel pipe arrays (DMS-PPA) based on analysis of energy dissipation. Build. Environ. **139**, 189–198 (2018). https://doi.org/10.1016/j.buildenv.2018.04.010
13. Shopayeva, A., Safiullin, R.: CFD-modeling of flow in confluence nodes of ventilation units of multi-storey buildings. In: IOP Conference Series: Materials Science and Engineering (2020).https://doi.org/10.1088/1757-899X/890/1/012157
14. Gao, R., et al.: A novel low-resistance duct tee emulating a river course. Build. Environ. **144**, 295–304 (2018). https://doi.org/10.1016/j.buildenv.2018.08.034
15. Gao, R., Liu, K., Li, A., Fang, Z., Yang, Z., Cong, B.: Biomimetic duct tee for reducing the local resistance of a ventilation and air-conditioning system. Build. Environ. **129**, 130–141 (2018). https://doi.org/10.1016/j.buildenv.2017.11.023
16. Zhou, M., Lian, H., Sigmund, O., Aage, N.: Shape morphing and topology optimization of fluid channels by explicit boundary tracking. Int. J. Numer. Methods Fluids. **88**, 296–313 (2018). https://doi.org/10.1002/fld.4667
17. Dilgen, C.B., Dilgen, S.B., Fuhrman, D.R., Sigmund, O., Lazarov, B.S.: Topology optimization of turbulent flows. Comput. Methods Appl. Mech. Eng. **331**, 363–393 (2018). https://doi.org/10.1016/j.cma.2017.11.029
18. Li, A., Chen, X., Chen, L.: Numerical investigations on effects of seven drag reduction components in elbow and T-junction close-coupled pipes. Build. Serv. Eng. Res. Technol. **36**, 295–310 (2015). https://doi.org/10.1177/0143624414541453

19. Ziganshin, A.M., Badykova, L.N.: Chislennoe modelirovanie techeniya v profilirovannom ventilyatsionnom troynike na sliyanie [Numerical investigation of flow in profiled ventilation tee at junction]. Izv. vuzov. Stroit. [News High. Educ. institutions. Constr.] **702**(6), 41–48 (2017)
20. Ziganshin, A., Logachev, K., Batrova, K.: Reducing the drag of midpoint lateral orifices of exhaust air ducts by shaping them along vortex zone outlines. Build. Environ. **188**, 107491 (2021). https://doi.org/10.1016/j.buildenv.2020.107491
21. Gauder, P., Karthick Selvam, P., Kulenovic, R., Laurien, E.: Large eddy simulation studies on the influence of turbulent inlet conditions on the flow behavior in a mixing tee. Nucl. Eng. Des. **298**, 51–63 (2016). https://doi.org/10.1016/j.nucengdes.2015.12.001
22. Grbčić, L., Kranjčević, L., Lučin, I., Sikirica, A.: Large eddy Simulation of turbulent fluid mixing in double-tee junctions. Ain Shams Eng. J. (2020). https://doi.org/10.1016/j.asej.2020.06.004
23. ANSYS® Academic Research Mechanical and CFD, Release 18.2, Help System, Coupled Field Analysis Guide, ANSYS, Inc. (2017)
24. Posohin, V.N., Ziganshin, A.M., Mudarisov, D.I.: O protyazhennosti zon vliyaniya vozmushchayushchih elementov truboprovodnyh sistem [The influence zones of disturbing elements in pipeline systems]. Izv. KazGASU [News KSUAE]. **28**, 121–126 (2014)
25. Ziganshin, A.M., Posokhin, V.N., Badykova, L.N., Gimadieva, G.A.: Numerical simulation of two-dimensional tee flow. Izv. VUZov. Constr. **677**, 89–95 (2015)
26. Logachev, K.I., Ziganshin, A.M., Averkova, O.A.: Simulations of dust dynamics around a cone hood in updraft conditions. J. Occup. Environ. Hyg. **15**, 715–731 (2018). https://doi.org/10.1080/15459624.2018.1492137
27. Ziganshin, A.M., Safiullina, G.R., Eremina, S.V., Gajfullin, A.A.: Validacija komp'juternoj modeli techenija v vytjazhnom i pritochnom simmetrichnyh ventiljacionnyh trojnikah [Validation of numerical model of a flow in an exhaust and supply symmetrical ventilation tees]. Izv. KGASU [News KSUAE]. **55**, 58–70 (2021). https://doi.org/10.52409/20731523_2021_1_58

Polymer Multilayer Films for Packaging of Building Materials

Zhanna Gerkina[1], Valentina Serova[1], and Viktor Stroganov[2(✉)]

[1] Kazan National Research Technological University, Kazan, Russian Federation
[2] Kazan State University of Architecture and Engineering, Kazan, Russian Federation

Annotation. One of the promising applications for thermoplastic polymer materials is their use as packaging shrink films. It is established that the leading direction in the field of transport (pallet) packaging of building materials is taken by polymer films of a multilayer structure. One of the main operational requirements for such films is the resistance to light aging during storage and transportation. The study was carried out when comparing two types of three-layer films based on high-pressure polyethylene obtained on the co-extrusion line for the production of Kiefel 1 bag film using the target components of the formulation, including those for protection against ultraviolet (UV) radiation. It was found that films containing an ethylene-butylene copolymer (SEB) in the inner layer have a higher degree of stability.

Keywords: Polymer films · Light aging · Ethylene-butylene copolymer · Films for packaging

1 Introduction

Polymer film materials are used as packaging in many sectors of the national economy [1–5]. At the same time, complex multilayer structures are most in demand [6–10]. Unlike traditional single-layer packaging films, the construction of polymer-polymer structures is the most effective way to provide the necessary complex of operational properties of the package. This is due to the unlimited possibilities of varying the components of a multilayer material, the number, thickness, and order of alternation of its layers [6, 11]. For their manufacture, the method of co-extrusion is most often used, which allows a single cycle to obtain a material that is a combination of several polymer layers [12–14].

The leading positions in the field of transport (pallet) packaging of building materials are occupied by polymer films of a multilayer heat-shrink structure [15]. They ensure the ability of oriented polymers to reduce linear dimensions: when heated above the softening temperature, they are able to shrink and tightly fit the packaged products, ensuring their safety, both during transportation and during intermediate storage at the place of application. This type of packaging is more economical in comparison with the traditionally used film materials due to the small volume of the package (tight fit), lower weight, lower material consumption and greater efficiency of the packaging design. The most economical and widespread variant of the three-layer structure is a film, the central

N. Vatin (Ed.): STCCE 2021, LNCE 169, pp. 223–229, 2021.
https://doi.org/10.1007/978-3-030-80103-8_24

layer of which is made from secondary polymer raw materials. This structure practically does not lose strength, compared to a single-layer film made of a mixture of secondary and primary raw materials and allows the manufacturer to reduce the cost of production.

The main consumers of shrink film are companies that produce building materials, such as bricks, cement, aerated concrete, rolled roofing materials, dry mixes, gypsum boards, etc. This takes into account the size and weight of the packaged products, the conditions of transportation and storage, as well as the requirements for the protective properties, appearance and packaging process of the package.

Strict requirements for modern packaging include high values of strength and performance indicators. Packaging for building materials must ensure their safety while transportation and storage. The most important thing is to protect the packaging from external influences, especially ultraviolet (UV) rays, i.e. it must have a good resistance to light aging [16].

The currently known publications usually consider the comparative properties of mono-and multilayer packaging films [17, 18], their production, barrier (protective) and some other indicators of multilayer polymer films for packaging both food and technical products [19–30], as well as the quality of the printing applied to them [31, 32]. There is insufficient data in the publications on the performance properties of polymer films for packaging construction and other materials, including their resistance to light aging.

The relevance of such studies is of both scientific and practical importance for predicting the service life and durability of this type of packaging films. From this point of view, the aim of this work is to determine the influence of the composite composition on the performance indicators, including the resistance to light aging of multilayer shrink films intended for pallet packaging of building materials.

2 Research Objects

Samples of two types (**I** and **II**) of three-layer heat-shrinkable films based on high-pressure polyethylene (PE) of the 15303–003 brand with a thickness of 150 microns, obtained in LLC "Kama Plant of Polymer Materials" on the co-extrusion line for the production of Kiefel 1 bag film, were selected for the study. The composition of film layers **I** and **II**, along with PE, includes a concentrate of mineral filler (talc) on the polymer basis of Camlen 4112 (KMN), PE waste, an ethylene-butylene copolymer(SEB), as well as a concentrate based on a photostabilizer of the HALS class and PE (KSS) (Table 1).

Table 1. Structure of three-layer PE-based shrink films

Composition of layers	
I	**II**
1) PE + KMN 2) PE + PE-waste + SEB 3) PE + KMN	1) PE + KMN 2) PE + KSS 3) PE + KMN

3 Methods

The strength and deformation of polymer films were determined under tension (rupture) according to GOST 14236 (ASTM D 882) on an automatic Shimadzu breaking machine model AGS-X with an extensometer SES-1000 [33]. The measurement error did not exceed $\pm$ 10%. Samples of each film were cut out with a special device in the longitudinal and transverse directions. During the test, the stretching diagrams of the samples were automatically recorded. The value of the breaking stress (σ) was determined by the formula:

$$\sigma = F/S,$$

where F is the breaking force, H; S is the cross-sectional area of the film sample. The arithmetic mean of at least 5 determinations in each direction was taken as the test result. The rapid method for assessing light resistance consisted in irradiating film samples with a UV light source – a high-pressure arc discharge mercury lamp DRT 240, the beam flux of which lies in the wavelength range of 240–320 nm, and its power is 24.6 W [34]. The film samples were placed on a substrate, their distance from the lamp was 30 cm.

4 Results and Discussion

The results of determining the physical and mechanical parameters of the films - the values of σ and the relative elongation (ε) are presented in Table 2. The different composition of the inner layer of the films does not lead to a noticeable difference in the values of these indicators, and regardless of the direction of their measurement, i.e. the films of structures **I** and **II** practically do not differ either in strength or in the amount of tensile deformation. The kinetics of light aging of films is illustrated by the corresponding dependences of σ and ε on the duration of their UV irradiation (t), shown in Figs. 1, 2, 3, 4.

According to the dependences shown in Figs. 1 and 2, in all cases, with an increase in the duration of UV irradiation t, a decrease in the values of σ and ε is observed to some extent. In order to compare the resistance of films **I** and **II** to light aging, the relative change in the parameters σ and ε after the test was determined – Δσ and Δε, respectively. The values of Δσ and Δε are also included in Table 2.

Based on the found values of Δσ and Δε, it can be noted that UV irradiation of the films of structures (**I** and **II**) for 20 h does not lead to a significant change in ε, since the values of Δε in all cases are within the measurement error.

It is determined that the strength of the films after the light resistance test, especially measured in the transverse direction, decreases much more noticeably. It was found that the degree of reduction in the strength of the films is also influenced by their structure. Thus, despite the presence of KSS in the inner layer, the structure **II** film has a more significant decrease in σ after the test, namely, the values of Δσ are almost 3 times greater than those of the structure **I** film. Therefore, film **I**, which contains SEB along with PE waste, is more resistant to light aging.

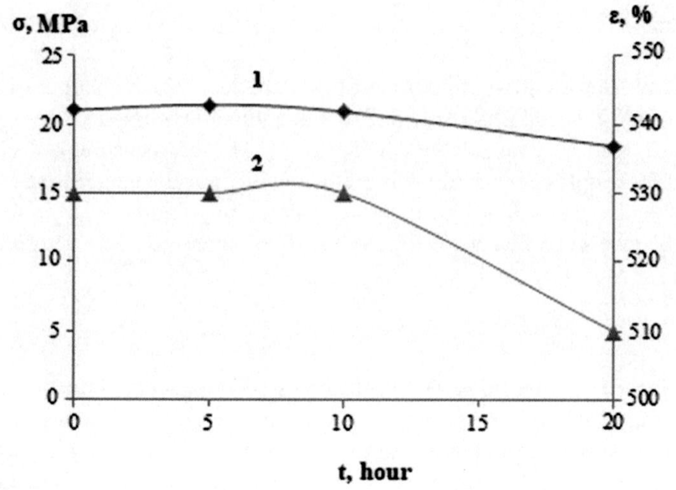

Fig. 1. Dependence of the breaking stress (1) and the relative elongation (2) of samples **I** measured in the longitudinal

Directions on the duration of UV irradiation.

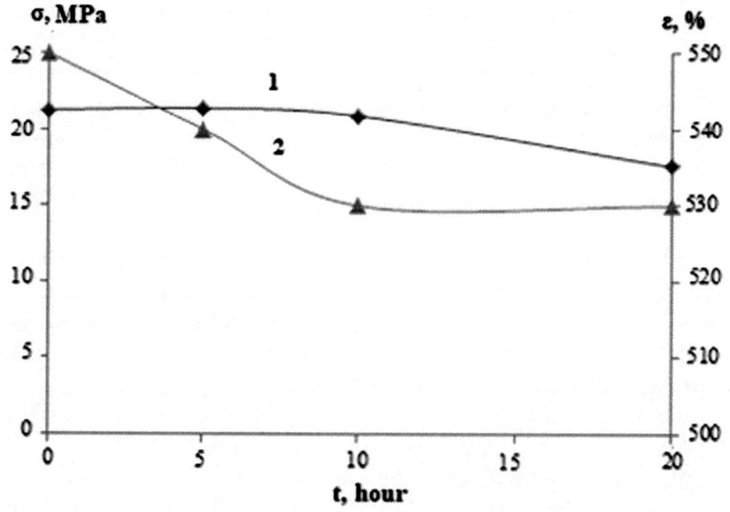

Fig. 2. Dependence of the breaking stress (1) and the relative elongation (2) of samples **I** measured in the transverse directions

On the duration of UV irradiation.

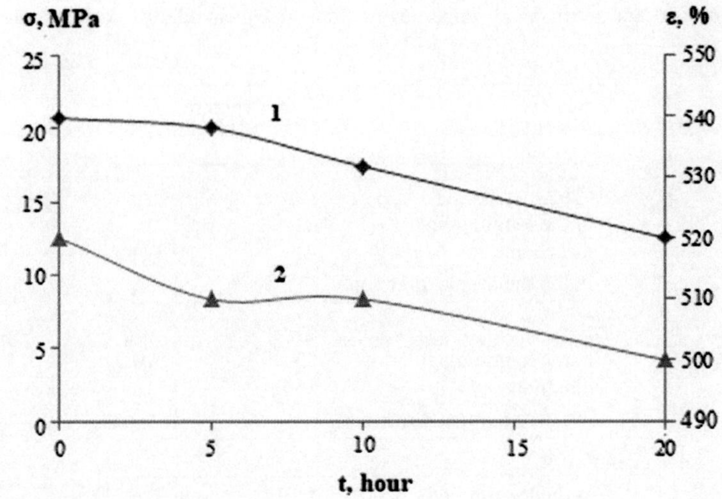

Fig. 3. Dependence of the breaking stress (1) and the relative elongation (2) of samples **II** measured in the longitudinal directions

On the duration of UV irradiation.

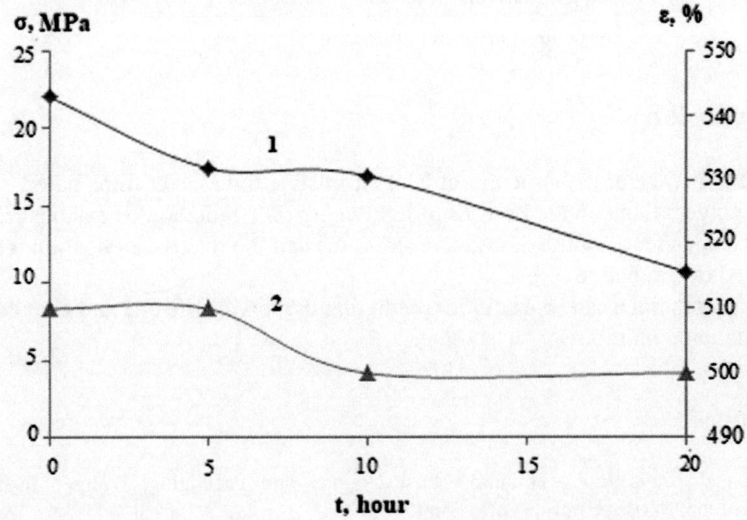

Fig. 4. Dependence of the breaking stress (1) and the relative elongation (2) of samples **II** measured in the transverse directions

On the duration of UV irradiation.

Table 2. Physical and mechanical properties of films and their relative change after the light resistance test

Indicator	Film structure	
	I	II
σ, MPa		
– in the longitudinal direction – in the transverse direction	21 21	21 22
ε, %		
– in the longitudinal direction – in the transverse direction	530 550	520 510
$\Delta\sigma^*$, %		
– in the longitudinal direction – in the transverse direction	11,9 16,6	33,3 43,2
$\Delta\varepsilon^*$, %		
– in the longitudinal direction – in the transverse direction	3,8 3,6	2,9 2,7

*Determined after 20 h of testing

5 Conclusion

Thus, the influence of the structure of heat-shrinkable three-layer films based on high-pressure polyethylene on the kinetics of light aging (UV radiation) is established in the form of a dependence of the destructive stress (σ) and the relative elongation (ε) on the duration (t) of irradiation.

The data obtained can be useful for predicting the service life of **I** and **II** under harsh UV irradiation conditions.

References

1. Hanlon, J., Kelsey, R., Forsinio, H.: Packaging and Packaging: Design, Technology, Application, St. Petersburg, Profession (2004)
2. Abbass, A.H.: Polymer Thin Films, In-Teh Olajnica 19/2, 32000 Vukovar, Croatia (2010). https://www.intechweb.org
3. Abdel-Bari, E.M.: Polymer films, Per. s angl. /Pod red. G.E. Zaikova. SPb, Professija (2006)
4. Zelke, S., Kutler, D., Hernandez, R.: Plastic Packaging, SPb, Profession (2011)
5. Brooks, D., Giles, J.: Production of Packaging from PET, Per. s Angl., SPb, Profession (2010)
6. Tretyakov, A.O.: Packaging, vol. 6, p. 20 (2006)
7. Lukanina, Y.K.: Packaging and packaging, vol. 3, p. 40 (2007)

8. Guillory, R., Deschaines, T., Henson, R.: Analysis of_multi-layer_polymer_films. Material Today **12**, 38 (2009). https://doi.org/10.1016/S1369-7021(09)70112-9
9. Dhawan, S.: Polymeric-Based Multilayer Food Packaging Films for Pressure-Assisted and Microwave-Assisted Thermal Sterilization, Washington state university, Washington (2013)
10. Langhe D., Ponting, M.: Manufacturing and Novel Applications of Multilayer Polymer Films, Elsevier Science, Amsterdam (2016)
11. Lyubeshkina, E.G.: Polymer materials, vol. 4, p. 4 (2009)
12. McKeen, L.W.: Production of films. In: The Film Properties of Plastics and Elastomers (2017)
13. Skopintsev, I.V.: Production of containers and packaging made of polymer materials, SPb, Lan (2018)
14. Tikhonov, N.N., Sheryshev, M.A.: Modern technologies and equipment for polymer extrusion, SPb, Profession (2019)
15. Packaging for building materials. https://www.fleimina.ru/inform/pack/dlya-stroitelnyh-materialov
16. Petrov, A.G.: Modern shrink films. https://plastinfo.ru/information/articles/234
17. Serova, V.N., Sugonyako, D.V., Verizhnikov, M.L., Tyuftin, A.A.: Plastic masses, vol. 5-6, p. 54 (2014)
18. Serova, V.N., Mirkhusainov, E.R.: Plastic masses, vol. 3-4, p. 56 (2018)
19. Garipov, R.M., Efremova, A.A., Gerkina, Z.Y.: Bulletin of Kazan. Technol. Un-ta, **18**, 174 (2015)
20. Zagidullin, A.I., Garipov, R.M., Khasanov, A.I.: Adhesives, sealants. Technologies. **12**, 18 (2020)
21. Serova, V.N., Khasanov, A.I., Gerkina, Z.Y., Efremova, A.A.: Abstract book of XX Mendeleev Congress on General and Applied Chemistry. Ekaterinburg. **3**, 282 (2016)
22. Garipov, R.M., Zagidullin, A.I., Khasanov A.I.: Pathways to Modern Physical Chemistry: An Engineering Approach with Multidisciplinary Applications, Apple Academic Press Inc. Ch. 17, 309 (2016)
23. Jang, W.-S., Rawson, I., Grunlan, J.C.: Thin Solid Films, vol. 516, p. 4819 (2008)
24. Jagadish, R.S., Raj, B., Asha, M.R.: Blending of low-density polyethylene with vanillin for improved barrier and aroma-releasing properties in food packaging. J. Appl. Polymer Sci. **113**(6), 3732–3741 (2009). https://doi.org/10.1002/app.30221
25. Rhim, J.W., Hong, S.-I., Ha, C.S.: LWT – Food Science and Technology, vol. 42, p. 612 (2009)
26. Mokwen, K.K., Tan, J., Dunn, C.P., Yan, T., Cho, E.: Journal of Food Engineering, vol. 92, p. 291 (2009)
27. Ali, J., Hakkaraine, M.: Journal of Applied Polymer Science, vol. 118, p. 1084 (2010)
28. Tihminliogl, F., Ati, I.D., \"{O}ze, B.: Journal of Food Engineering, vol. 96, p. 342 (2010)
29. Galikhanov, M.F., Guzhova, A.A., Efremova, A.A., Nazmieva, A.I.: Effect of aluminum oxide coating on structural, barrier and electret properties of polyethylene terephthalate films. IEEE Trans. Dielect. Electr. Insul. **22**(3), 1492–1496 (2015). https://doi.org/10.1109/TDEI.2015.7116342
30. Muratov, I., Garipov, R., Efremova, A., Khasanov, A.: Key Engineering Materials (KEM), 869 KEM, p. 394 (2020)
31. Serova, V.N., Mirkhusainov, E.R., Gerkina, Z.Y., Khasanov, A.I.: Adhesives, sealants. Technologies **8**, 42 (2017)
32. Serova, V.N., Zagidullin, A.I.: Adhesives, sealants. Technologies **1**, 37 (2019)
33. Grellman, V., Seidler, S.: Testing of Plastics, SPb, Professija (2010)
34. Rabek, J.F.: Experimental Methods in Photochemistry and Photophysics, Part 2. Wiley, Chichester (1982)

Air Diffuser for Ventilation and Air Conditioning Systems with Quantitative Control

Rinat Safiullin(✉) and Vladimir Posokhin

Kazan State University of Architecture and Engineering, 420043 Kazan, Russia

Abstract. This paper proposes a new design of the variable geometry air diffuser for energy-efficient ventilation and air conditioning systems with variable airflow. The device enables to adjust the outflow parameters, and therefore the jet properties (range, expansion angle, decay rate). The kinematic characteristics of the formed jet are studied for different ratios of the inlet and outlet diameters of the air diffuser. In the course of the experiments, the jet axis velocities were registered at different distances from the outflow plane. It was shown that at the jet formation section ($x \leq 0.25$ m), the axial velocity first decreases and then increases intensively, which indicates the presence of a rarefaction zone behind the baffle. It was found that the change in the relative jet velocity as a function of the distance to the air diffuser is practically the same for all the considered outflow velocities at a fixed angle of the case petals' expansion. The change in the velocity coefficient with an increase in the expansion angle of the air diffuser plates was plotted. It was found that for small expansion angles the kinematic coefficient equaled 5.1, which is typical for a compact jet. At angles of 45° and 80° it equaled 3.9 and 2.9, respectively, which is typical for conical and fan jets. Thus, the proposed air diffuser allows adjusting the parameters of the outflow and, accordingly, the kinematic coefficient in a wide range, and can be used for ventilation and air conditioning systems with variable air supply.

Keywords: Variable geometry air diffuser · Jet · Kinematic · Characteristics · Axial velocity · Change in the velocity coefficient

1 Introduction

When designing ventilated rooms, for which the main type of harmful emissions is heat, there are two ways to determine air exchange. The first option is to assess it by the heat balance of summer and to consider it unchanged in all periods of the year. The second way is to consider it variable and assess it following the predicted change in heat balance. In the first case, partial air recirculation is provided for the winter and transition periods. In the second case, the supply systems must be provided with a quite complex system of automatic air supply control. Note that a similar problem arises in all engineering systems with a variable consumption of working fluid (heat and gas supply systems, water supply systems, etc.).

The experience shows that the energy savings achieved in the second case can be quite significant, especially for large systems. However, the most popular is the first

N. Vatin (Ed.): STCCE 2021, LNCE 169, pp. 230–238, 2021.
https://doi.org/10.1007/978-3-030-80103-8_25

option, since in this case, the automatic control system is simpler and cheaper, therefore, the requirements for the qualifications of the personnel serving the ventilation systems are lower.

Modern trends in designing the engineering systems of buildings are aimed at reducing energy supply and energy-saving [1]. In particular, one should search for the principles of constructing intelligent ventilation and air conditioning systems with variable flow rates. Such systems should provide for the possibility of deep regulation of air exchange in individual zones of the premises and in time, depending on the actual demand (for example, the degree of occupancy of the premises by people). Here the annual operating costs can be reduced by 15–20% or more as compared with quality control systems [2–5].

An important factor restraining the widespread use of ventilation systems with variable flow rates is the unsteadiness of the air circulation regime in the room, and hence the complicacy of maintaining normalized conditions in the habitable zone. When the outflow conditions change, the jet velocity distribution changes. If the jet is non-isothermal, the ratio of gravitational and inertial forces at the outlet changes, which can result in a change in the jet shape, the place of separation of the jet from the ceiling when cooled air is supplied with a spreading jet [6, 7].

Although the modern level of automation systems makes it possible to solve the issues of automatic airflow control according to demand (to maintain the required temperature, CO_2 concentration, humidity, and other parameters in the room), nowadays the task of creating intelligent ventilation and air-conditioning systems has not been solved. The practical implementation of such systems faces several technological problems [8–10]. The main one is the absence of special air diffusers (AD) for these systems. Air diffusers are generally designed to operate with a constant air volume since there is a need to form a given geometry of the supply jet [11, 12]. A change in the airflow rate affects the jet angle, which changes the flow energy and leads to a deviation of the actual air parameters in the working area of the room from the calculated ones [7]. If the air outflow from the AD is too slow, an uncontrolled fall of the jet into the area where people are located may occur and a draft will be created in the area served by this diffuser [13, 14].

A change in airflow from a conventional AD causes a change in the ratio of gravitational and inertial forces in non-isothermal supply jets. In this case, the jet trajectory and the place where the spreading jet separates from the ceiling or floor surface change, and the intensity of attenuation of excess temperatures in the supply jets increases or decreases. AD for systems with a variable flow must ensure the constancy of parameters at the point where the jet enters the working area, and it is rational to supply cooled air with a horizontal fan-shaped jet, and warm air with a vertical direct-flow one. Therefore, AD must have an automatically variable geometry of the outlet section to control the shape and range of the supply jet, for example, with a thermal drive [15–17]. However, this greatly complicates its design and cost.

So, the task of developing simple and cheap ADs with automatically changing geometry is urgent. They will enable regulating the shape and properties of the jet (distribution of velocity and excess temperature over sections, long-range action, etc.). Here we propose a design of adjustable AD possesses such qualities [18], the operating principles are shown in Fig. 1.

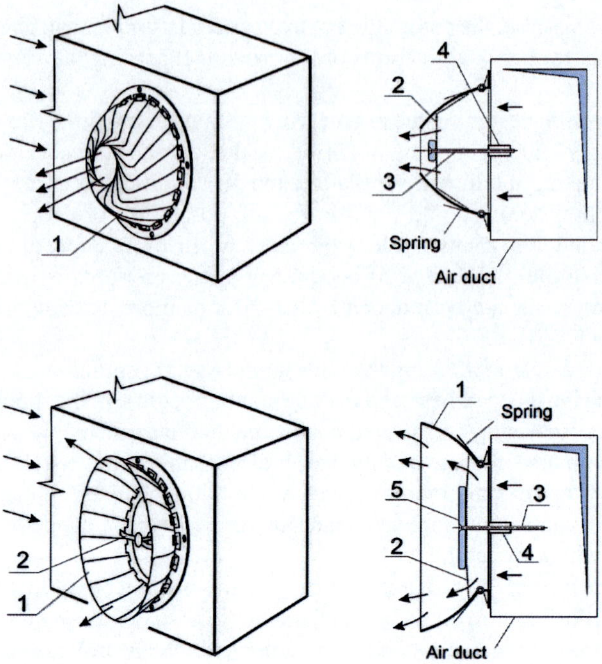

Fig. 1. AD schemes with adjustable geometry to create: *a*) compact jet; *b*) fan jet.

The air diffuser consists of a case with an adjustable outlet profile and a mechanism for adjusting the shape and parameters of the supply jet. The case is made of a set of overlapping and spring-loaded petal plates 1, supported from the inside by a control thrust plate 2. The plate 2 can move on the rod 3 along the axis driven by the magnetic coil 4, the voltage on which is controlled by the automation of the supply unit. The air diffuser can be equipped with a perforated or impermeable baffle 5 to create fan-shaped supply jets.

So, the air diffuser has a spring-loaded petal structure, inside which a control thrust plate is located on the movable axis. By changing the plate position, one can choose one of two options for air supply – a fan jet (expanding conical bell with a baffle) or a compact jet (a tapering conical bell formed when the petals are pulled together).

Figure 2 shows its structural elements: a support frame fixing the magnetic coil 4, hinged assemblies for attaching the petal plates 2 to the air diffuser flange.

The air diffuser works as follows. With an increase in the airflow rate, the automation of the supply unit gives a control signal, and the rod 3 is drawn into the magnetic coil 4. The ends of the plate 2 abut against the bases of the case petals 1 and they spread apart. With a decrease in airflow, the rod 3 moves towards the outlet, the thrust plate 2 stops pressing on the spring-loaded petals 1 and they close. By changing the position of the rod 3 and plate 2, one of two options for air release is achieved – a conical or compact jet. A conical jet is formed when the petal plates 1 are spread apart, when the case takes the form of an expanding conical bell. At large expansion angles, a perforated or impermeable baffle 5 is installed on the rod 3. A compact jet is formed by a tapering conical bell, which is formed when petals 1 are closed.

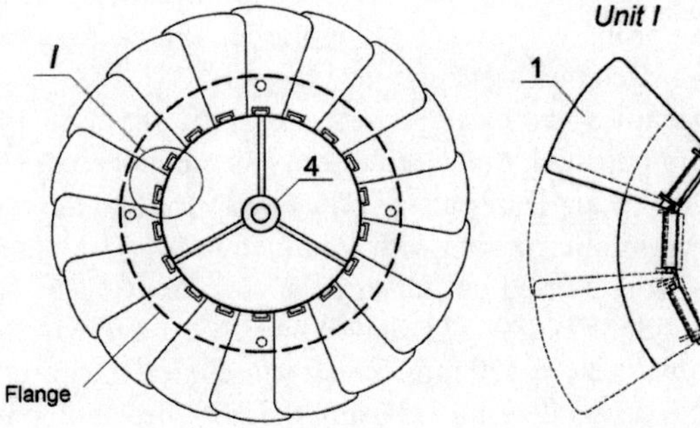

Fig. 2. The scheme for fixing the plates to the air diffuser flange.

Thus, in the proposed air diffuser the jet parameters are regulated in an automatic mode. The degree of outlet expansion is controlled by the air diffuser actuator (thrust plate 2 movable on the rod 3), connected with the regulating elements of the supply unit. The similarly designed ADs can be used in various rooms, equipped with systems for automatic control of air parameters. For its wide use, it is necessary to provide designers with information on the kinematic characteristics of the jet at different angles of petal opening.

2 Methods

The developed prototype of the air diffuser was tested on an experimental bench, the scheme of which is shown in Fig. 3. In the course of the experiments, the kinematic characteristics of the jet and the AD velocity coefficient were determined.

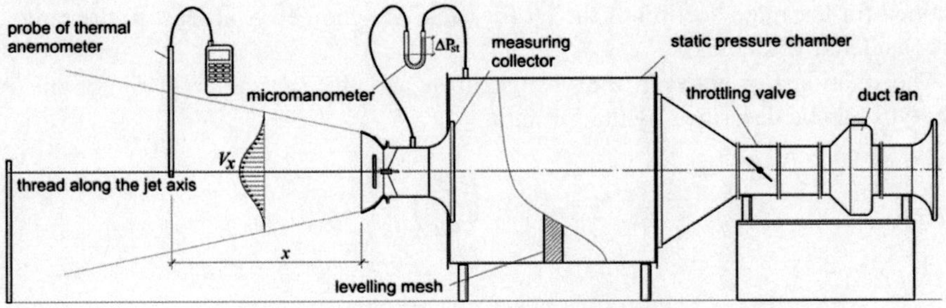

Fig. 3. The scheme of the experimental setup.

The stand consists of a high-pressure centrifugal fan, pipeline, plenum box with a leveling mesh inside, and a measuring manifold with a diameter of 125 mm. The tested

AD was connected to the manifold. The air from the room was sucked in by a fan and then pumped through the pipeline to the manifold and AD. The average air velocity in AD was determined from the difference in static pressures in the pressure chamber and in the manifold pipe. The air velocity in the jet was measured with a hot-wire anemometer. The hot-wire anemometer probe was moved along the jet axis using a two-axis coordinate system along a thread stretched along the jet axis.

The tested AD samples had case diameters of 115, 185, and 140 mm. The rod 3 position was manually adjusted, the width of the gap δ between the baffle 5 and case 1 was changed (Fig. 1). The angle of petals' expansion varied in the range $\alpha = 0° \div 85°$. The chosen baffle 5 was impermeable with a diameter $D_1 = 100$ mm.

The experiments were carried out for three angles $\alpha = 0°$, 45° and 80° for the AD airflow rates of 181, 258, 387, 489, 538, and 553 m^3/h. At a fixed velocity v_0 in the AD outlet pipe with a diameter $d_0 = 120$ mm, the air velocity on the jet axis $v_x{}^{ja}$ at distances x from the outlet section was determined using the hot-wire anemometer. The distance x between the measurement points was taken uniformly with a step of 100 mm to obtain $v_x{}^{ja}$ velocities comparable to the natural air mobility in the room.

3 Results and Discussions

The obtained plots of the axial velocity drop for various airflow rates are shown in Fig. 4. It is seen that in the jet formation section ($x \leq 0.25$ m), the axial velocity first decreases and then increases rapidly, which indicates the presence of a rarefaction zone behind the baffle. This characteristic property of the jet is shown in Fig. 5.

Figure 6 shows the experimentally obtained axial jet velocity for three values of the AD geometric parameter $D/d_0 = 0.97$, 1.17 and 1.54 in dimensionless coordinates $\overline{v} = v_x^{ja}/v_0$ and $\overline{x} = x/\sqrt{F_0}$, where $F_0 = \pi d_0{}^2/4$ is the area of the AD inlet section. Here the distance x is related to the area of the outlet section, and the axial speed v_x is related to the average AD outflow rate.

Figure 6 shows that the change in relative velocity v_x/v_0 as a function of distance to air diffuser $x/\sqrt{F_0}$ is almost the same for all the considered velocities v_0 for the constant angle of petals' expansion α. Thus, a unified velocity coefficient $m = \frac{v_x}{v_0} \cdot \frac{x}{\sqrt{F_0}}$ can be obtained for the main section of the jet for each AD geometry (at least in the range of the considered velocities).

The relationship between the change in m and the parameter D/d_0 for the AD prototype can be described by the formula:

$$m = 4.83 \cdot \left(\frac{D}{d_0}\right)^{-1.16}.$$

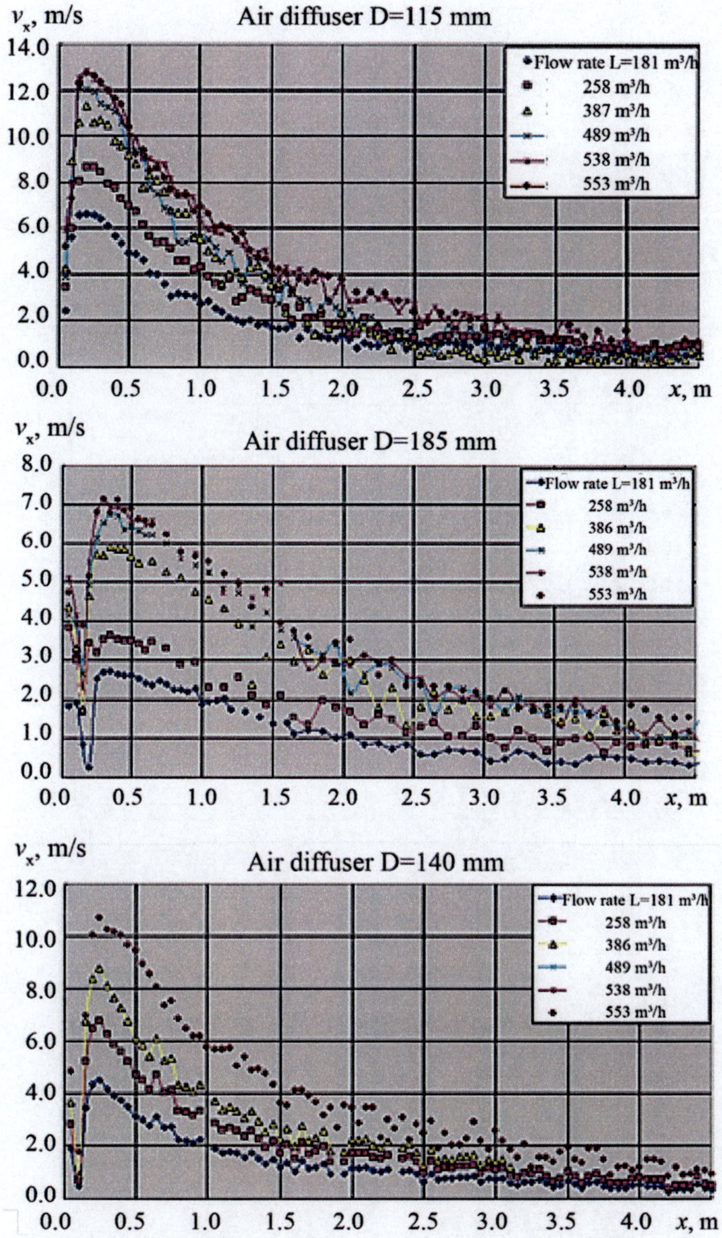

Fig. 4. Axial velocity drop for various airflow rates through AD.

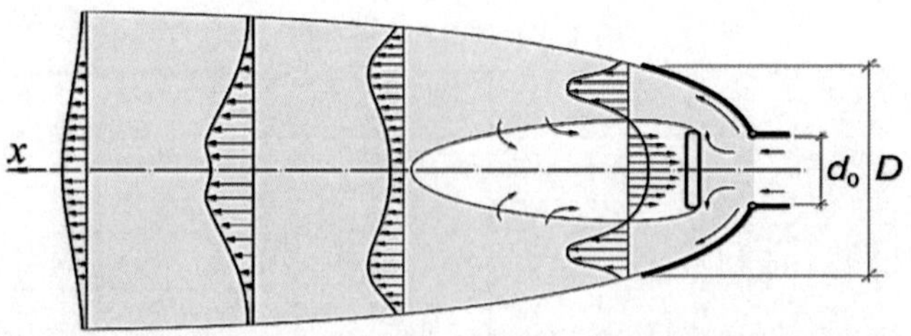

Fig. 5. Scheme of jet formation in the presence of an impenetrable baffle.

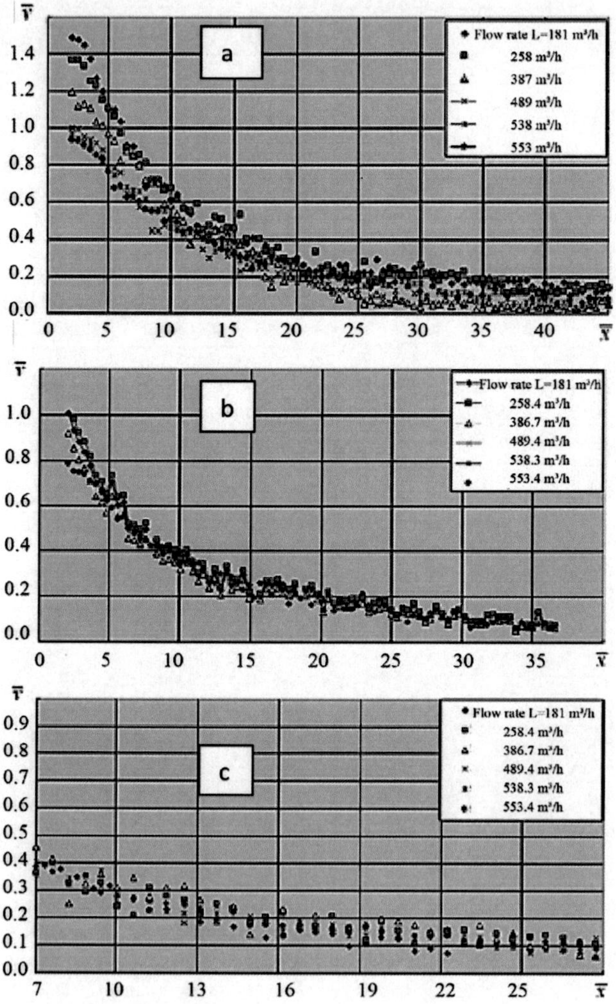

Fig. 6. Change in the axial jet velocity of the AD prototype for various geometric parameters D/d_0: a – 0.97; b – 1.17; c – 1.54.

For $D/d_0 = 0.96$, m equals 5.1, which is typical for a compact jet. For $D/d_0 = 1.17$ and 1.54 m equals 3.9 and 2.9, which are typical for conic and fan jets, respectively. Thus, the proposed AD allows adjusting the outflow parameters and, accordingly, the kinematic coefficients in a wide range, and can be used for systems with variable air supply. The change in m with an increase in the expansion angle of petals α is shown in Fig. 7.

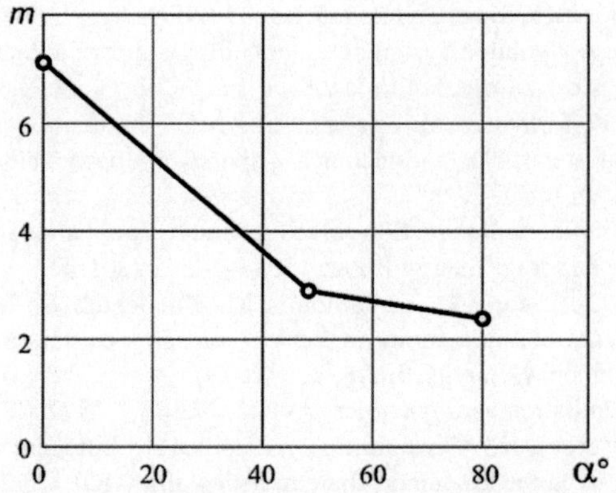

Fig. 7. Change in velocity coefficient m as a function of the expansion angle AD.

4 Conclusion

The results obtained showed that the proposed air diffuser can form jets of different geometries, which is important when used in ventilation systems with variable airflow. A conical jet is formed when the petal plates move apart, and the case takes the form of an expanding conical bell with a baffle. The compact jet is formed by a tapering conical bell, which is formed when the petals are closed.

The following characteristics of air diffuser were determined experimentally:

- At the jet formation section ($x \leq 0.25$ m), the axial velocity first decreases and then increases rapidly, which indicates the presence of a rarefaction zone behind the baffle;
- The change in the relative jet velocity as a function of distance to the air diffuser is practically the same for all considered outflow velocities at a fixed angle of the case petals' expansion;
- A plot showing the relationship between the changes in velocity coefficient m with an increase in the angle the petal plates' expansion α was obtained;
- The relationship between m and the parameter D/d_0 of the air diffuser prototype was obtained for the main section of the jet for each geometry (at least in the range of the considered velocities). For $D/d_0 = 0.96$, $m = 5.1$, which is typical for a compact jet; for $D/d_0 = 1.17$ and 1.54, m is equal to 3.9 and 2.9, respectively, which is typical for conical and fan jets.

Thus, the proposed air diffuser allows adjusting the outflow parameters and, accordingly, the kinematic coefficient in a wide range, and can be used for ventilation and air conditioning systems with variable air supply.

References

1. Goshka, L.L.: On the issue of the need to introduce effective use efficient systems for air conditioning of buildings. Mag. Civ. Eng. **7**, 33–37 (2009)
2. Advanced Variable Air Volume System. Design Guide. California Energy Comission (2003). https://newbuildings.org/sites/default/files/A-11_LG_VAV_Guide_3.6.2.pdf
3. Introducing High Performance Air System. An AMCA International White Paper. https://www.amca.org/educate/articles-and-technical-papers/whitepapers/air-systems/high-performance-air-systems.html
4. Hurnik, M.: Novel cylindrical induction controller and its application in VAV air conditioning system in an office building. Energy Build. **130**, 341–349 (2016)
5. Anand, P., Sekhar, C., Cheong, D., Santamouris, M., Kondepudi, S.: Occupancy-based zone-level VAV system control implications on thermal comfort, ventilation, indoor air quality and building energy efficiency. Energy Build., 204 (2019)
6. Posokhin, V.N.: Ventilation aerodynamics. AVOK-PRESS, 175 (2008)
7. Kareeva, J., Bliznyakova, K., Ashadullina, D., Zakieva, R.: Influence of geometrical parameters of air inlet hole on the kinematic characteristics of jet. IOP Conf. Ser. Mater. Sci. Eng. **890**(1), 012164 (2020)
8. Li, Z.H., Zhang, J.S., Zhivov, A.M., Christianson, L.L.: Characteristics of diffuser air jets and airflow in the occupied regions of mechanically ventilated rooms – a literature review. USA, ASHRAE Transactions 99(Pt. 1) (1993)
9. Yau, Y.H., Poh, K.S., Badarudin, A.: A numerical airflow pattern study of a floor swirl diffuser for UFAD system. Energ. Buildings **158**, 525–535 (2018)
10. Industrial Ventilation Design Guidebook 1: Fundamentals. Academic Press. 2nd edition, 550 (2020)
11. Hurnik, M., Kaczmarczyk, J., Popiolek, Z.: Study of radial wall jets from ceiling diffusers at variable air volume. Energies **14**, 240 (2021)
12. Hagström, K., Sirén, K., Zhivov, A.: Calculation methods for air supply design in industrial facilities. Literature review. Helsinki University of Technology, Laboratory of Heating, Ventilating and Air Conditioning B60, 205 (1999)
13. Gad, M.A., El. Shahat, F.A., Mohammed, F.A., Ramy, H.M.: Experimental and numerical study of influence of air ceiling diffusers on room airflow characteristics. Energ. Buildings **55**, 738–746 (2012)
14. Dozier, J., Hunka, R.S., Kline, J.R.: U.S. Patent № 5860592. Variable-air-volume diffuser with independent ventilation air assembly and method, 19 January (1999)
15. Subban, R., Raj, R.S.: U.S. Patent № US10670285 B2. Personal comfort variable air volume diffuser, 2 June (2020)
16. Kim, S.Y.: U.S. Patent № US 201102.75306A1. Motorized diffuser, 10 November (2011)
17. Zhivov, A.M.: Variable-air-volume ventilation systems for industrial buildings// USA, ASHRAE Transactions 96 Part 2, 367–372 (1990)
18. Safiullin, R.G., Khafizova, M.R.: Patent № 2490562 (Russia). Air diffuser. Bulletin of inventions № 23, 20 August 2013

Model of Degradation of Composite Materials of Building Structure's Load-Bearing Elements

Rashit Kayumov[1,2], Alfred Sulejmanov[1], and Dmitry Strakhov[1](✉)

[1] Kazan State University of Architecture and Engineering, Zelenaya Street, 1, 420043 Kazan, Russia

[2] Kazan National Research Technical University named after A.N. Tupolev, K. Marks Street, 10, 420111 Kazan, Russia

Abstract. The study of the degradation of flat extended elements made of polymer composite material reinforced with glass fiber has been carried out. The stresses in the elements exposed to UV radiation were determined to assess the long-term strength by the level of permanent deformations. The dependence of the strength of the composite element on the degree and time of irradiation is offered. Based on experimental data on the residual strength of specimens subjected to buckling and UV irradiation, a relation is proposed that reflects the change in the strength of the test material on the stress level, irradiation power, and exposure time of these factors. A comparison of the experimental and calculated values of the breaking loads was carried out, which showed the consistency of the results. A linear relationship was established with short exposure time.

Keywords: Composite materials · Polymer compositions · Temperature-time analogy · Degradation of polymer materials · Behavior in time · Aging of composites

1 Introduction

The use of load-bearing structural elements made of polymer composite materials is often constrained by insufficient knowledge of the mechanisms of damage development that affect the durability and load-bearing capacity of elements and their joints.

The variety of compositions of modified composites, with a variety of reinforcement and different operational properties, leads to the lack of accurate experimental data on certain products when performing calculations.

Evaluation of long-term fracture characteristics, when calculating polymer composite materials for strength and durability, is undeniably an interesting and urgent task today.

The most versatile and technologically advanced production of various fiberglass elements of load-bearing structures of the constant cross-section is the method of pulling highly fiber-filled polymer compositions through a system of dies with a gradually decreasing cross-section. The reinforcement of such elements occurs in the direction in which the composite mass is stretched.

N. Vatin (Ed.): STCCE 2021, LNCE 169, pp. 239–249, 2021.
https://doi.org/10.1007/978-3-030-80103-8_26

Under complex influences during operation on a polymer composite element of temperature, water, alkaline or acidic environment under the atmosphere and various biological damage, climatic and natural aging of materials occurs [1–10]. With the simultaneous action of UV radiation and the action of the applied load, the influence of solar radiation on the growth of cracks was revealed [11]. The study of a polymer composite material in the form of flat extended elements reinforced with glass fiber in various corrosive environments is considered [12]. The deformation of the test specimens was carried out in the form of three-hinged (Fig. 1) fastening, by analogy with the buckling method [13–28].

Heat resistance, strength, and durability characterizing the performance of composite material are determined by a set of physical constants, whereby changing one performance parameter, it is possible to achieve a change in the other two [29]. The degradation process of a polymer composite material is an integral part of engineering calculations that affect the durability of a structure. Thus, by simulating the aging process, it is possible to predict the durability of the material with sufficient accuracy.

The proposed study makes it possible to develop a calculation method that allows predicting the service life of load-bearing structures made of polymeric materials that provide strength reliability, considering aggressive factors affecting the decrease in the characteristics of the material under study.

2 Materials and Methods

2.1 Simulation of the Test Process

In this paper, we investigate flat extended elements made of polymer composite material, and polymer composite reinforced with glass fiber. The problem of predicting the deformation in time of polymers at different temperatures is investigated. The combination of aliphatic and aromatic fragments during the formation of epoxy polymers was investigated. As a component, including aliphatic (elastic) fragments, we used a reactive epoxy rubber-oligoepoxyurethane, since the introduction of rubbers has a significant effect on their elastic-deformation and relaxation properties. As a curing system, an amine curing agent, diethylenetriaminomethylphenol, was used, combining aliphatic and aromatic fragments in its structure.

The tests were carried out on a tensile testing machine under uniaxial tension at various temperatures and stresses. The samples were heated in a vertical muffle chamber with three heater sections and holes for capturing samples with the smallest temperature gradient along the height of the working space. Observation of the samples and measurement of their deformation was carried out through the observation windows provided in the body of the heat chamber.

The deformations of the samples were measured using two cathetometers intended for measuring vertical segments, in the same plane in places inaccessible for direct measurement, with a basic permissible error limit of $\pm$ 0.0015 mm.

When taking measurements, the measured length (the distance between two points of the object) was compared with the scale of the device by sequentially sighting the telescope at the boundary points of the measured segment.

The time dependences of deformation at different temperatures and loads obtained during the test are shown in Fig. 1

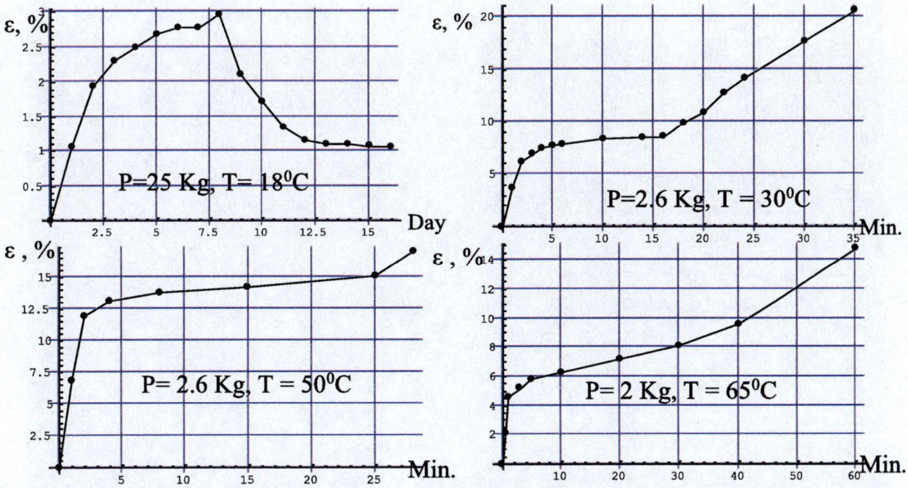

Fig. 1. Deformation of samples in time at different loads and temperatures.

To develop the deformation model, only the first two sections of the diagram were used - the hardening section and the section of steady-state creep.

To predict long-term creep, the method of analogies (superpositions) was chosen, based on the use of factors that accelerate relaxation processes. According to the principle of temperature-time analogy, temperature and time of deformation are interrelated and mutually equivalent.

The deformation of the studied group of samples was carried out (Fig. 2) to obtain various stress levels with exposure to UV radiation to assess the long-term strength by the level of residual deformations with periodic measurements of residual deflections.

It is known that after eighteen days the residual deflection practically ceases to change [12]. Therefore, as a margin of safety, it can be assumed that the following value can be taken as the elastic part of the deflection:

$$f^{elastic} = f - f^{residual} \quad (1)$$

Then, using the Blaznov or Euler formulas, one can find the maximum deformation in the middle of the sample, which remains constant at any time:

$$\varepsilon_{max} = \pm \frac{100b}{2\rho}$$

$$\rho = \frac{L}{\pi\sqrt{\frac{1}{2} \cdot \frac{3+0.252\delta^2+0.077\delta^3+0.079\delta^4}{0.125-0.015\delta-0.008\delta^2}}} \quad (2)$$

where L and b are the length and thickness of the sample [11].

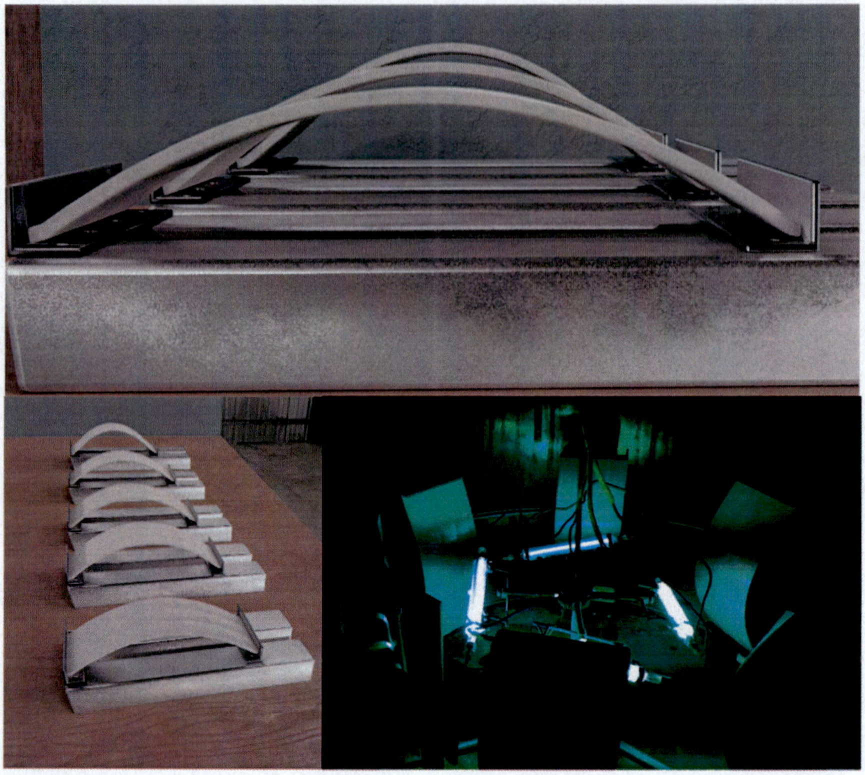

Fig. 2. General views of the studied samples, and samples in the process of UV radiation (authors' illustration).

The stress is determined according to Hooke's law:

$$\sigma_{max} = E.\varepsilon_{max}^{\text{elastic}} = E.\left(\varepsilon_{max} - \varepsilon^{\text{residual}}\right) \tag{3}$$

To find the value of $\varepsilon_{max}^{\text{elastic}}$ we use the proportion:

$$\frac{f^{\text{elastic}}}{f} = \frac{\varepsilon^{\text{elastic}}}{\varepsilon_{max}} \tag{4}$$

Now you can find the true values of the compressive force F, for example, using the Blaznov formula:

$$\sigma_{max} = \frac{F.f^{\text{elastic}}}{W} \tag{5}$$

As a result of the calculation, we get the following values (Table 1):

Table 1. Stiffness and strength parameters.

№ exp.	F	F residual	F elastic	F/F break	σ/σ break
1	38.61	8.53	30.08	0.156	0.156
2	48.1	9.61	38.44	0.240	0.240
3	52.71	11.13	41.58	0.394	0.394
4	63.7	13.04	50.46	0.477	0.477

Here and in (5), F is understood as the real compressive force acting on the sample after stress relaxation.

It follows from (5) that

$$\frac{\sigma}{\sigma_{break}} = \frac{F}{F_{break}} \tag{6}$$

In what follows, σ_{break} will be called the ultimate strength and also denoted by the letter R.

To simulate the aging process of the material, let us take into account that UV in the first moments of time leads to an increase in its strength.

Let us choose the following approximation for ultimate strength as a function of time t and intensity (power) of irradiation γ in the absence of stress:

$$R_0(t,\ \gamma) = R_{00}\frac{1 + t\gamma/k_0}{1 + c(t\gamma/k_0)^2} \tag{7}$$

Here $k_0,\ c$ are the material constants sought from the experiment;

R_{00}– initial strength;

$R_0(t,\ \gamma)$ – is the strength that an unstressed material acquires after a time t under UV irradiation.

It is clear that the breaking loads are proportional to $R_0(t,\ \gamma)$ and R_{00}:

$$m.F_0^{break}(t,\ \gamma) = R_0(t,\ \gamma) \tag{8}$$

$$m.F_{00}^{break}(0,\ 0) = R_{00}, \tag{9}$$

where m is the coefficient of proportionality.

Substituting (8) and (9) in (7), we get:

$$F_0^{break}(t,\ \gamma) = \frac{1 + t\gamma/k_0}{1 + c(t\gamma/k_0)^2}.F_{00}^{break}(0,\ 0) \tag{10}$$

From the experiment, the value $F_0^{brea}(t_{max},\ \gamma)$ and the value $F_{00}^{break}(0,\ 0)$ are known:

$$F_{00}^{break}(0,\ 0) = 57\ \text{kgf} \tag{11}$$

$$F_0^{break}(t_{max},\ \gamma) = 63\ kgf \tag{12}$$

Here

$$t_{max} = 56\ days, \qquad \gamma = 2,0\ kW/m^2 \tag{13}$$

To obtain an additional condition, we assume that for

$$t > t_{max} = 56\ days$$

Further hardening cannot be expected, this goes into the margin of safety. Then this condition can be written as:

$$F_{max}^{break} = F_0^{break}(t_{max},\ \gamma) = 63\ \text{kgf} \tag{14}$$

By Fermat's theorem we get:

$$\left.\frac{\partial F_0^{break}(t,\ \gamma)}{\partial t}\right|_{t=t_{max}} = 0 \tag{15}$$

Solving together the system of Eqs. (13), (10) at $t = t_{max}$ we find $k_0,\ c$:

$$c = \frac{[F_{00}^{break}]^2}{4\left[F_{max}^{break} - F_{00}^{break}\right].F_{max}^{break}} \tag{16}$$

$$k_0 = \frac{F_{00}^{break}.t_{max}.\gamma}{2\left[F_{max}^{break} - F_{00}^{break}\right]} \tag{17}$$

Note: Eq. (15) after differentiation and simplifications has the following form:

$$k_0^2 - c.t.\gamma.(t.\gamma + 2k_0) = 0 \tag{18}$$

After substituting the values (11)–(13) into the system and solving it, we obtain:

$$k_0 = 532\frac{kJ}{m2}, \qquad c = 2.15 \tag{19}$$

The form of function (10) is shown in Fig. 3. It characterizes the change in strength. $R_0(t,\ \gamma) = m.F_0^{break}(t,\ \gamma)$ from UV exposure.

Next, we introduce an approximation for the residual strength $R(t, \gamma, \sigma)$, which changes due to stress, namely in the form:

$$R(t, \gamma, \sigma) = R_0(t, \gamma).e^{-\beta.|\sigma|.t} \tag{20}$$

That is, the strength of the stress σ is assumed to decrease exponentially.

To obtain the value of the constant β, you need to have one experiment. Again we believe that:

$$\frac{R(t, \gamma, \sigma)}{R_0(t, \gamma)} = \frac{F^{break}(t, \gamma, \sigma)}{F_0^{break}(t, \gamma)} \tag{21}$$

Then from (20) and (21) the expression for β follows:

$$\beta = \frac{1}{|\sigma|t}\left\{ln\left[F_0^{break}(t,\gamma)\right] - ln\left[F^{break}(t,\gamma,\sigma)\right]\right\} \quad (22)$$

As $F^{break}(t,\gamma,\sigma)$ we take the value obtained from the experiment at the maximum deflection value:

$$F^{break}\left(t_{max},\gamma,\sigma^{break}.0,477\right) = 0,477.F_{00}^{break} \quad (23)$$

$$F_0^{break}(t_{max},\gamma) = 63\ kgf \quad (24)$$

The calculation results give the value:

$$\beta = 0,00647$$

Figure 4 shows the dependence of the breaking load on time in the interval [0–56 days] for the last specimen with a deflection $f = 63,7$ mm.

Figure 5 shows the experimental values of the breaking load depending on the ratio $\sigma/\sigma_{pa3p} = 0.156; 0.240; 0.394; 0.477$ at exposure $t_{max} = 56\ days$ and the calculated dependence of $F^{break}(t_{max},\gamma,\sigma)$ on stresses.

Note that this dependence is close to linear due to the small value of the β coefficient and the small time interval. If you increase the time interval of irradiation, then the nonlinearity of the dependence of $F^{break}(t_{max},\gamma,\sigma)$ on the relative stress σ/σ_{break} will manifest itself more strongly. This is also seen from Fig. 6, while the exposure time under irradiation $\gamma = 2.0$ $\gamma = 2,0\ \mathrm{kW/m^2}$ was taken as 250 days.

3 Results

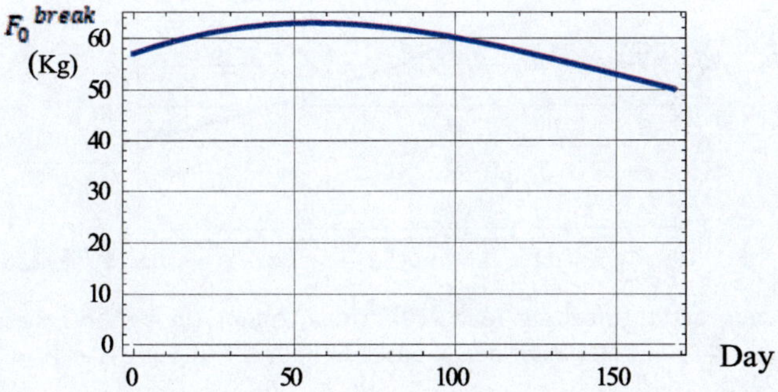

Fig. 3. Calculated dependence of the breaking load $F_0^{break}(t,\ \gamma)$ (in kgf) on time (in days) for an unstressed sample when exposed only to UV radiation with a power of $\gamma = 2,0\ kW/m^2$ (authors' illustration).

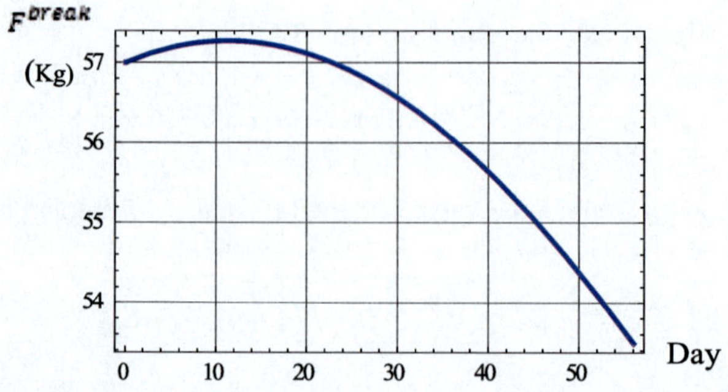

Fig. 4. Dependence of the breaking load $F^{break}(t, \gamma, \sigma)$ (in kgf) on time (in days) for the last sample when exposed to both UV and stress $\sigma/\sigma_{break} = 0.47$ (authors' illustration).

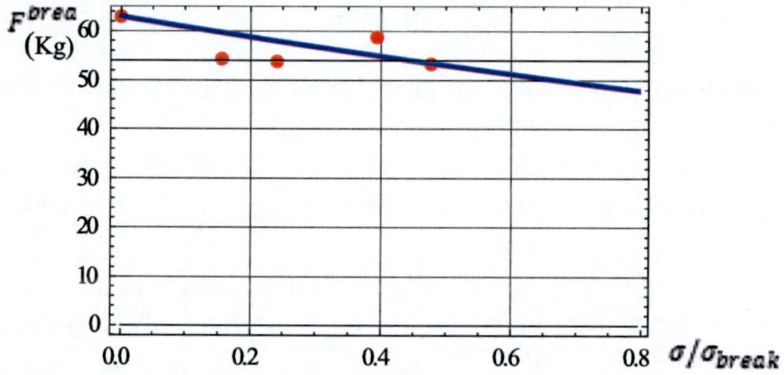

Fig. 5. Dependence of the breaking load $F^{break}(t_{max}, \gamma, \sigma)$ (in kgf) on the relative stress σ/σ_{break} in $t_{max} = 56$ days under exposure to both UV and stress. Markers indicate experimental, solid line - calculated values (authors' illustration).

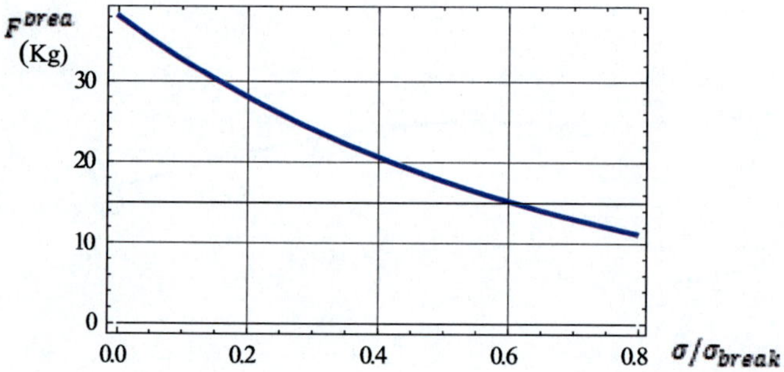

Fig. 6. Dependence of the breaking load $F^{break}(t_{max}, \gamma, \sigma)$ (in kgf) on the relative stress σ/σ_{break} through $t_{max} = 250$ days when exposed to both UV and stress at $\gamma = 2,0\frac{kW}{m^2}$.

4 Discussions

The experiments and the assumptions make it possible to determine the approximate values of stresses in the sample, at which aging (degradation) of the material occurs, in the margin of safety.

Based on the analysis of test results of unloaded samples, the dependence of the strength of fiberglass on the degree and time of irradiation is proposed (Eq. (7)).

Based on the results of experimental data on the residual strength of specimens subjected to buckling and UV irradiation, a relation has been proposed that reflects the change in the strength of the fiberglass under study on the stress level, irradiation power, and exposure time of these factors (relation (20)).

A comparison of the experimental and calculated values of the breaking load obtained from this ratio was carried out, which showed their satisfactory consistency (Fig. 6). It was found that at small times this dependence can be taken to be linear.

Acknowledgements. The work was performed within the framework of the Russian Foundation for Basic Research (project No. 19–08-00349, sections 1, 2) and the Russian Science Foundation (project No. 19–19-00059).

References

1. Blaznov, A.N., Savin, V.F., Atyasova, E.V., Babenko, F.I., Fedorov, Y.: Effect of temperature on the strength of basalt and fiberglass plastics. Polzunovskii Vestnik. **4**(2), 154–159 (2014)
2. Fedorov, Y., Babenko, F.I., Gerasimov, A.A., Lapiy, G.P.: Influence of cold climate factors on the mechanical properties of composite rods made of glass and basalt plastics. Metal Proc. (Technol., Equip., Tools). **2**, 56–61 (2012)
3. Blaznov, A.N., Savin, V.F., Krasnov, A.A.: Predicting the strength of composite rods under the influence of an alkaline medium. South Siberian Scientific Bulletin. **4**(8), 12–14 (2014)
4. Sen, R., Mullins, G., Salem, T.: Durability of e-glass/vinylester reinforcement in alkaline solution. ACI Struct. J. **3**(99), 369–375 (2002)
5. Cabral-Fonseka, S., Correia, J.R., Rodrigues, M.P., Branco, F.A.: Artificial accelerated aging of GFRP pultruded profiles made of polyester and vinylester resins: characterization of physical-chemical and mechanical damage. Strain **48**, 162–173 (2012). https://doi.org/10.1111/j.1475-1305.2011.00810
6. Koretskaya, L., Aleksandrova, T.: Influence of water on the properties of fiberglass. Technologies and materials. **1**(31), 38–41 (2011)
7. Nikolaev, E.V., Barbotko, S.L., Andreeva, N.P., Pavlov, M.R., Grashchenkov, D.V.: Comprehensive study of the impact of climatic and operational factors on a new generation of epoxy binder and polymer composite materials based on it. Part 4. Field climatic tests of polymer composite materials on based on epoxy matrix, Proceedings of VIAM, 6(42) (2016)
8. Kablov, E.N., Startsev, O.V., Krotov, A.S., Kirillov, V.N.: Climatic aging of composite materials for aviation. III. Significant Aging Factors. Deformation and Fracture of Materials. Electronic version: 1 (2011). www.viam.ru/public
9. Efimov, V.A., Shvedkova, A.K., Korenkova, T.G., Kirillov, V.N.: Research of polymer structural materials under the influence of climatic factors and loads in laboratory and field conditions, Aviation materials and technologies, 68–73 (2013)

10. Karapetyan, K.A.: Resistance to deformation and destruction of reinforced plastics. NAS RA: Gitutiun Yerevan, Armeniya (2014)
11. Yartsev, V.P., Kiseleva, O.A.: Predicting the behavior of building materials under unfavorable operating conditions. Tutorial (2009)
12. Kayumov, R., Strakhov, D.: The study of the rheological properties of composite materials of power elements of building structures. In: IOP Conference Series: Materials Science and Engineering, 890 (2020). https://doi.org/10.1088/1757-899X/890/1/012096
13. Arnautov, A.K., Tarnopolskiy, Y.: Buckling as a method for determining the flexural strength of composite materials. Mech. Compos. Mater. **1**(40), 25–42 (2004)
14. Savin, V.F., Blaznov, A.N.: Method for determining the long-term strength of fiberglass reinforcement. Meas. Autom. Model. Ind. Sci. Res. Interuniversity Collect. 214–219 (2001)
15. Savin, V.F., Volkov, Y.P., Lugovoi, A.N., Blaznov, A.N., Khe, A.I.: Buckling as a means of controlling mechanical characteristics of composite structural materials. Meas. Autom. Model. Ind. Res. Interuniversity Collect. 167–172 (2002)
16. Kiselev, N.M., Savin, V.F., Blaznov, A.N.: Fatigue strength of rods made of composite materials in buckling and tension. In: Reports of the VII All-Russian scientific-practical conference Technique and Technology for the Production of Thermal Insulation Materials from Mineral Raw Materials, 21–23 May, 2008, BTI AltSTU Biysk, 137–142 (2008)
17. Blaznov, A.N., Volkov, Y.P., Lugovoi, A.N., Savin, V.F., Khe, A.I.: Investigation of the deformation of fiberglass rods during buckling. Meas., Autom. Model. Ind. Sci. Res. Interuniversity Collect. 180–185 (2003)
18. Petrov, M.G.: Analysis of strength and durability of unidirectional fiberglass from the standpoint of the kinetic concept of fracture. Mech. Compos. Mater. Struct. **3**(9), 376–397 (2003)
19. Volkov, Y.P., Lugovoy, A.N., Savin, V.F., Khe, A.I., Blaznov, A.N.: Method for determining the mechanical characteristics of rods based on the results of buckling test. Factory Lab. **9**(70), 266–268 (2004)
20. Ruslantsev, A.N., Dumansky, A.M.: Model of nonlinear deformation and accumulation of damage in polymer composites, Electronic scientific and technical publication Science and Education, 2 (2014). https://doi.org/10.7463/0214.0687567
21. Kayumov, R.A., Strakhov, D.E.: Prediction of deformation in time of polymeric materials with shape memory at different temperatures. Izvestija KazGASU. **2**(16), 195–199 (2011)
22. Khozin, V., Khokhryakov, O., Nizamov, R.: A carbon footprint of low water demand cements and cement-based concrete. In: IOP Conference Series: Materials Science and Engineering, 890(1) (2020). https://doi.org/10.1088/1757-899X/890/1/012105
23. Vdovin, E.A., Stroganov, V.F.: Optimization of complex frost-resisting additives in line with operational requirements towards freezing temperature in road dressing constructions. In: IOP Conference Series: Materials Science and Engineering. 832(1) (2020). https://doi.org/10.1088/1757-899X/832/1/012034
24. Salakhutdinov, M.A., Gimranov, L.R., Kuznetsov, I.L.,Fakhrutdinov, A.E., Nurgaleeva, L.M.: PFRP structures under the predominately short term load. Mag. Civil Eng. Vol. **96**(4), 3–14. https://doi.org/10.18720/MCE.96.1 ISSN: 20714726 PFRP structures under the predominately short term load
25. Nizameev, V.G., Basharov, F.F, Nizameeva, L.V.: Load carrying ability and optimal characteristics of the roof slab made of profiled sheeting. In: IOP Conference Series: Materials Science and Engineering, 890, 012049 (2020)
26. Gimazetdinov, A., Kayumov, R., Shmelev G.: Method of calculating the shear stiffness of the flooring under the action of horizontal loads. In: IOP Conference Series. Materials Science and Engineering, 012056 (2020). https://doi.org/10.1088/1757-899X/890/1/012056

27. Kayumov, R.A., Tazyukov, B.F., Mukhamedova, I.Z.: Identification of mechanical characteristics of a nonlinear-viscoelastic composite by results of tests on shells of revolution. Mech. Compos. Mater. **55**(2), 171–180 (2019). https://doi.org/10.1007/s11029-019-09802-3
28. Kayumov, R.A., Mukhamedova, I.Z., Tuisina, E.B.: Analysis of a beam with a crack under irregular cyclic load. Russ. Aeronaut. **59**(4), 452–459 (2016). https://doi.org/10.3103/S1068799816040036
29. Ratner, S.B., Yartsev, V.P.: Physical mechanics of plastics. How performance is predicted (1992)

Design Characteristics of a Horizontal Light Tube Construction Considering Requirements of Lighting and Thermal Protection of Enclosing Structures

Artem Petrov(✉) and Maksim Kolchin

Kazan State University of Architecture and Engineering, 420043 Kazan, Russia

Abstract. Light pipes are an effective solution for providing natural light to rooms in those areas of the building where traditional natural lighting systems are not available. The use of vertical pipes may not be optimal in a multi-storey building due to a significant lengthening of the tube and a decrease in their efficiency, as well as the possible need for holes in the ceilings of each floor. In this case, the use of horizontal light pipes passing through the outer walls into the room can be an alternative to vertical ones. However, these systems are less studied and tested. There is ambiguity in their luminous efficiency and thermotechnical qualities. These openings can lead to heat loss and condensation. This article presents the results of an experimental and calculated analysis of the luminous efficiency of hollow horizontal light pipe, and also a solution has been developed for the junction of the light receiver with the outer wall, taking into account the requirements of thermal protection. The regularity of the efficiency of the pipe on the ratio of the section height to the length of the tube and the angle of inclination of the reflecting walls is revealed.

Keywords: Light pipe · Natural light · Light receiver · Thermal protective · Qualities · Heat loss · Condensation · Enclosing structures

1 Introduction

Natural lighting has a significant advantage over artificial lighting. Natural light has an erythemal effect, that is, a healing effect (has a beneficial effect on metabolism, blood composition, immunity, psyche, etc.). Light tube structures are an effective solution for providing natural light to rooms in those areas of the building where traditional natural lighting systems are not possible. For this reason, these systems are widely used in architectural and construction practice. For example, during the reconstruction of the Kazan NCC building (Russia), a system of light guides was implemented to illuminate the premises of the National Library, Fig. 1. It should be noted that in this project, tubular vertical light tubes were used, passing through the roofing of the building. The use of vertical light tubes is one of the most effective, since their entrance holes face the zenith part of the sky, which has the greatest brightness and is not obscured by buildings.

N. Vatin (Ed.): STCCE 2021, LNCE 169, pp. 250–260, 2021.
https://doi.org/10.1007/978-3-030-80103-8_27

However, their use may not be optimal in a multi-storey building due to a significant lengthening of the light pipe and a decrease in their efficiency, as well as the possible need for holes in the ceilings of each floor [1–3].

Fig. 1. Application of light pipes at a national cultural center, Kazan.

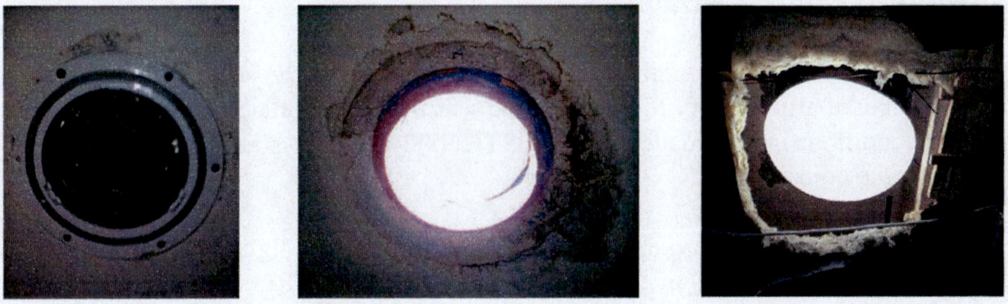

Fig. 2. Condensation and destruction of the building envelope.

In this case, the use of horizontal light tubes passing through the outer walls into the room can be an alternative to vertical ones, Fig. 3.

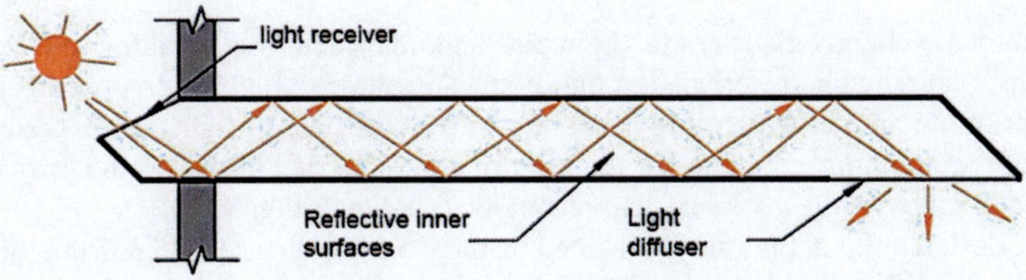

Fig. 3. Schematic diagram of the operation of horizontal light tube.

However, these systems are less studied [4–8, 8]. In [9–15], the economic efficiency of hollow tubular light pipe systems for natural lighting of premises and their rational use from the point of view of the comfort of a person's stay in the premises was assessed. It should be noted that the number of studies devoted to the assessment of their thermo technical characteristics is much less [16–21]. At the same time, it is logical to assume that the presence of a hollow structure in the outer enclosure of a building can lead to an increase in its thermal heterogeneity and requires an assessment of compliance with the thermal protection requirements. Cases of condensation formation and destruction of the enclosing structure unit due to incorrect accounting of thermos-physical processes occurring at the junction of the light pipe fixture and the wall are not uncommon. An example of a destruction process is shown in Fig. 2.

For these reasons, the study sets the following tasks:

- experimental study of the luminous efficiency of a wedge-shaped light tube for various geometric parameters of the structure,
- comparison of experimental results with calculated ones based on computer simulation,
- development of the joint of the light receiver unit and the enclosing structure, taking into account the requirements of thermal protection.

2 Materials and Methods

The experimental research method was based on the use of the scale model approach [22]. A standard artificial sky (dome) method was used to simulate real outdoor natural lighting conditions in accordance with GOST 266024–99 (Russian standard).

Test setup consisting of:

- a diffuse light source of type A (artificial skies of reflected light, painted with white diffusely reflecting paint) in accordance with GOST 7721 (Russian standard).;
- a light-measuring chamber painted with a matte white diffusely reflecting paint, separated by a horizontal partition with an opening and a support grid in it for the installation of the test specimen;
- light dimmer in accordance with GOST 15543 (Russian standard).

2.1 Experimental Study of a Horizontal Wedge-Shaped Light Pipe

The geometric characteristics of the light pipe should be determined taking into account the space-planning characteristics of the room. The choice of its length depends on the location of the area at which it is necessary to provide natural light. An experimental study of the luminous efficiency of the light pipe was carried out using the example of a typical room with the geometric characteristics shown in Fig. 4.

The design point in the study is located on the working surface 0.75 m from the level of the finished floor and at a distance of 9 m from the inner surface of the outer wall. The length from the inner edge of the wall to the beginning of the diffuser is 7.8 m. The length of the diffuser is 2.4 m.

Model of the light pipe for the experimental study was made on a reduced scale of 1:10, taking into account the scale model approach. In this case, the layout of the light tube is made in the form of a transformable structure, which allows to change the angles of the reflecting walls, Fig. 5.

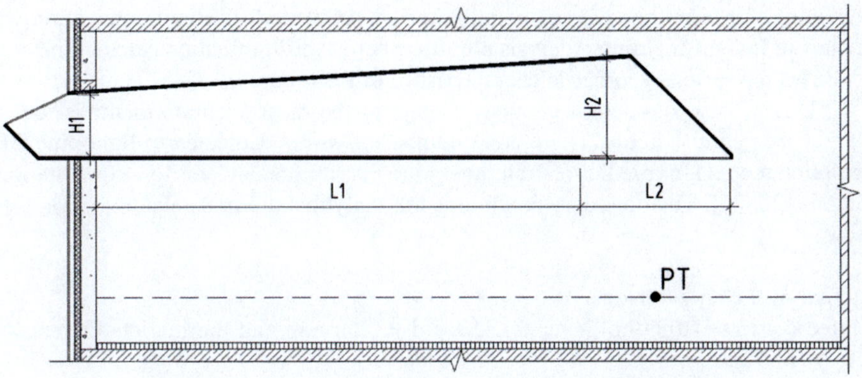

Fig. 4. Model of the investigated hollow horizontal wedge-shaped light pipe.

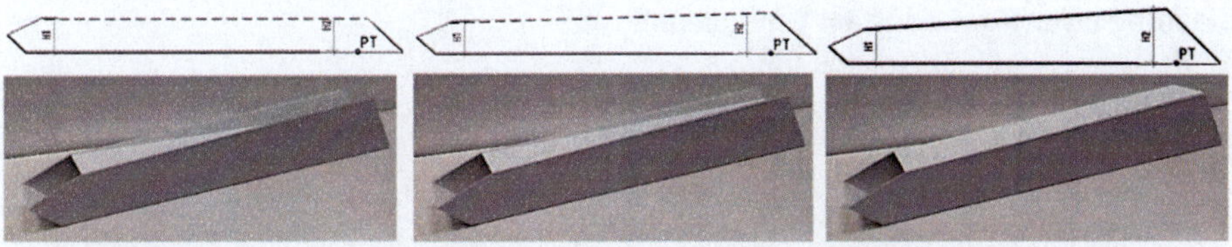

Fig. 5. Layout of the light pipe with transformation into a wedge-shaped type.

The model was placed in an artificial sky dome, where an illumination equal to 1000 lx was formed. A light meter is placed under the diffuser of the model, and the illumination values are recorded at different angles of inclination of the upper surface of the pipe, Fig. 6.

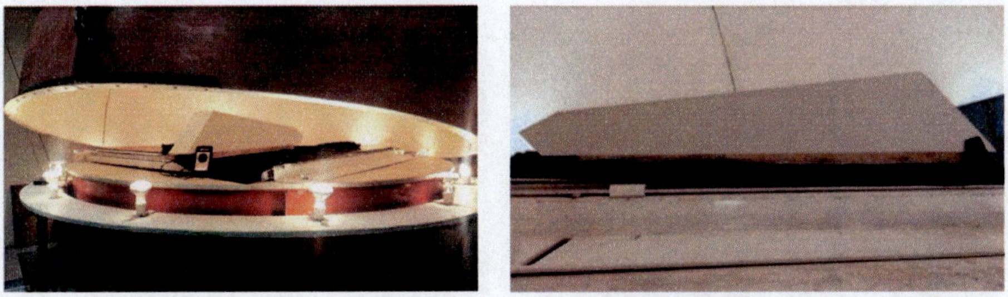

Fig. 6. The investigated model of the light pipe in the artificial sky.

2.2 Daylight Factor and Thermal Calculation Boundary Conditions

There are various analytical methods for calculating illumination, including using computer calculations [23–25]. The analytical calculation method is based on computer simulation of a room and a pipe model based on the Autodesk Revit application with the built-in Insight Lighting Analysis plug-in, a series of illumination calculations were performed for various geometric characteristics of the tube, Fig. 7.

CFD heat flow simulation is currently one of the most accurate methods for calculating temperature fields. There are a number of works dedicated to light pipe CFD simulation models for prediction of temperature distribution and air flow computational methods [20, 21]. The boundary conditions for analytical thermal calculation are listed below.

External wall characteristics:
Aerated concrete structural layer thickness d = 250 mm and thermal coefficient λ = 0.26 W/(m·°C).
Monolithic reinforced concrete bulkhead – d = 250 mm, λ = 2.04 W/(m·°C).
Heat-insulating stand made of mineral wool insulation – d = 100 mm, λ = 0.044 W/(m·°C).
Cladding layer from a curtain ventilated facade.
Characteristics of a horizontal tube:
Outer transparent organic glass – d = 10 mm, λ = 0.2 W/(m·°C).

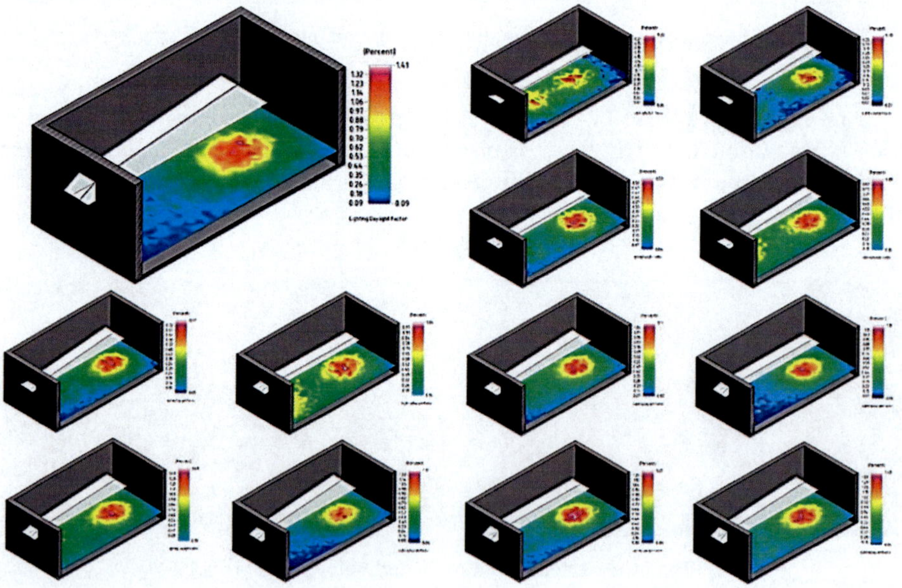

Fig. 7. Calculation and analysis of illumination by computer simulation (Autodesk Revit + Insight plugin).

The supporting frame is made of steel, which ensures the rigidity of the structure. Anodized aluminium inner layer with 99.8% reflectivity. The heat-insulating layer is made of hard mineral wool insulation. General parameters of the tube walls – d = 50 mm, $\lambda = 18$ W/(m·°C).

Light diffuser made of organic glass – d = 10 mm, $\lambda = 0.2$ W/(m·°C).
Air inside the tube – $\lambda = 0.02563$ W/(m·°C).
Climate characteristics:
Outside air temperature of the coldest five-day period – 19° C,
Heat transfer coefficient of the outer surface $a_{ext} = 23$ W/(m^2·°C).
Microclimate characteristics:
Internal air temperature + 20 °C,
Indoor air humidity 55%,
Dew point temperature + 10.7 °C,
Heat transfer coefficient of the inner surface $a_{int} = 8.7$ W/(m^2·°C).

3 Results and Discussion

3.1 Experimental study results

The data obtained from the experimental study are summarized in Table 1 and graph in Fig. 8. The graph shows that with an increase in the height of the cross-section of the inlet H1 to the length of the pipe L1, the daylight coefficient increases. An increase in the H2 height significantly increases the efficiency of the pipe, which is explained by a decrease in the number of reflections of light from its inner surfaces.

Table 1. Dependence of the daylight factor level, % on the geometry of the tube.

H1, mm	H2, mm									
	100	110	120	130	140	150	160	170	180	190
100	0.04	0.047	0.054	0.067	0.074	0.08	0.087	0.36	0.094	0.1

3.2 Analytical Calculation Results

The results of the analytical calculation are shown in Table 2 and the graph in Fig. 9.

The graph shows that the results are in good agreement with the experimental values. For example, for a pipe with a cross section of 400 mm, an increase in the H2 height from 400 mm to 1000 mm increases the daylight factor value from 0.53% to 0.88%, respectively, that is, more than 1.5 times. Thus, in terms of illumination, the most efficient pipe type is the one with the maximum H1/L1 and H2/L1 ratios. However, this ratio requires an increase in the size of the hole in the enclosing structure (H1), which negatively affects the thermo technical qualities of the enclosure.

It is logical to assume that within the framework of the design it is necessary to find a balance between the illumination and the amount of heat loss, taking into account the

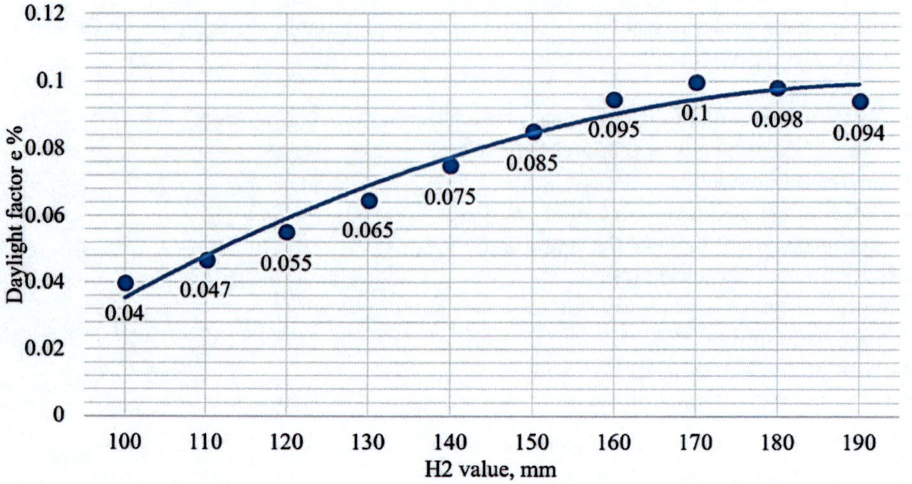

Fig. 8. Change in the daylight factor at the design point for different geometric characteristics of the pipe layout (H1value is fixed and equals to 100 mm).

Table 2. Dependence of the Daylight factor level, % on the geometry of the tube.

H1, m	H2, m											
	0.3	0.4	0.5	0.6	0.7	0.8	0.9	1.0	1.1	1.2	1.3	1.4
0.2	0.25	0.28	0.32	0.34	0.36	0.38	0.36	0.35	0.34			
0.4		0.53	0.61	0.7	0.78	0.81	0.85	0.88	0.84	0.81	0.77	
0.6				1.04	1.09	1.15	1.2	1.17	1.14	1.11	1.09	1.06
0.8						1.18	1.24	1.31	1.37	1.41	1.45	1.49
1.0								1.29	1.34	1.39	1.44	1.45

sanitary and hygienic requirements of thermal protection, namely the risk of condensation in places of thermal heterogeneity. As an example, a public space was considered with the required daylight factor value of 1% (according to SP 52.13330.2016 «Natural and artificial lighting»). According to the data obtained (Fig. 9), the minimum height of the pipe section required for DF = 1% is 600 mm.

3.3 Thermal Fields Analysis

Analytical calculation of three-dimensional temperature fields was carried out taking into account the processes of convection inside the tube. It can be seen that the bottom surface of the tube has significantly lower temperatures than the upper surface. This is explained by the process of convection inside the tube that has a through hole through the wall. Based on the results of the calculation, it can be concluded that condensation occurs at the junction in the lower part of the pipe with the inner surface of the wall.

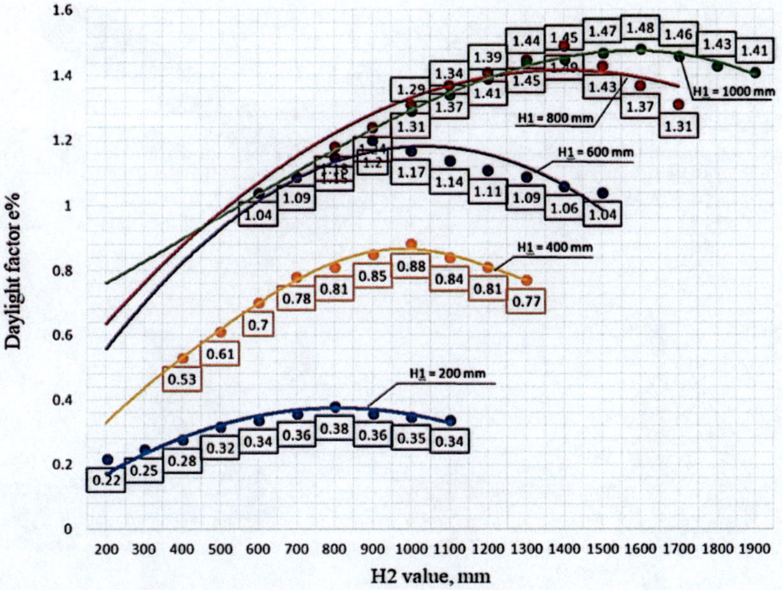

Fig. 9. Change in the daylight factor at the design point (PT) for different geometric characteristics of the pipe layout.

Since the temperature at the junction point t = 7.1 °C is lower than the dew point temperature t_{dew} = 10.7 °C. In order to eliminate condensation, various types of this unit were analysed. It was established that in order to eliminate condensation, it is enough to insulate the outer unit from the end sides of it with a layer of insulating material. However, calculations show the presence of condensation inside the pipe which cannot be eliminated by simple solutions without reducing the daylight efficiency. This problem requires further research. Also, for the independent descent of atmospheric precipitation, the outer glass of the pipe is turned at 60 degrees to the horizon (Fig. 10).

Based on the results of the calculation, it can be concluded that no condensation occurs at the junction in the lower part of the pipe with the inner surface of the wall, since the temperature at the junction point t = 13.4 °C is higher than the dew point temperature t_{dew} = 10.7 °C. Thermal protection requirements are met.

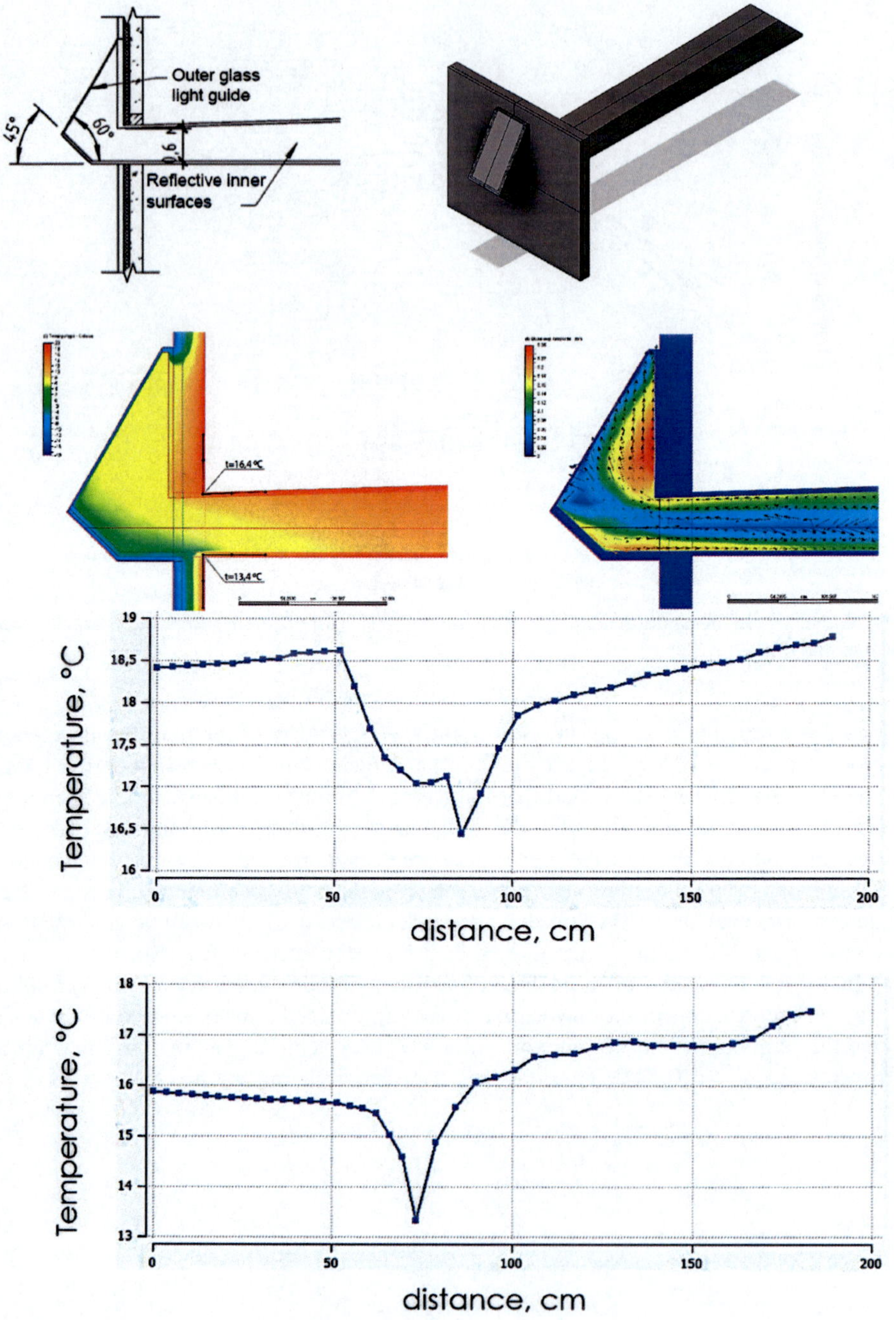

Fig. 10. Temperature plots at the junction of the pipe and the outer wall.

4 Conclusion

The relationship between the luminous efficiency of a light pipe and its geometric characteristics is obtained. The most effective is the wedge-shaped type with the maximum opening angle;

The lower surface of the tube has significantly lower temperatures than the upper surface. This is explained by the process of convection inside the tube that has a through hole through the wall;

It is shown that in a horizontal light pipe with a through hole through the wall, condensation will occur inside the tube. This problem requires further research;

Light pipe assembly has been developed, which ensures that condensation does not fall out in the area of the joint with the outer wall, which is achieved due to additional insulation of the outer part of the light receiver.

References

1. Veronica, G.-H.: Ian, edmonds: methods for the illumination of multilevel buildings with vertical light pipes. Sol. Energy **117**, 74–88 (2015). https://doi.org/10.1016/j.solener.2015.04.017
2. Sterkhov, A.I., Palagin, A.V., Loshkarev, I.Y.: Study of lighting systems with extended hollow light guide. Light Eng. **28**(2), 54–60 (2020)
3. Sterhov, A.I., Petrov, K.A., Loshkarev, I.Y., Engineering design of solar concentrator for transporting sunlight through optical fiber. J. Phys. Conf. Ser. Int. Conf. HIRM High-Tech Innovations Res. Manuf. 012004 (2019)
4. Heng, C.Y.S., Lim, Y.-W., Ossen, D.R.: Horizontal light pipe transporter for deep plan high-rise office daylighting in tropical climate. Build. Environ. 171, 106645 (2020).https://doi.org/10.1016/j.buildenv.2020.106645
5. Obradovic, B., Matusiak, B.: A customised method for estimating light transmission efficiency of the horizontal light pipe via a temporal parameter with an example application using laser-cut panels as a collector, MethodsX, 101339 (2021). https://doi.org/10.1016/j.mex.2021.101339
6. Kwok, C., Chung, T.: Computer simulation study of a horizontal light pipe integrated with laser cut panels in a dense urban environment. Light Res. Technol. **40**(4), 287–305 (2008)
7. Biljana, O.: Barbara, szybinska matusiak: daylight autonomy improvement in buildings at high latitudes using horizontal light pipes and light-deflecting panels. Sol. Energy **208**, 493–514 (2020)
8. Obradovic, B., Matusiak, B.S.: Daylight transport systems for buildings at high latitudes. J. Daylight. **6**(2), 60–79 (2019)
9. Elsiana, F., Soehartono, F., Kristanto, L.: Collaboration of two optical daylighting systems at office building in the tropics. Int. Jt. Conf. SENVAR-iNTA-AVAN, 223–236 (2015)
10. Sharma, L., Ali, S.F., Rakshit, D.: Performance evaluation of a top lighting light-pipe in buildings and estimating energy saving potential. Energ. Build. **179**, 57–72 (2018)
11. Chi, D.A., Moreno, D., Navarro, J.: Correlating daylight availability metric with lighting, heating and cooling energy consumptions. Build. Environ. **132**, 170–180 (2018)
12. Wong, I.L.: A review of daylighting design and implementation in buildings. Renew. Sustain. Energ. Rev. **74**, 959–968 (2017)
13. Beltrán, L.O.: Using sunlight as a light source in building cores. In: The 9th International Energy Forum on Advanced Building Skins (2014)

14. Lechner, N.: Heating, cooling, lighting Sustainable Design Methods for Architects. John, New Jersey (2015)
15. Galatioto, A., Beccali, M.: Aspects and issues of daylighting assessment: a review study. Renew. Sust. Energ. Rev. **66**, 852–860 (2016)
16. Alrubaih, M.S., Zain, M.F.M., Alghoul, M.A., Ibrahim, N.L.N., Shameri, M.A., Elayeb, O.: Research and development on aspects of daylighting fundamentals. Renew. Sustain. Energy Rev. **21**, 494–505 (2013)
17. Kraynov, D.V., Maslennikov, I.A.: Calculation of the reduced heat transfer resistance of translucent structures made of aluminum profiles. Izvestija KGASU **4**(54), 35–43 (2020)
18. Ondřej, Š: Jitka, Mohelníková, Josef, Plášek: thermal analysis of light pipes for insulated flat roofs. Energ. Build. **85**, 436–444 (2014)
19. Harrison, S.J., McCurdy, G.G., Cooke, R.: Preliminary evaluation of the daylighting and thermal performance of cylindrical skylights. In: Proceedings of International Daylight Conference, pp. 205–212 (1998)
20. Callow, J.M.: Daylighting Using Tubular Light Guide Systems. Ph.D. Thesis, University of Nottingham. Access from the University of Nottingham repository. http://eprints.nottingham.ac.uk/10026/1/Thesis_-_Joel_Callow.pdf
21. Šikula, O., Mohelníková, J., Plášek, J.: Thermal CFD analysis of tubular light guides. Energies **6**(12), 6304–6321 (2013). https://doi.org/10.3390/en6126304
22. Sikula, O., Mohelnikova, J.: CFD simulation of thermal behaviour of tubular light guides. In: SOLARIS, Brno University of Technology: Brno, Czech Republic, 252–258 (2011)
23. Boccia, O., Zazzini, P.: Daylight in buildings equipped with traditional or innovative sources: a critical analysis on the use of the scale model approach. Energy Build. **86**, 376–393 (2015)
24. Taengchum, T., Chirarattananon, S.: Ray tracing method of light through rectangular light pipe with bends. Energy Procedia **79**, 791–798 (2015)
25. Malet-Damour, B., Guichard, D.B.S., Boyer, H.: Photometrical analysis of mirrored light pipe: From state-of-the-art on experimental results (1990–2019) to the proposition of new experimental observations in high solar potential climates. Sol. Energy **193**, 637–653 (2019)
26. Petržala, J.: Simple analytic formula for the light-tube optical efficiency under overcast sky conditions. Sol. Energy **194**, 47–50 (2019)

Foamed Wood-Polymer Composites Based on Polyvinyl Chloride

Anvar Islamov(✉) and Venera Fakhrutdinova

Kazan State University of Architecture and Engineering, 420043 Kazan, Russia

Abstract. The work investigates the possibility of obtaining foamed wood-polymer composites based on polyvinyl chloride (PVC), since they have higher physical and mechanical properties, fire resistance compared to other thermoplastic polymers (PE, PP, PS). Nevertheless, they have disadvantages, one of which is a high density (1300–1400 kg/m^3) when compared to the products from natural wood (500–600 kg/m^3) and wood-polymer composites based on polyolefins, PS (900–1000 kg/m^3). The influence of various chemical blowing agents (azodicarbonamide, sodium bicarbonate) and nucleating agents on physical and mechanical properties of the received materials was studied. Azodicarbonamide in comparison to sodium bicarbonate is more effective as a blowing agent since it leads to the formation of uniform close-meshed structure. It has been found that the introduction of a small amount of wood-filler into the polymer melt prevents a decrease in the density of specimens even in the presence of gassing nucleators, which presumably occurs due to a decrease in the compliance of melt with the blowing gases released. The authors achieved a decrease in the density of composites by 10–15% depending on the degree of filling, without a significant deterioration in strength properties.

Keywords: Wood-polymer composite · Polyvinylchloride · Foaming · Blowing agent · Azodicarbonamide · Sodium bicarbonate

1 Introduction

Modern building materials science is impossible to imagine without the use of composite materials due to the valuable combination of technological and operational properties [1]. Composites with polymer matrix occupy a special place among them.

In this paper the research is devoted to wood-polymer composite (WPC) – relatively recently appeared modern high-performance building material, which is a thermoplastic polymer filled with wood flour (WF). It combinins the positive properties of both plastics and organic filler. Compared to natural wood products, WPC is distinguished by high water and biological resistance, resistance to external atmospheric actions, isotropy of physical and mechanical properties.

WPC based on PVC has higher physical and mechanical properties, incombustibility compared to composites based on other thermoplastic polymers (polyethylene, polypropylene, polystyrene) [2, 3]. However, they have disadvantages, one of which

N. Vatin (Ed.): STCCE 2021, LNCE 169, pp. 261–268, 2021.
https://doi.org/10.1007/978-3-030-80103-8_28

is high density of this material (1300–1400 kg/m^3) compared to products from natural wood (500–600 kg/m^3) and WPC based on polyolefins, polystyrene (900–1000 kg/m^3) [4–6]. This disadvantage limits the use of PVC-based WPC for decorative elements, exterior siding, window and door frames. A possible way to solve this problem is to foam materials to density values at which the decrease in the strength and performance properties of the composite will be insignificant.

Foaming of polymers depends on technological parameters of equipment, the properties of the initial components and physical properties of melt of processed compositions (viscosity, diffusion coefficient of gases, modulus of elasticity) [7]. The final objective of this process is to obtain a gas-filled material with a uniform fine-porous structure with mainly closed pores. This structure provides the most rational combination of physical and mechanical properties [8].

Foaming of polymeric materials is carried out with blowing agents, subdivided into physical (PBA) and chemical (CBA) ones. PBA are compounds that release gas as a result of physical processes (evaporation, boiling, desorption) when increasing temperature or decreasing pressure, without undergoing chemical transformations. Solid substances (coal, clay, silica gel), low-boiling volatile liquids (aromatic, aliphatic, cyclic, halogenated hydrocarbons, ethers, ketones, alcohols) and gases (air, N_2, CO_2, Ar, He, NH_3) can be used as PBA [5, 9]. The use of PBA in the production of foamed plastics is increasing every year. The disadvantages of this type of blowing agents include the low quality of the resulting foam structure (interconnected pores), the need for large capital expenditures on special equipment, compliance with strict safety requirements for the environment and production from the point of view of explosion and fire safety [5, 7].

Chemical blowing agents (CBA) have found the greatest practical application, emitting gases as a result of irreversible thermal decomposition (porophores), reversible (equilibrium) thermal decomposition and chemical interaction with any ingredient of the compositions or between the components making up the blowing agents [5]. This is due to the simplicity of introducing blowing agents into the composition and the possibility of their processing on existing equipment without any significant changes. For the production of foamed PVC, compounds of various nature and their mixtures are used as CBA. The most commonly available and cheapest are inorganic CBA. These include various salts that release carbon dioxide and ammonia during decomposition (sodium, potassium, ammonium bicarbonates and carbonates), hydrogen (sodium borohydride), as well as water of crystallization (gypsum dihydrate, sodium perborate, metal phosphates). Their disadvantages are: low decomposition temperature (below the melting point of polymers), high diffusion of the resulting CO_2 and NH_3 gases through the cell walls, leading to the formation of a non-uniform open-cell structure [5, 7, 10].

Among CBA of organic nature (azo and diazo compounds, nitroso compounds, sulfohydrazides, urea derivatives, etc.), azodicarbonamide (ADCA) is the most widely used for most thermoplastics and elastomers [10]. The nitrogen released during decomposition is an effective blowing gas due to the lower diffusion rate through the polymer melt [11]. The evolving gas of ADCA is 220–260 cm^3/g, which can be increased to 420 cm^3/g in the presence of catalysts [12].

There is a number of publications [13–21], in which the authors obtained foamed composites based on PVC with a density close to natural wood, characterized by a small degree of filling (40–50 phr). The analysis of these works let us conclude that it is difficult to obtain these materials. This is due to the fact that the introduction of wood flour leads to an increase in the viscosity of the PVC melt, which is already highly viscous. An increase in the processing temperature (over 190 °C) is unacceptable, which is associated with the occurrence of destructive processes, both in PVC and in WF.

Based on this, the purpose of our work was to obtain foamed wood-polymer composites based on PVC of reduced density with a high degree of filling without a significant change in strength properties.

2 Materials and Methods

In the work, foaming of PVC-composition was carried out, the composition of which is presented in Table 1.

Table 1. Formulation of the wood-filled PVC composite.

Component	Content (phr)
Suspension PVC (K-value 70)	100
Heat stabilizer Interstab	5.5
Impact modifier FM-22	7
Wood flour modified with 1.72% (by weight) coupling agent (silica)	0–100
Blowing agent	0–3
Dioctyl phthalate	0–40

Azodicarbonamide (ADCA) and sodium bicarbonate (SB) were used as blowing agents.

The coupling agent silica is a colloidal solution containing nanosized micelles of amorphous silica of spherical shape (average particle size 5–9.5 nm). Modification of wood flour with a coupling agent was carried out by preliminary mechanical mixing of wood filler with silica in a paddle mixer and subsequent drying at 103 ± 2 °C until constant weight.

Mixing of compositions was carried out using a laboratory roller mill at a temperature below the decomposition of blowing agent. Compositions with plasticizer content were heated for 40 min at 80°C before processing with the roller mill to accelerate plasticization.

Foamed specimens from the prepared compositions were obtained using Gotech gt-7100-mi extrusion plastometer at 190°C and load of 21.6 kg. The melted composition at the outlet of the die expanded due to a sharp decrease in pressure to form a foamed extrudate.

The following characteristics were determined for the obtained specimens: density according to GOST 15139–69 (Russian standard); tensile strength according to GOST 11262–2017 (ISO 527–2:2012); water absorption during 24 h according to GOST 4650–2014 (ISO 62:2008).

The structure of foamed specimens was studied using a laboratory polarization microscope Axioskop 40 Pol in reflected light at a magnification of 100 times.

3 Results and Discussion

At the initial stage of the work, the efficiency of the selected blowing agents (ADCA, SB) for foaming PVC-compositions without wood filler was investigated. In addition, carbamide and a mixture of sodium bicarbonate/citric acid (ratio 1:1; 1.3:1; 3:1) were considered as blowing agents, but due to their low foaming ability, they were not considered in the paper. The obtained foam specimens were measured for density, tensile strength, water absorption (Fig. 1).

The figures show that the optimum concentration of ADCA is 0.4 phr, at which the density is decreased from 1.4 to 0.95 g/cm^3 (by 32%). Further adding of blowing agent has practically no effect on the density. Optimal concentration of SB is 2.2 phr, at which the density is decreased from 1.4 to 0.85 g/cm^3 (by 39%).

Addition of blowing agent lead to a consequent decrease in strength of PVC specimens – by 54% for ADCA, and by 74% for SB, and an increase in water absorption by 2 and 6 times, respectively. The sharp increase in water absorption after passing the optimal concentration is associated with the formation of open porosity.

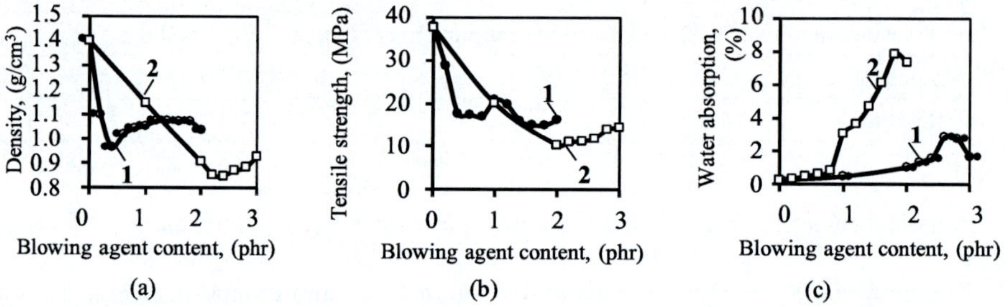

Fig. 1. Influence of blowing agents (1 – ADCA, 2 – SB) on density (a), water absorption (b) and tensile strength (c) of foamed PVC composites.

The cellular structure of PVC specimens (without wood flour content) formed by the introduction of blowing agent was investigated with an optical microscope (magnification 100x). From the microphotographs (Fig. 2) it is shown that ADCA, compared to SB, contributes to the formation of a more homogeneous fine-cell structure, and is more effective as a blowing agent.

It is known that in the interfacial areas the wood filler is not wetted by polymers (especially for hydrophobic polymers), so the phase interface can form channels for rapid migration of gases from composites. Therefore, the addition of WF pre-modified with silica was carried out in PVC composites containing blowing agent. WF modification consisted in its preliminary impregnation with an aqueous solution of silica, careful mixing and subsequent drying in order to remove water. The concentration of silica in the WF was 1.72% (in terms of dry residue). This modifier was used as a coupling agent to increase the degree of interaction between the polymer and the filler, to reduce the migration of foaming gases from the composite.

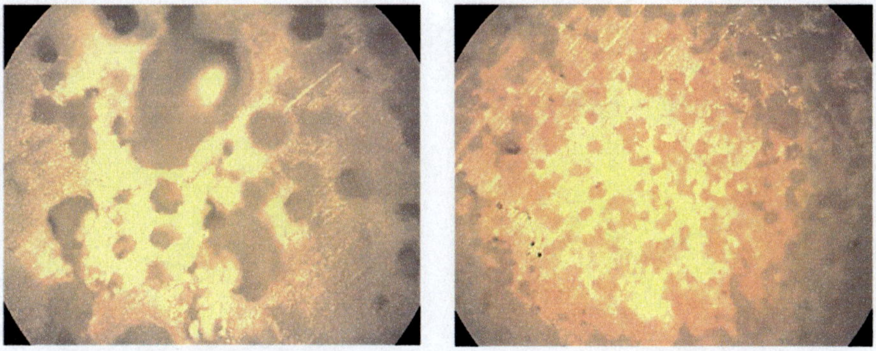

(a) (b)

Fig. 2. Micrographs of the cellular structure of PVC specimens containing SB (a) and ADCA (b).

It was shown (Fig. 3) that the addition of even a small amount of filler (10–20 phr) led to a decrease in the foaming efficiency of the material. At filling more than 30 phr the density of composites changes insignificantly and practically is similar to the density of wood-polymer composites not containing blowing agent. The tensile strength decreases (after passing the extremum) and the water absorption increases.

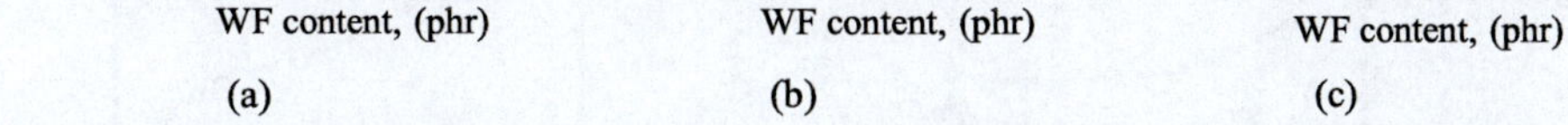

Fig. 3. Effect of WF on density (a), water absorption (b) and tensile strength (c) of wood-filled PVC specimens foamed by 0.4 phr ADCA (1); 2.2 phr SB (2).

During processing the WF remains in the solid phase and is not softened, which causes a significant increase in the melt viscosity to values at which the expanding force of the foaming gases is insufficient for the pore-forming process. Also the WF particles themselves inevitably collide with the growing bubbles and destroy the cell walls [22, 23].

To avoid this and to increase the foaming ability of blowing agent, attempts to add nucleating agents to PVC compositions without WF have been made. As it is known

[2], they serve as embryos for the formation of primary cells, activate the optimal gas release and contribute to the development of a fine-pored cellular structure. The following nucleating agents were tested: titanium dioxide, hydrophobized chalk, microsilica, talc, polyethylene wax, stearic acid, zinc stearate, carbamide, and dioctyl phthalate (DOP). DOP proved to be the most effective (Fig. 4).

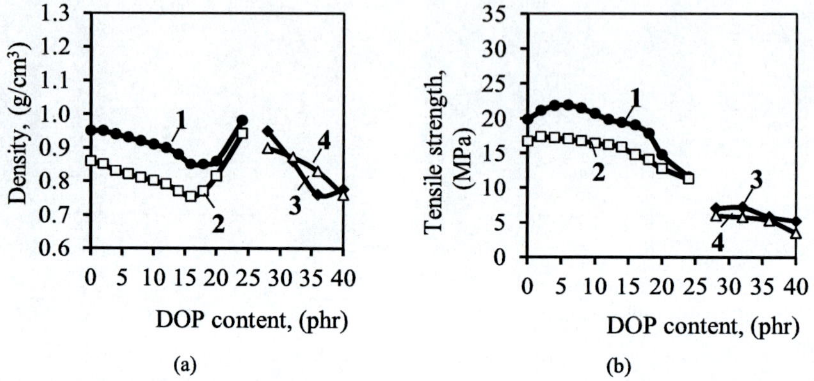

Fig. 4. Effect of DOP on density (a) and tensile strength (b) of wood-filled PVC specimens foamed by 0.4 phr ADCA (1 and 3); 2.2 phr SB (2 and 4). Extrusion load 21.6 kg (1 and 2), 10 kg (3 and 4).

The optimal concentration of DOP in PVC compositions without WF is the value of 16 phr for both blowing agents, at which the reduction of strength of composites is insignificant. Its use as a nucleating agent reduces the density of the PVC matrix by another 10–12%. The interruption of the curves is due to the fact that the specimens with plasticizer content higher than 24 phr were extruded at a lower load (10 kg) as plasticized PVC material.

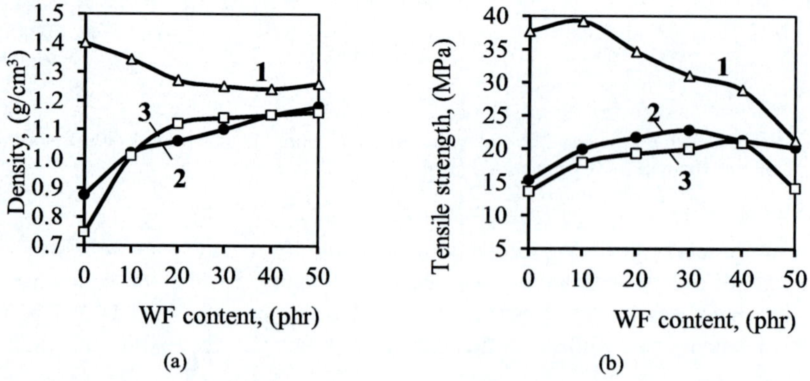

Fig. 5. Effect of WF on density (a) and tensile strength (b) for the virgin PVC specimens (1); specimens containing ADCA and 16 phr DOP (2), specimens containing SB and 16 phr DOP (3).

Thus, the use of plasticizer (16 phr DOP) allows reducing slightly the density of the obtained wood-polymer composites (Fig. 5). However, as it has been revealed earlier, the increase in the amount of the WF strongly reduces the effect of blowing agents, due to the increase in the viscosity of the melt. It was found that the order of addition of the components into the composition, as well as temperature conditions do not affect the efficiency of foaming.

4 Conclusions

1. The possibility of obtaining wood-polymer composites on the basis of PVC with reduced density as a result of using chemical blowing agents – ADCA or SB is shown.
2. For PVC compositions unfilled with WF, optimal concentrations of blowing agents are 0.4 phr ADCA, or 2.2 phr SB. Their use leads to a decrease in density by 32% and 39%, respectively, and a decrease in tensile strength by 54% and 74%, respectively.
3. The method of optical microscopy revealed that ADCA, compared to SB, contributes to the formation of smaller and more closed cells in the PVC matrix, as well as their uniform distribution.
4. Adding WF to PVC compositions leads to a sharp decrease in the foaming ability of blowing agents, as a result of an increase in the viscosity of the melt during processing.
5. The use of the nucleating agent (16 phr DOP) together with blowing agents ADCA or SB allows achieving the density decrease of wood-polymer composites by 15% with the addition of 20–30 phr WF, and by 10% with the addition of 40–50 phr WF with preservation of strength properties.

References

1. Vdovin, E.A., Stroganov, V.F.: Properties of cement-bound mixes depending on technological factors. Mag. Civ. Eng. **93**, 147–155 (2020). https://doi.org/10.18720/MCE.93.12
2. Galeev, R., Nizamov, R., Abdrakhmanova, L.: Influence of dispersed fillers on properties and processability of polyvinyl chloride compositions, In: E3S Web of Conferences 164, p. 14018 (2020). https://doi.org/10.1051/e3sconf/202016414018
3. Galeev, R., Nizamov, R., Abdrakhmanova, L., Khozin, V.: Resource-saving polymer compositions for construction purposes. In: IOP Conference Series: Materials Science and Engineering 890, 012111 (2020). https://doi.org/10.1088/1757-899X/890/1/012111
4. Patterson, J.: New opportunities with wood-flour-foamed PVC. J. Vinyl Addit. Technol. **7**, 138–141 (2001). https://doi.org/10.1002/vnl.10281
5. Klempener, D., Sendijarevic, V.: Polymeric Foams and Foam Technology. Hanser Publications, 2nd edition, 603 (2004)
6. Jiang, H., Kamdem, D.P.: Development of poly(vinyl chloride)/wood composites. Lit. Rev. J. Vinyl Addit. Technol. **10**, 59–69 (2004)
7. Lee, S.-T., Park, C.B., Ramesh, N.S.: Polymeric Foams Science and Technology. CRC Press, Taylor & Francis Group, Boca Raton, 202 (2007). https://doi.org/10.1201/9781420004625

8. Jin, F.L., Zhao, M., Park, M., Park, S.J.: Recent trends of foaming in polymer processing: a review. Polymers **11**, 953 (2019). https://doi.org/10.3390/polym11060953
9. Gendron, R.: Thermoplastic foam processing: Principles and development. CRC Press, Boca Raton, 1st edition, 320 (2004)
10. Grossman, R.F.: Handbook of Vinyl Formulating: Second Edition. Wiley-Interscience; 2nd edition, 544 (2007). https://doi.org/10.1002/9780470253595
11. Heck, R.L.: A Review of Commercially Used Chemical Foaming Agents for Thermoplastic Foams. J. Vinyl Addit. Technol. **4**, 113–116 (1998). https://doi.org/10.1002/vnl.10027
12. Chanda, M.: Plastics Technology Handbook, Fifth Edition. CRC Press; 5th edition, 1044 (2017). https://doi.org/10.1201/9781315155876
13. Feng, A., Wu, G., Wang, Y., Pan, C.: Synthesis, preparation and mechanical property of wood fiber-reinforced poly(vinyl chloride) composites. J. Nanosci. Nanotechnol. **17**, 3859–3863 (2017). https://doi.org/10.1166/jnn.2017.13987
14. Matuana, L.M., Park, C.B., Balatinecz, J.J.: Processing and cell morphology relationships for microcellular foamed PVC/wood-fiber composites. Polym. Eng. Sci. **37**, 1137–1147 (1997). https://doi.org/10.1002/pen.11758
15. Stucky, D.J., Elinski, R.: Foamed polymer-fiber composite. United States Patent US6344268B1 (2002)
16. Zehner, B.E.: Foam composite wood replacement material. United States Patent US6590004B1 (2003)
17. Finley, M.D.: Foamed thermoplastic polymer and wood fiber profile and member. United States Patent US6054207 (2000)
18. Atas, C., Sevim, C.: On the impact response of sandwich composites with cores of balsa wood and PVC foam. Compos. Struct. **93**, 40–48 (2010). https://doi.org/10.1016/j.compstruct.2010.06.018
19. Cope, C.W.: Polymer and wood flour composite extrusion. United States Patent US5847016 (1998)
20. Matuana, L.M., Mengeloglu, F.: Manufacture of rigid PVC/wood-flour composite foams using moisture contained in wood as foaming agent. J. Vinyl Addit. Technol. **8**, 264–270 (2002). https://doi.org/10.1002/vnl.10373
21. Petchwattana, N., Covavisaruch, S.: Influences of particle sizes and contents of chemical blowing agents on foaming wood plastic composites prepared from poly(vinyl chloride) and rice hull. Mater. Des. **32**, 2844–2850 (2011). https://doi.org/10.1016/j.matdes.2010.12.044
22. Niska, K.O., Sain, M.: Wood-Polymer Composites. CRC Press Inc. 366, (2008). https://doi.org/10.1533/9781845694579
23. Guo, G., Rizvi, G.M., Park, C.B.: Wood-polymer composite foams. IWood-Polym. Compos. 227–256 (2008). https://doi.org/10.1533/9781845694579.227

Modification of Styrene-Acrylic Coatings with Carbon-Containing Shungite Filler

Victor Stroganov[1], Maxim Amel'chenko[1], Evgenii Vdovin[1](✉), Radmilla Tabaeva[1], and Eduard Kraus[2]

[1] Kazan State University of Architecture and Engineering, 420043 Kazan, Russian Federation
[2] German Plastics Centre (SKZ), 97082 Wuerzburg, Germany

Abstract. The use of styrene-acrylic paints is an affordable and reliable way to protect buildings and premises from external factors, including electromagnetic radiation. It is known that shungite is used as a filler in radioprotective polymer compositions (polyurethane, polyethylene, rubbers). However, there are few data to determine its effect on the durability and resistance of acrylic-based protective coatings. It was found that the application of a protective filler allows maintaining an increased level of adhesive strength of coatings to concrete (over 3.0 MPa) and steel (over 1.5 MPa) substrates. It is shown that the exposure of styrene-acrylic free films in aggressive media for 7 days causes a decrease in their elasticity, a change in color characteristics, assessed by the CIE Lab method, which is accompanied by the formation of a white bloom on the coating surface, detected by the microphotographic method. The obtained result can be explained by the chemical interaction of aggressive media on the components of the polymer composition (film). The absence of the formation of defects and the uniformity of free films after their exposure in aggressive media was also noted.

Keywords: Shungite · Styrene-acrylic copolymer · Protective coatings · Carbon-containing filler · Building materials · Adhesion

1 Introduction

Water-dispersion styrene-acrylic paints have proven to be a reliable and affordable way to protect buildings, structures and premises from external factors [1, 2]. One of the important problems is the protection of humans and the environment from electromagnetic radiation (EMR), which has a negative impact [3–6]. It is known that fillers are used for protection against EMR: metal powders, nanotubes, graphite in radioprotective polymer compositions based on polyethylene, polyurethanes, styrene-butadiene latexes, etc. [7–10]. However, the high cost of the listed fillers makes it necessary to replace them with more affordable ones, among which should be noted the carbon-containing filler - shungite, which is used in the composition of structural polymer composites [9, 11, 12] for various purposes. The durability of polymeric protective coatings depends on such important indicators as adhesion strength to substrates and resistance to the action of aggressive media. These indicators are largely determined by the type and dispersion

N. Vatin (Ed.): STCCE 2021, LNCE 169, pp. 269–278, 2021.
https://doi.org/10.1007/978-3-030-80103-8_29

of fillers. However, data on the use of shungite as a filler in the composition of water-dispersion styrene-acrylic paints for construction purposes are very few. In this regard, research on the modification of styrene-acrylic coatings with a shungite filler and the determination of their properties is relevant, which is the purpose of the work.

The object of research is shungite-filled styrene-acrylic protective coatings.
The subject of research is shungite filler.

2 Methods

2.1 Obtaining a Paint Composition and Fillers

The paint material is a polymer composition in which other components are allocated in the volume of the polymer binder: fillers, pigments and functional additives. As a film-forming agent, we used an aqueous styrene-acrylic dispersion of the "Lakroten® E-21" brand (LLC PCC "Orgchimprom"), which is most often used as a binder in polymer protective compositions for constructions [13–15]. In order to ensure the protective properties of the coatings, mineral fillers were introduced into the formulation - kaolin, chalk, radioprotective filler - shungite, as well as some functional additives such as dispersant, antifoam, coalescent and thickener.

The preparation of the paint composition was carried out on a dissolver according to the method [16, 17], consisting of the following stages:

- dry components were premixed in specified proportions, then distilled water and functional additives were added;
- the resulting mass was placed in a dissolver and stirred at a speed of n = 2000 rpm for 30 min to obtain a pigment paste;
- the resulting pigment paste was mixed with the styrene-acrylic binder in a ratio of 1:0.5 (pigment paste: binder).

The mixing (~ 30 min) was carried out until there was obtained a homogeneous composition that did not contain aggregates and air bubbles.

The preparation of the shungite, the marketable form of which is sand with a fraction of 0.5–1 mm, was carried out in a vortex layer apparatus (VLA) CMD5 (297). Shungite sand was loaded into a sealed container of the apparatus together with the grinding bodies (cylindrical ferromagnetic particles with a diameter of 2 mm) and dispersed for 5–7 min.

2.2 Characterization of Free Paint Films and Filler

The dispersion of the radioprotective mineral shungite was determined using a HORIBA LA-950 device, the principle of which is based on the scattering and detection of reflected / refracted laser light of red and blue spectra (650 and 405 nm), in accordance with ISO 13320–1.

The adhesive strength of paint coatings was determined using a PSO-10MG4S adhesion tester. A paint composition was applied on previously degreased, cleaned surfaces until they were completely covered. After that this composition was dried within 24 h. Next, a steel plate of 50x50 mm in size with a ball shank was attached to the formed coating with the help of the epoxy glue consisting of ED-20 epoxy resin (10 parts by weight) and diethylenetriamine hardener (1 part by weight). After the adhesive layer had cured, the shank was fixed in the forks of the adhesion tester, and then a force was applied using the handle, the vector of which was directed perpendicular to the surface. When the handle was loaded, the strain gauge transducer of the device started to generate an electrical signal that changed in proportion to the applied load, which was recorded by the electronic unit and converted into force. The bond strength was calculated as the ratio of the force expended on the separation of the steel plate to its area.

The determination of changes in the quality indicators of coatings was carried out on free films, and tests of their resistance to the action of aggressive media were carried out in acidic, alkaline, and neutral media, which are 3% sulfuric acid solution, 0.1 N sodium hydroxide solution, and distilled water.

Free films were obtained by applying a paint composition on a fluoroplastic substrate (coating formation time is up to 7 days at (20 ± 2) °C). Their formation was controlled by the changing of the state of the surface to "tack" and the subsequent removal of the films.

The presence of possible microcracks and defects on the surface of paint coatings before and after their exposure in aggressive environments was determined with the help of the Levenhuk DTX90 digital microscope (x280 magnification).

The change in the color characteristics of free films was carried out in accordance with GOST 52490–2005. In the graphic light model, "lightness" is set by the L coordinate, and the chromatic component of the color is set by two Cartesian coordinates a and b (Fig. 1).

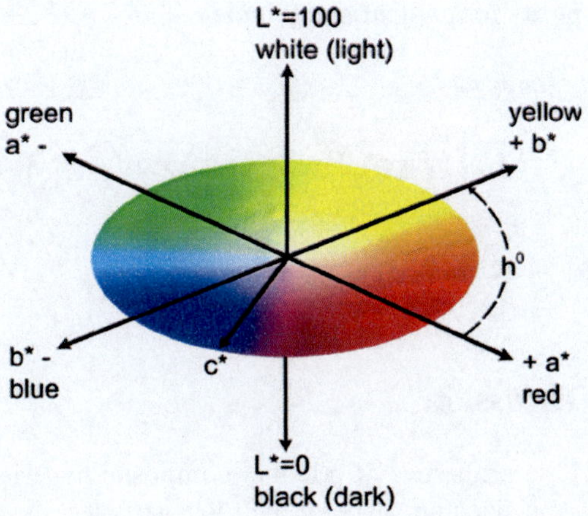

Fig. 1. Graphical representation of the Lab color model

The processing of the results obtained by measuring the data of the coordinates of the comparison samples and the tested paint coatings, and the subsequent calculation of the parameters of the difference in their color characteristics, was carried out according to the formulas (1–8):

- total color difference:

$$\Delta E^*_{ab} = \left[(\Delta L^*)^2 + (\Delta a^*)^2 + (\Delta b^*)^2\right]^{1/2} \quad (1)$$

$$\Delta L^* = L^*_T - L^*_R \quad (2)$$

$$\Delta a^* = a^*_T - a^*_R \quad (3)$$

$$\Delta b^* = b^*_T - b^*_R \quad (4)$$

ΔL^*_T, Δa^*_T, Δb^*_T – color coordinates of the test sample;
ΔL^*_R, Δa^*_R, Δb^*_R – color coordinates of the comparison sample.

- color purity differences:

$$\Delta C^*_{ab} = C^*_{ab,\,T} - C^*_{ab,\,R} \quad (5)$$

$$\Delta C^*_{ab,\,T} = (a^{*2}_T + b^{*2}_T)^{1/2} \quad (6)$$

$$\Delta C^*_{ab,\,R} = (a^{*2}_R + b^{*2}_R)^{1/2} \quad (7)$$

$\Delta C^*_{ab,T}$ – color purity for the test sample;
$\Delta C^*_{ab,R}$ – color purity for comparison sample.

- differences in color tone:

$$\Delta H^*_{ab} = k_H \left\{\left[\left(\Delta E^*_{ab}\right)^2 - \left(\Delta L^*_{ab}\right)^2 - \left(\Delta C^*_{ab}\right)^2\right]^{1/2}\right\} \quad (8)$$

$k_H = +1$ if $(a^*_R b^*_T - a^*_T b^*_R) \geq 0$,
$k_H = -1$ -if$(a^*_R b^*_T - a^*_T b^*_R) < 0$.

3 Results and Discussion

As we know [18–21] the properties of polymer composite materials are largely determined by the filling - the size and shape of the filler particles, its chemical nature, etc. Shungite of the Zazhoginsky occurence was a sandy material with a fraction size of 0.5–1 mm, which caused a low level of sedimentation stability of the polymer composition. In this regard, before the injection of shungite into the composition, it was pre-dispersed in a vortex layer apparatus, the use of which depends on the results of studies confirming its high grinding efficiency in comparison with the traditional methods [22, 23].

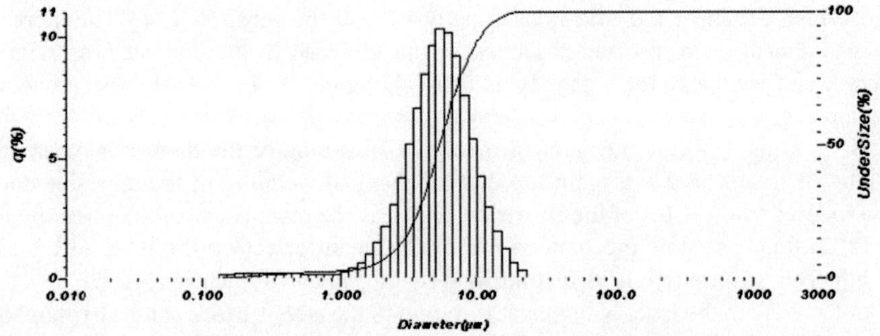

Fig. 2. The results of determining the dispersion of shungite milled in vortex layer apparatus

The size of shungite particles after processing in the vortex layer apparatus was 6 μm (Fig. 2), which indicates its high dispersion in comparison with other fillers of the paint composition such as kaolin and chalk.

The most important characteristic of protective paint coatings is their durability, the level of which can be indirectly conditioned by the indicators of adhesive strength. Widely used building materials are steel and concrete, so, the determination of adhesive strength was carried out on the steel grade S245 and the concrete grade B25 surfaces.

Analysis of the results of adhesive strength (Fig. 3) showed that filling with shungite provides a high level adhesion of the protective coating to the substrate - over 1.5 MPa for S245 and over 3 MPa for the concrete surfaces, in comparison with unmodified coatings (grey area of the histogram on Fig. 3).

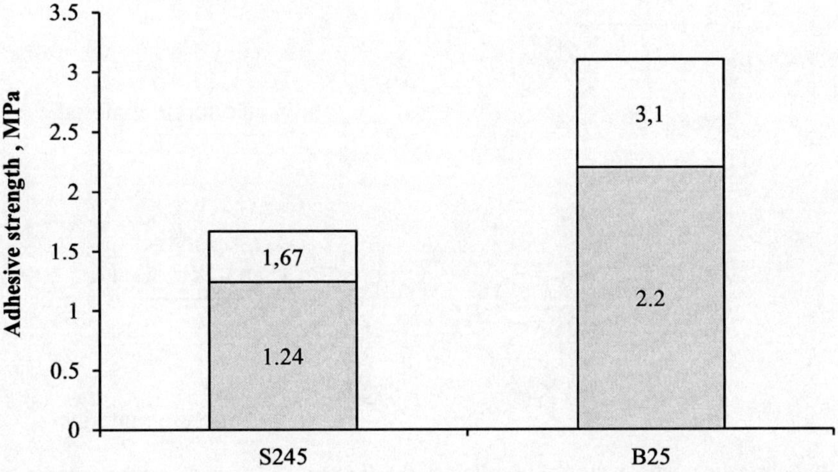

Fig. 3. The adhesive strength of shungite-containing styrene-acrylic coatings to the steel and concrete surfaces

The high values of the adhesive strength to the concrete surface can be explained by both a sufficiently high polarity and an increased surface roughness in the submicron range, which enhances the adhesion of the coating to the substrate [24]. The increased level of adhesion can also be explained by an increase in the contact area with the protected surface due to the high dispersion of shungite [25, 26], which is schematically shown in Fig. 4.

As we know [27] the mechanical theory of adhesion, or the theory of mechanical wedging, is based on the assumption that mechanical wedging of the adhesive occurs in microdefects and pores of the substrate, which is the main cause of the phenomenon. Indeed, in this case, this mechanism has a significant effect on the level of adhesive strength. This assumption is also confirmed by the cohesive fracture (Fig. 5).

An increase in the level of adhesive strength to the steel surface can be explained by an increase in the intermolecular interaction between the paint composition and metal atoms on the steel surface, caused by the increased contact area of the substrate with the coating.

The main function of building paints is to protect against external factors, including aggressive environments. In this regard, the color characteristics of free films exposed for 7 days in solutions of 3% sulfuric acid, 0.1 N sodium hydroxide solution and water were determined in order to establish the change in the level of protective characteristics caused by chemical reactions with the environment.

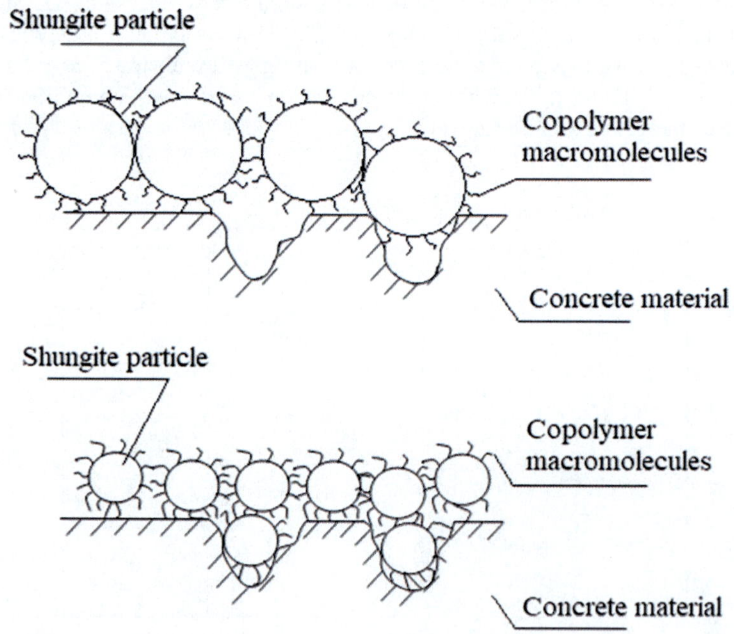

Fig. 4. Schematic representation of the effect of the dispersion of shungite on the increase in the contact area between the substrate and the composition a - before dispersing b - after dispersing the vortex layer in the apparatus

Fig. 5. The cohesive fracture of styrene-acrylic coatings filled by shungite

Analysis of the results showed that the sample exposed to water has the minimum change in color coordinates (Table 1), which can be explained by the weak chemical interaction between the coating and the liquid.

Table 1. Color coordinates of the source and exposed films

Exposure media	Coordinate value		
	L	A	b
Initial sample	2	3	0
H_2SO_4	2	– 1	1
NaOH	4	0	1
H_2O	1	1	– 1

The most noticeable changes occur in the samples after the exposure to the acidic and the alkaline media, which is proved by the calculated indices of the total color difference, differences in color purity and color tone (Table 2). The appearance on the surface of a white bloom is noted (Fig. 6), and it can be explained both by the precipitation from the solution of the basic substance and by the chemical interaction between the components of the coating and the media. In particular, a number of authors do not exclude the possibility of the reaction of the acrylic component of the polymer binder agent on carboxyl groups [28, 29].

It should be noted that the uniformity of the coatings (the formation of micro-cracks) does not occur during the exposure in the aggressive media (Fig. 6), but there is a slight decrease in the elasticity of free films, apparently due to the chemical interaction between the polymer binder and the aggressive media.

Table 2. The results of calculating the parameters of the difference in color characteristics

Exposure media	Parameter values					
	ΔL^*	Δa^*	Δb^*	ΔE_{ab}^*	ΔC_{ab}^*	ΔH_{ab}^*
H_2SO_4	0	– 4	1	4,12	– 1,58	– 3,08
NaOH	2	– 3	1	3,74	– 2	– 2,45
H_2O	– 1	– 2	– 1	2,45	– 1,58	– 1,58

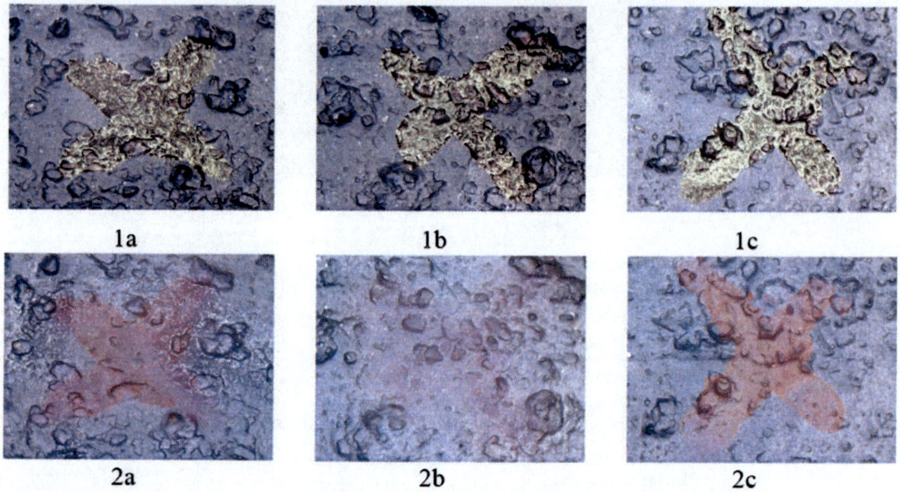

Fig. 6. Microphotographs of the surface of the samples before (1) and after the exposure (2) in the aggressive media: a - water; b - 0.1 N sodium hydroxide; c - 3% sulfuric acid solution

4 Conclusion

To sum it up it should be said that the filling of the paint coatings with shungite in order to increase their multifunctionality provides a high level of the adhesive strength - over 1.5 MPa - to concrete and steel surfaces. It was found that during the exposing of shungite-containing styrene-acrylic coatings in the aggressive media the acidic and the alkaline media have the most significant effect. It manifests the change of the color characteristics of coatings, the formation of white bloom on the surface and the loss of their elasticity, which can be explained by the chemical interaction between the polymer binder and the aggressive media. It should be also noted that the uniformity of the coatings retains, there are no microcracks, swellings, etc.

The results obtained confirmed the effectiveness of the use of shungite in the composition of styrene-acrylic paints and the expediency of continuing research on the study of their radioprotective properties.

References

1. Lin, H., Wang, C.: Influences of electromagnetic radiation distribution on chaotic dynamics of a neural network. Appl. Math. and Comp. 369 (2020). https://doi.org/10.1016/j.amc.2019.124840
2. Przystupa, K., Pohrebennyk, V., Mitryasova, O., Kochan, O.: The Influence of Electromagnetic Field on Human Health, in Applications of Electromagnetics in Modern Engineering and Medicine, PTZE, 9–12 June, 2019. Janow Podlaski, Poland (2019)
3. Redlarski, G., et al.: The influence of electromagnetic pollution on living organisms: Historical trends and forecasting changes. BioMed Res. Int., **2015** (2015). https://doi.org/10.1155/2015/234098
4. Abdulhadi, H.D.A., Belousova, A.S., Prudnik, A.M., Lynkou, L.M.: Mechanical and electromagnetic pro perties of shields on the base of needle-punched material with conductive and dielectric coatings. In: Proceedings of the National Academy of Sciences of Belarus, Physical-Technical Series (2019). https://doi.org/10.29235/1561-8358-2019-64-1-44-50
5. Guan, H., Liu, S., Duan, Y., Cheng, J.: Cement based electromagnetic shielding and absorbing building materials. Cem. Concr. Compos. **28**, 468–474 (2006)
6. Majcher, K., et al.: Methods of protecting buildings against hpm radiation—a review of materials absorbing the energy of electromagnetic waves. Materials (Basel). **13**, 1–21 (2020)
7. Barsukov, V., Senyk, I., Kryukova, O., Butenko, O.: Composite carbon-polymer materials for electromagnetic radiation shielding. Mater. Today Proc. **5**, 15909–15914 (2018)
8. Mukhametrakhimov, R., Shafigullin, R.I., Kuprijanov, V.N.: Development of radioprotective schungite-containing gypsum-fiber facing sheets. Izv. KGASU. **3**(41), 224–231 (2017)
9. Terukov, E.I., Babaev, A.A., Tkachev, A.G., Zhilina, D.V.: Radio-Wave absorbing properties of polymer composites on the basis of shungite and carbon nanomaterial taunit-M. Tech. Phys. **63**(7), 1044–1048 (2018). https://doi.org/10.1134/S1063784218070289
10. Solovieva, A.B., Neschadina, L.E., Rojkova, N.N., Gorbatkina, J.A., Kolbanev, I.V., Wolfson, S.A.: The Shungite effect on the physico-mechanical properties of elastomers and polypropylene. Int. J. Polym. Mater. Polym. Biomater. **36**, 11–17 (1997). https://doi.org/10.1080/00914039708044132
11. Khanjani, J., Hanifpour, A., Pazokifard, S., Zohuriaan-Mehr, M.J.: Waterborne acrylic-styrene/PDMS coatings formulated by different particle sizes of PDMS emulsions for outdoor applications. Prog. Org. Coat. 141 (2020)
12. Sbardella, F., Pronti, L., Santarelli, M.L., Gonzàlez, J.-M.A., Bracciale, M.P.: Waterborne acrylate-based hybrid coatings with enhanced resistance properties on stone surfaces. Coatings **8**(8), 283 (2018). https://doi.org/10.3390/coatings8080283
13. Anikina, N.A., Smirnov, V.F.: A study of resistance of some acrylic polymers to microscopic fungi. Vest. Nizhegor. Univ. Im. N.I. Lobachevskogo. **6**(1), 142–145 (2013)
14. Stroganov, V.F., Amel'chenko, M.O.: Effect of silicate fillers on the properties of styrene-acrylic coatings. Izv. KGASU. **3**(45), 196–202 (2018)
15. Rodionov, D.A., Nikolaeva, T.V.: Wall penetrating primers based on LACROTEN nano latex. Lakokras. Mater. i Ikh Primen. **10**, 18–21 (2006)
16. Talbert, R.: Paint Technology Handbook. Paint Technology Handbook. CRC Press, Boca Raton (2007). https://doi.org/10.1201/9781420017786
17. Strehmel, B., Groteklaes, M., Mischke, P.: Lehrbuch der Lacktechnologie Vincentz Network (2019). https://doi.org/10.1515/9783748600220
18. Gysau, D.: Fillers for Paints. Vincentz Network (2019). https://doi.org/10.1515/9783748600312
19. Rothon, R. (ed.): Fillers for Polymer Applications. PPCARS, Springer, Cham (2017). https://doi.org/10.1007/978-3-319-28117-9

20. Nizamov, R., Abdrakhmanova, L.: Influence of Fillers on Polyvinyl Chloride Materials Thermal Resistance (2016)
21. Galeev, R.R., Nizamov, R. K., Abdrakhmanova, L. A.: Filling of Epoxy Polymers with Chemically Precipitated Chalk from Chemical Water Treatment Sludge (2021)
22. Voitovich, V.A., Shvarev, R.R., Zacharichev, E.A. et al.: The effectiveness of the vortex layer apparatus in the processes of grinding powder materials. Nov. Ogneup. 10, 48–53 (2017)
23. Vdovin, E., Stroganov, V., Konovalov, N.: Modification of road soil cement with activated fillers. In: Vatin, N., Borodinecs, A., Teltayev, B. (eds.) EECE 2020. LNCE, vol. 150, pp. 335–345. Springer, Cham (2021). https://doi.org/10.1007/978-3-030-72404-7_33
24. Starostina, I.A., Stoyanov, O.V., Kraus, E.: Development of wetting methods for assessing surface conditions Publishing House KSTU (2019)
25. Kuznetsova, V.A., Silaeva, A.A., Emelyanov, V.V., Marchenko, S.A.: Influence of manufacturing techniques of the disperse reinforced paint and varnish materials on operational properties of coverings on their basis. Trudi VIAM. **6**(78), 75–83 (2019)
26. Mayer, P., Dmitruk, A., Lubecki, M., Stosiak, M.: Adhesion of epoxy coatings with micrometer size fillers to steel substrate. J. Theo. Appl. Mech. (Poland) **58**, 437–443 (2020)
27. Habenicht, G.: Kleben: Grundlagen, Technologien, Anwendungen. Springer Berlin Heidelberg, Berlin, Heidelberg (2009). https://doi.org/10.1007/978-3-540-85266-7
28. Prokopchuk, N.R., Krut'ko, E.T., Globa, A.I.: Chemical modification of polymer binders. BGTU (2012)
29. Stoye, D., Freitag, W.: Paints. WILEY-VCH Verlag GmbH, Coatings And Solvents (1998)

Effect of the Viscogel Additive on the Rheological Parameters of Bitumen

Maxim Lashin[1], Marina Vysotskaya[1], and Evgeniy Vdovin[2](✉)

[1] Belgorod State Technological University named after V.G. Shukhov, Kostyukov Street, 46, 308012 Belgorod, Russia

[2] Kazan State University of Architecture and Engineering, Zelenaya Street, 1, 420043 Kazan, Russia

Abstract. Bitumen is a complex multicomponent colloidal dispersed system based on the interaction of asphaltenes (phase) and maltenes (medium). In the absence of external influence, this system is in a relatively stable state. Any external action shifts the equilibrium state of the colloid and initiates phase aggregation with a general transition of the binder from the state of gel or sol-gel to sol, with a qualitative change in properties. Thus, a quantitative change of the colloid phase entails a qualitative change in its state. Rheological additives are some of the promising ways to control the structure formation and characteristics of bitumen. This paper considers road bitumen modified by Viscogel additive in a concentration range of 1–5%. The influence of the additive on the rheological characteristics of bitumen was determined. The optimal amount of additive which positively affects the properties of bitumen was established. This was confirmed by the determined contact angles for bitumen samples with different concentrations of additives subjected to the aging process. Test data indicated system stability over 48 h of exposure to high temperatures. A model of interaction of a rheological additive within the asphaltenes-maltenes system is proposed based on the available data on the structure and dispersion of bitumen.

Keywords: Rheology · Bitumen · Viscosity · Aging

1 Introduction

Bitumen is a complex colloidal system comprised of paraffinic and naphthenic hydrocarbons, cyclic systems, and aromatic hydrocarbons connected by aliphatic chains. In this colloid, the maltenic medium acts as a dispersant for asphaltenes, which are contained in bitumen as a dispersed phase in the form of micelles [1]. The uniqueness of the bitumen structure determines its features, including an imbalance in the colloidal stability of the system under the influence of external factors. The most aggressive of them are high temperatures and pressure. They give rise to the process of aggregation of asphaltene complexes and contribute to a change in the properties of bituminous binder and its aging during the technological cycle of the binder existence and the entire life cycle of the composite based on it.

N. Vatin (Ed.): STCCE 2021, LNCE 169, pp. 279–288, 2021.
https://doi.org/10.1007/978-3-030-80103-8_30

From the point of view of colloidal chemistry, aging process of bituminous binder manifests itself unevenly, by changing its dispersed structure, rheological, and physical characteristics.

Asphaltenes are the most condensed polar centers of this system, which accumulate the maltenic part around the nuclei. Maltenes consist of oils and resins, the main difference of which is the ratio of hydrogen and carbon [2]. Thus, the maltenic part of bitumen is a kind of buffer that protects polar asphalt-resinous components (ARC) from premature aggregation. As a rule, asphaltenes are high-molecular, condensed hetero-organic compounds, consisting of carbon (80–84%), hydrogen (7.5–8.3%), sulfur (4.6–8.3%), oxygen (up to 6%), and nitrogen (0.4–1%). Resins are hydrocarbons composed of aromatic, naphthenic, and heterocyclic rings connected by short aliphatic bridges. Oils are composed of carbon (79–87%), hydrogen (8.5–9.5%), oxygen (1–10%), sulfur (up to 2%). Oils are paraffinic, paraffinic-naphthenic, and aromatic hydrocarbons, consisting of several polycyclic groups [3, 4].

Nowadays the following theories of bitumen structure are considered [5–8]:

- Bitumen as an oil dispersed system (ODS) [9]
- Bitumen as a colloidal system with a dispersed phase (asphaltenes) and a dispersed medium (maltenes) [10].

Within the framework of ODS [9, 11], conglomerates of asphaltenes and maltenes are usually called a complex structural unit (CSU), consisting of a core and a solvate shell (Fig. 1).

Within the framework of the colloidal theory [10, 12, 13], bitumen particles are called a micelle, which consists of a core, a layer of counterions, a colloidal particle and a diffuse layer (Fig. 2).

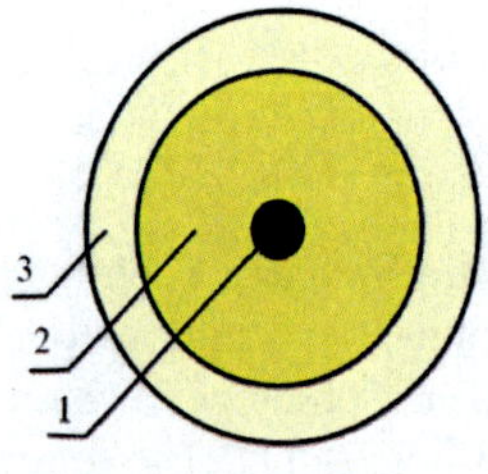

1-Core, 2-Solvate shell, 3- Intermediate layer;

Fig. 1. A complex structural unit [9]

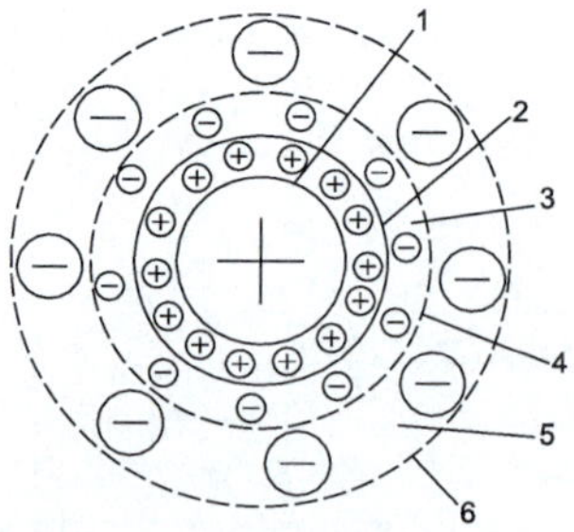

1-Aggregate, 2-Core, 3- Layer of counterions,
4-Colloidal particle, 5- Diffuse layer, 6-Micelle

Fig. 2. Micelle structure [12]

By analysing the particle structure, and the nature of resins and asphaltenes [5–14], we state the following:

- Asphaltenes act as the aggregate core possessing the system charge.
- Resins, adsorbed on the surface of asphaltenes, form a layer of counterions and enter the diffuse layer of the system environment.
- Oils are a dispersed medium in which interactions between particles take place.

Based on these models, Unger [14] proposed a description of the molecules' interaction in a dispersed system according to the principle of charge and spin (Fig. 3).

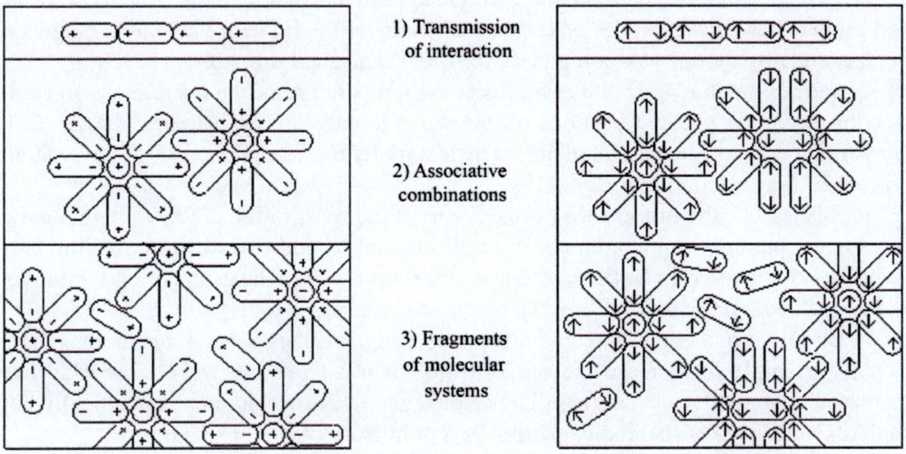

Fig. 3. Charge and spin models of molecular systems interactions [14]

According to the theory [14], the systems of charge principle can transfer charge over a long distance, and an associative combination model can be built based on the excess or missing charge. The uniform distribution of micelles in the system can be explained by the presence of a similar charge at the outer ends of molecules, as shown in Fig. 2. Based on the magnetic properties, the system comes to equilibrium.

In the absence of external influence, the spin model of interaction works according to the same principle as the charge one: a molecule is capable of transmitting an exchange effect, thus spreading it to the entire system. The concentration of the exchange action in a large volume leads to a multiplicity of spin-polarized layers, just as charge polarization leads to multiple electric layers. Since the tails of spin-polarized molecules are parallel, mutual repulsion occurs, which makes the system rather strong and stable [14].

Thus, it can be assumed that the charge interaction affects all processes occurring in the ODS, however, the influence of high temperatures or pressure affects the magnetic properties of the system, provoking a shift in equilibrium with a change in the properties of the system. Depending on the intensity of external factors, one can observe an increase in magnetic particles as a result of their agglomeration (an increase in the number and size of asphaltene aggregates) [15].

The most detrimental effect on the dispersed system of bitumen is exerted by the aging process, which occurs during both the mixture preparation and the road surface operation [16]. In the process of mixture preparation at the asphalt-concrete mixing plant, bitumen is distributed on the surface of the stone material in a thin film, which increases the negative effect of temperature and pressure, and leads to irreversible changes in the dispersed structure of bitumen by depleting the maltenic part of bitumen. As a result, the binder in the mixture has unpredictable properties [17].

To improve the rheological characteristics, the bitumen structure can be treated by magnetic field [18], sonication [19], or the introduction of rheological additives [5, 20–22].

The influence of magnetic field on the dispersed system of bitumen contributes to an increase in saturated and aromatic compounds and a decrease in the content of resins and asphaltenes. However, the system, which is under the influence of only a magnetic field, eventually partially or completely restores its original properties [18, 23].

Ultrasonic treatment of bitumen affects the temperature of the crystallization start, slowing down the phase transitions of the maltenic part into asphaltenes [19, 24, 25]. However, these changes cannot maintain their stability for a long time, and the relaxation process of the system is 2–7 days [19].

In this regard, the introduction of additives is the most rational way to regulate the rheological parameters of bitumen. A small amount of additive can improve the performance characteristics (softening temperature, brittleness, plasticity) of both bitumen and asphalt concrete mixture [26, 27].

At present, organoclay-based additives are used in various fields of technology [28–31], for example to increase the durability of paints. Concrete with organoclay has increased strength. However, the influence of organoclays on bitumen is practically not studied, which makes rheological additives a promising object of research.

This work aims at studying the influence of the rheological additive Viscogel on the rheological, physical and structural characteristics of bitumen.

2 Materials and Methods

The rheological additive Viscogel is a finely dispersed white powder. Its general chemical analysis showed a significant presence of carbon and oxygen, and a smaller amount of various impurities (Table 1). It also contains a small amount of Na, P, S, Cl, K, Ca.

Table1. Chemical content of the modifying additive

Chemical element	C	O	Al	Mg	Si	Fe
Mass,%	33.3	34.52	7.59	1.36	20.13	1.50

The powder structure was studied using the scanning electron microscope. It showed that the additive consists of microdispersed grains with an average size of 20–40 microns (Fig. 4).

The additive influence on the bitumen structure was studied by assessing the dynamic viscosity, which was chosen as the fundamental rheological factor. If the additive has some influence on the dispersed structure of bitumen, this will primarily affect the change

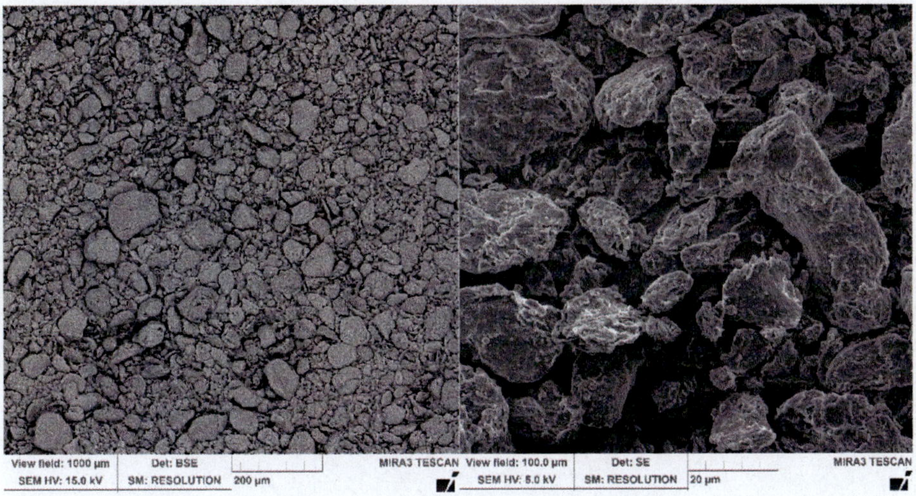

Fig. 4. Microphotographs of the viscogel additive

in the sample viscosity [32, 33]. The BND 50/70 road bitumen from the Moscow Oil Refinery was studied (Table 2).

Table 2. Main characteristics of bitumen

Depth of needle penetration, mm		Softening temperature, °C
0 °C	25 °C	53
25	55	

An additive in the amount of 1–5% was introduced into the prepared samples of bitumen. Further, the material was mixed using the laboratory mixer Silverson L5T for 30–40 min. Immediately after mixing, the samples were placed in cuvettes for further testing using the viscometer.

The dynamic viscosity was measured using the Brookfield DV2T rotational viscometer. The test temperature was varied from 120 to 200 °C with a step of 20 °C. The spindle rotation speed was selected individually for each test temperature, based on the testing instructions "Broorfield DV2T. Operating Instructions. Manual No. M13–167".

Also, the contact angle was determined. It is an important characteristic of the adhesion properties of bitumen and stone material, which is important for the major bitumen purpose, which consists in gluing the framework of the mixture [34, 35].

For this, the bitumen samples were modified with an additive in the amount of 0.5%, 1%, 2.5% of the bitumen mass and mixed according to the same principle as when determining the viscosity. The samples obtained were subjected to the aging process in a drying oven for 24 and 48 h at a temperature of 165 °C, having a bitumen contact with the air. The droplet was deposited on a prepared glass at a bitumen temperature of

120 °C. At this temperature the samples have the largest discrepancies in the viscosity and, as a consequence, their rheological properties are more pronounced.

3 Results and Discussion

Taking into account the colloidal structure of bitumen and the nature of the interaction of its components, it can be assumed that the additive particles, interacting with the maltenic medium, swell and are attracted to the most polar bitumen particles - asphaltenes, create a stable halo that prevents further interactions. Thus, it can be assumed that the separation of the phase (asphaltenes) and the maltenic medium occurs (Fig. 5).

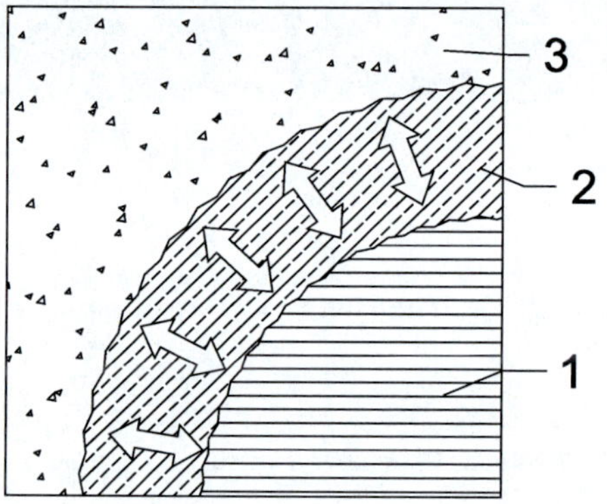

1-Asphaltene phase, 2-Halo of additive in maltenic medium, 3-Maltenic medium

Fig. 5. Scheme of interaction of the additive in the dispersed structure of bitumen

It can be assumed that such an interaction will prevent the aggregation of asphaltene conglomerates and phase transitions of maltenes into asphaltenes, keeping the system stable over time under the influence of external factors, making such a system less prone to aging processes.

The measured dynamic viscosity (Fig. 6) shows that an increase in viscosity is observed in samples with an additive concentration of 2–5%.

For the sample with an additive content of 1%, the viscosity curve was lower than that for the original bitumen sample throughout the entire test cycle in the temperature range of 120–200 °C. These results confirm the assumptions on the influence of Viscogel on the rheological properties of bitumen. It can be assumed that the decrease in the colloid viscosity is associated with the effect of the additive on the structural bonds within the system. In addition, the obtained viscosity allows us to assume that in the future, bitumen modified with 1% Viscogel will have more stable rheological properties.

The determined contact angles of the bitumen samples confirm the viscosity data on the stability of the rheological properties of the sample modified with 1% Viscogel (Table 3).

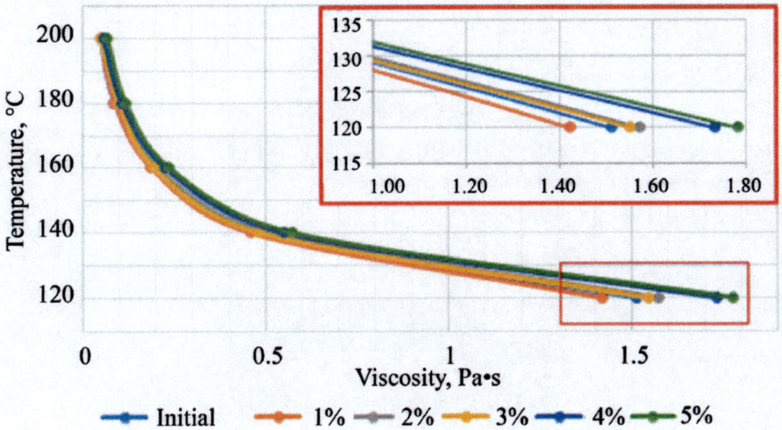

Fig. 6. Viscosity of bitumen modified by the additive

Table 3. Contact angle of the modified bitumen samples

Aging time, hours	Modifier concentration, %			
	0	0,5	1.0	2,5
0	95.52	101.89	94.00	96.52
24	103.37	102.25	94.24	101.23
48	115.77	108.78	96.61	110.31

The presented data show that the sample with 1% of the modifier, there was practically no change in the contact angle after 24 h of testing. After 48 h the increase in the indicator is much less than in all other cases.

According to studies [36], the contact angle may indicate the quality of the bituminous binder used in the asphalt concrete mixture (Fig. 7).

Wetting describes the behavior of a liquid upon contact with a solid surface when the system behavior is determined by the surface tension. In our case, this liquid is the bituminous colloid. It can be assumed that if a 1% rheological additive is introduced into bitumen, stable cohesive bonds are formed, which can keep the drop stable under the influence of high temperatures over time.

Obviously, due to an increase in cohesion inside the original bitumen and binders modified with a non-optimal additive content, the surface wetting deteriorates and the system rigidity increases due to an imbalance in the colloidal system manifested in the depletion of the maltenic medium. In this case the main characteristic of the state is the mechanical coupling of the phases involved, caused by the strengthening of the molecular interaction in the boundary layer. According to the theory of adsorption and wetting, the binder modified with the 1% additive maintains thermodynamic equilibrium during the entire process of thermostating. If the boundary layer of the bituminous colloid is thermodynamically unstable, the drop changes.

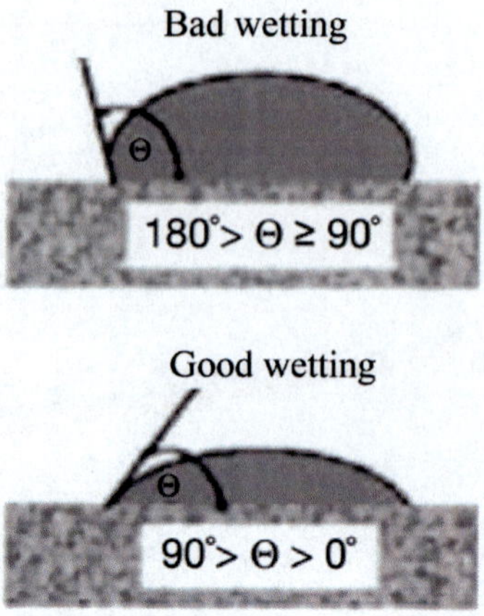

Fig. 7. Contact angle of a bituminous drop [34]

Further we introduce the Δ coefficient, denoting the ratio of the characteristics of bitumen samples before and after aging, and obtain the following data (Fig. 8).

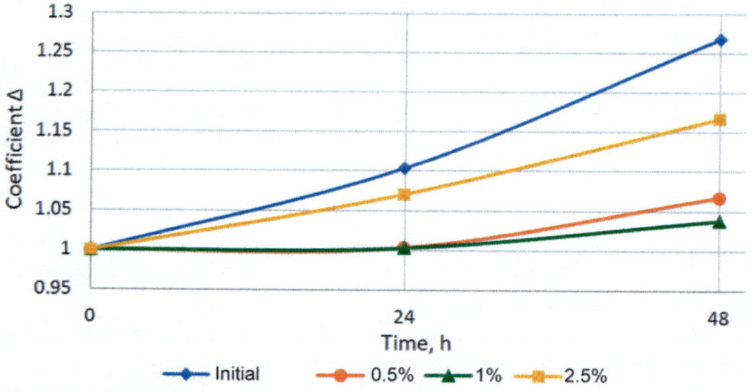

Fig. 8. Dynamics of change in the bitumen viscosity during aging

The graph shows that the sample modified with 1% rheological additive has the smallest deviation from the initial bitumen parameters.

4 Conclusions

A scheme for the interaction of a rheological additive with a dispersed colloidal structure of bitumen is proposed.

Viscosity and contact angle of wetting were determined. The rational content of the additive was established, which makes it possible to control the rheological and technological parameters of bitumen. It can be concluded that the introduction of the

1% Viscogel additive has the best effect on the bitumen performance, reducing the viscosity, but at the same time maintaining the cohesive properties of the bitumen.

The efficiency of the Viscogel additive as a tool maintaining the colloidal system stability during thermostating was proven. This modifier enables to protect bitumen from the harmful effects of aging processes while maintaining the stable properties of the asphaltenes-maltenes system.

The results of the study indicate the need for further investigation of the effect of the rheological additive not only on bitumen but also on the properties of the asphalt concrete mixture.

References

1. Savitskaya, T.A.: Manual for independent work on the lecture course Colloidal chemistry: questions, answers and exercises. (2009)
2. Zolotarev, V.A.: Road bituminous binders and asphalt concretes (2014)
3. Syunyaeva, Z.I.: Chemistry of oil (1934)
4. Safieva, R.Z.: Oil dispersed systems: composition and properties (part 1) (2004)
5. Dolomatov, M.Y.: Investigation of the structure of nanoparticles of petroleum asphaltenes. Bashkir Chem. J. **18**(3), 18–21 (2011)
6. Ryskulova, G.R., Shiryaeva, R.N., Serebrennikov, D.V.: Investigation of the composition of asphaltenes of high-viscosity oils by IR spectroscopy. Bashkiria **21**(4), 928–929 (2016)
7. Dolomatov, M., Shutkova, S.A., Dezortsev, S.V.: Investigation of the characteristics of the electronic structure of petroleum resins and asphaltenes. Bashkir Chem. J. **17**(3), 211–218 (2010)
8. Shutkova, S.A.: Investigation of the supramolecular structure of nanoparticles of petroleum asphaltenes. Bashkir Chem. J. **19**(4), 220–225 (2012)
9. Sunyaev, Z.I., Safieva, R.Z., Syunyaev, R.Z.: Oil disperse system (1990)
10. Baranov, V.Y., Frolov, V. I.: Electrokinetic phenomena. Educational manual. RSU of Oil and Gas (2002)
11. Tumanyan, B.P.: Scientific and applied aspects of the theory of oil dispersed systems (2000)
12. Savitskaya, T.A.: Manual for independent work on the lecture course Colloidal chemistry: questions, answers and exercises (2009)
13. Lefedova, O.V., Nemtseva, M.P.: Basic concepts and definitions of the courses Physical chemistry and Colloidal chemistry (2017)
14. Unger, F.G.: Nanosystems, dispersed systems, quantum mechanics, spin chemistry (2010)
15. Lesin, V.I., Koksharov, Y.: Khomutov. G. B: Magn. Nanoparticles in oil. **50**(2), 114–117 (2010)
16. Rybachuk, N.A.: Aging of bituminous binder, 2(97) (2015)
17. Skripkin, A.D., Starkov, G.B., Kolesnik, D.A.: Evaluation of bitumen aging in thin films using the Iatroscan Mk-5 thin chromatography analyzer. **40**, 32–35 (2008)
18. Musina, N.S., Maryutina, T.A.: Application of magnetic processing for changing the composition and physico-chemical properties of oil and petroleum products. 71(1), 29–36 (2016)
19. Anufriev, R.V.: Influence of ultrasonic treatment on the structural and mechanical properties of oil dispersed systems (2017)
20. Belyaev, K.V., Chulkova, I.L: Modification of bitumen with technical carbon. 4.(68) (2019)
21. Ilyin, S.O., Arinina, M.P., Mamulat, Y.S., Malkin, A.Y., Kulichikhin, V.G.: Rheological properties of road bitumen modified with polymer and nanoscale solid additives. Colloidal j. **76**(4), 461–471 (2014)

22. Galeev, R., Abdrakhmanova, L., Nizamov, R.: Nanomodified organic-inorganic polymeric binders for polymer building materials. Solid State Phenom. **276**, 223–228 (2018). https://doi.org/10.4028/www.scientific.net/SSP.276.223
23. Loskutova, Y.V.: Influence of the magnetic field on the structural and rheological properties of oils. **309**(4), 104–109 (2006)
24. Mullakaev, M.S.: Investigation of the influence of ultrasonic exposure and chemical reagents on the rheological properties of viscous oils. 5, 31–34 (2010)
25. Anufriev, R.V., Volkova, G.I.: Influence of ultrasonic treatment conditions on the properties of high-tar paraffin oil. In: Collection of scientific papers of the X International Conference of Students and Young Scientists, 242–244 (2013)
26. Shestakov, N.I., Urkhanova, LA., Buyantuev, S.L., Semenov, A.P., Smirnyagina, N.N.: Asphalt concrete using carbon nanomodifiers, **6**, 21–24 (2015)
27. Inozemtsev, S.S., Korolev, E.V.: Development of nanomodifiers and study of their influence on the properties of bituminous binders. 10, 131–139 (2013)
28. Begieva, M.B., Soblirova, A.A., Balashov, A.V., Amshokova, D.B., Kharaev, A.M.: Organoglina modified with n, N-diallylamino isopentanoic acid and investigation of its structure, (53), 58–2 (2016)
29. Loganina, V.I., Petukhova, N.A.: Increasing the resistance of polystyrene paints when introducing an organomineral additive into the formulation. EEJET. 6(63) (2013)
30. Loganina, V.I., Akzhigitova, E.R.: Dry building mixes with the use of additives based on mixed-layer clays (2014)
31. Bulanov, P.E., Mavliev, L.F., Vdovin, E.A., Yagund, E.M.: The interaction between the kaolinite or bentonite clay and plasticizing surface-Active agents. Mag. Civ. Eng. **75**(7), 171–179 (2017). https://doi.org/10.18720/MCE.75.17
32. Helal, E., Sherif, E., Alaa, G., Saaid, Z.: Evaluation of asphalt enhanced with locally made nanomaterials. Nanotechnol. Constr. 8(4) (2016). https://doi.org/10.15828/2075-8545-2016-8-4-42-67
33. Sarsam, S.: Improving asphalt cement properties by digestion with nano materials. Res. Appl. Mater. J. (RAM) **1**(6), 61–64 (2013). https://doi.org/10.12966/ram.09.01.2013
34. Arkhipov, V.A.: Determination of the edge angle of wetting of the coal surface. Phys. Tech. Dev. Miner. **5**, 22–27 (2011)
35. Abdullin, A.I., Emelyanycheva, E.A., Diyarov, I.N.: Assessment of bitumen adhesion to mineral material in asphalt concrete on the basis of its wetting properties 4 (2009)
36. Huchenreuther, Y.: Asphalt in road construction (2013)

Biostability of Road Cement-Mineral Materials Reinforced with Dispersed Fibre

Evgenii Vdovin[1(✉)], Victor Stroganov[1], Maxim Amel'chenko[1], Eduard Kraus[2], and Ilshat Fazleev[3]

[1] Kazan State University of Architecture and Engineering, 420043 Kazan, Russia
[2] German Plastics Centre (SKZ), 97082 Wuerzburg, Germany
[3] JSC «Tatavtodor», 420012 Kazan, Russia

Abstract. Modification of building and road materials with fibres is a promising area of construction material science. This type of dispersed reinforcement can increase several performance indicators. However, a small number of works consider biodamage of such materials. This work studies biostability of road cement-sand materials modified with cellulose (a natural fibre) in the conditions simulating culture fluids of molds of the Aspergillus niger and Penicillium genus. The relationships between the bending strength loss of fibre-cement-sand materials and the holding time in the tested media of carboxylic acids were obtained. It was found that the maximum decrease in strength after biodamage reaches 40% (from the initial). It was shown that the mixture of acids simulating the culture fluid of molds of the Aspergillus niger genus has the greatest ability to biodegrade road cement-sand materials modified with cellulose fibre. It is recommended to use this method of determining the hardness of the test materials when designing and constructing the regional and local highways, especially those connecting agricultural, livestock, and other complexes.

Keywords: Bending tensile strength · Road cement-mineral materials · Dispersed reinforcement · Fibre · Biodamage · Biostability · Model media of carboxylic acids

1 Introduction

Road cement-mineral materials (RCMM) based on cement-crushed stone-gravel-sand or cement-soil mixtures are composite building materials used for the construction of foundations and pavements. Their application is technically and economically efficient, especially in regions with an absence of durable stone materials [1–3]. However, it was shown that in certain cases utilization of cement-mineral materials in the structural layers of road pavements could cause low deformability of layers, and, hence, increased cracking of pavement surfaces [4, 5].

Dispersed reinforcement of cement-mineral road-building materials by artificial (steel, glass, basalt) [6, 7], natural (cellulose, wood) [8, 9], and synthetic (polypropylene, polyester, polyamide) [10, 11] fibres is one of the most acute directions of their

N. Vatin (Ed.): STCCE 2021, LNCE 169, pp. 289–296, 2021.
https://doi.org/10.1007/978-3-030-80103-8_31

modification. This solution is used to improve the material construction and technical properties and durability, as well as to increase its deformation in the elastic stage, with no cracking [12, 13].

It is generally accepted that hardening during fibre dispersed reinforcement of RCMM is based on the fact that the matrix of the composite material transfers the applied load to the fibres uniformly distributed in it due to the tangential forces acting at the interface. If the elastic modulus of the fibre exceeds that of the concrete matrix, most of the stress arising in the material is absorbed by fibres, therefore the total strength of the composite increases proportionally to the distributed load [14–19]. In addition, when the material is loaded, the fibres distributed over its volume cause the processes of crack growth deceleration (energy dissipation), a decrease in the concentration and distribution of stresses [20–26].

The issues of RCMM biostability are most relevant for regional and local automobile roads, including those connecting objects of agricultural, livestock, and other complexes [27]. Natural fibres, of all types of materials for dispersed reinforcement, can be subject to the greatest bio damage as their biostability is much lower than that of artificial and synthetic ones. Thus, the application of natural cellulose fibre as a reinforcing additive in road materials for such roads necessitates an assessment of their biostability, and this is the purpose of the present work.

2 Materials and Methods

2.1 Preparation of Fibre-Cement-Sand Materials

The studied RCMM modified with fibre were of 100 × 100 × 10 mm size. They had the following content (by weight):

- Portland cement Cem I 42.5 N – 20.0%;
- Natural fine-grained quartz sand with a fineness modulus of 1.8 (Table 1) – 74.5%;
- Unbleached softwood sulphate cellulose with water-cement ratio W/C = 0.4 – 5.5%;

Table 1. Granulometric composition of sand.

Residue	8.000	4.000	2.000	1.000	0.500	0.250	0.125	0.063	<s0.063
Total residue, %	0.0	0.0	0.2	1.6	14.4	79.2	98.2	99.7	100.0
Total out siftings, %	100.0	100.0	99.8	98.5	85.7	20.9	1.9	0.3	0.0

The RCMM was prepared as follows:

1. Fluff fibres were mixed with cement and sand to form a fibre-cement-sand mass (FCSM) with water-cement ratio W/C = 0.4;

2. The obtained FCSM was packed into molds and put under pressure using the following mode: pressure rise to 600 N/cm^2 at a speed of 10 kgf/s for 10 min, and holding under pressure for 10 min;
3. The obtained samples were subjected to thermal and moisture treatment in a steaming chamber. The following processing mode was used: temperature rise to 65 °C for 2 h, isothermal holding for 4 h, cooling for 2 h. Additionally, the samples were subjected to autoclaving, i.e., rising to a temperature of 175–180 °C for 3 h, isothermal holding under the pressure of 0.9–1 MPa for 6 h, cooling for 3 h.

2.2 Testing

The prepared RCMM samples were tested for bending strength in accordance with GOST.

The biostability of RCMM samples was determined using the laboratory kinetic setup [26]. The tests were carried out for 7 days at an ambient temperature of 40 ± 2 °C, which is equivalent to 28 days at 20 ± 2 °C according to Van't Hoff rule, with automatic regulation of the main medium parameters – temperature and pH.

The laboratory setup [29] consists of a test container (in which the test samples were loaded, immersed in the model media); make-up and drain tanks; measuring container for assessing the model media pH and calibrating the electrode pH; an electronic control unit, measuring pH, temperature of the model media and maintaining them at a given level.

Citric acid (model media 1) and a mixture of carboxylic acids: lactic, oxalic, malic, and citric acids (model media 2) were used as model media. Their choice is specified by their presence in the metabolites of microorganisms isolated from building materials, affected sections of buildings and road structures [30, 31].

3 Results and Discussion

It is advisable to use fibres of plant origin in RCMM as their fibrous structure can form a three-dimensional skeleton of fibres, providing a minimum degree of delamination and maximum stability of RCMM technological properties.

It is known that biological damage to materials is caused by various types of bacteria and, first of all, mold fungi of the Aspergillus and Penicillium genus, the natural reservoir of which are soils. This work studies their influence on road cement-mineral materials reinforced with cellulose fibre. The studies were carried out in model media – aqueous solutions of carboxylic acids and citric acid, and in a mixture of lactic, oxalic, malic, and citric acids. The composition of these mixtures corresponds to the culture fluid of molds of the Penicillium genus and the Aspergillus niger genus [30, 31]. The culture fluid of the mold fungus Aspergillus niger has a more complex chemical composition [30], which is confirmed by liquid chromatography data (Fig. 1).

The qualitative and quantitative composition of the culture fluid analogue was determined using the chromatographic analysis (Table 2).

It is known that the culture fluids under study have pH = 3 [32], which was used to simulate the processes of biological damage. These conditions ensured that the action of the model media corresponded to the culture fluids of carboxylic acids.

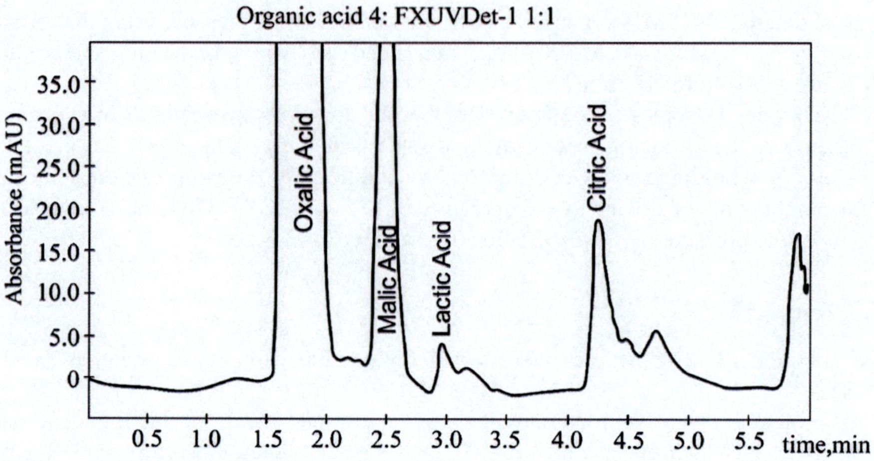

Fig. 1. Chromatogram of the culture fluid of the mold fungus of the Aspergillus niger genus.

Table 2. Chemical composition of the culture fluid produced by molds of the Aspergillus niger genus.

Mixture components	Acid concentration (weight %)
Oxalic acid	10.0
Malic acid	70.9
Lactic acid	6.6
Citric acid	12.5

The biostability of RCMM samples was determined by the coefficient of chemical resistance, which characterizes the change in the strength characteristics of the samples after testing them for ultimate bending strength (K_{bend}) before and after holding to model media (Tables 3 and 4).

Table 3. Results of biostability testing in a citric acid medium at 40 °C for samples № 1.

Time (τ), days	Bending strength $R_{bend}1$, MPa	$K_{bend}1$
0	25.3	1.0
2	23.1	0.91
4	20.8	0.82
6	18.5	0.73
8	16.3	0.64

Table 4. Results of biostability testing in a mixture of carboxilyc acids at 40 °C for samples № 2.

Time (τ), days	Bending strength $R_{bend}2$, MPa	$K_{bend}2$
0	25.3	1.0
2	20.1	0.79
4	18.5	0.73
6	16.8	0.66
8	15.2	0.60

The established dependencies are described by linear equations:

$$R_{bend}1 = 25.4 - 1.14 \cdot \tau, \quad (1)$$

where $R_{bend}1$ is the ultimate bending strength after holding in citric acid, MPa, with the correlation coefficient $r = -0.999$ and standard deviation $S_d = 0.038$; τ is the holding time in a model medium, in days;

$$R_{bend}2 = 21.8 - 0.82 \cdot \tau, \quad (2)$$

where $R_{bend}2$ is the ultimate bending strength after holding in a mixture of carboxylic acids, MPa, with the correlation coefficient $r = -0.999$ and standard deviation $S_d = 0.031$; τ is the holding time in a model medium, in days.

Experimental data on the change in ultimate strength R_{bend} of RCMM samples in citric acid and in a mixture of carboxylic acids are shown in Fig. 2.

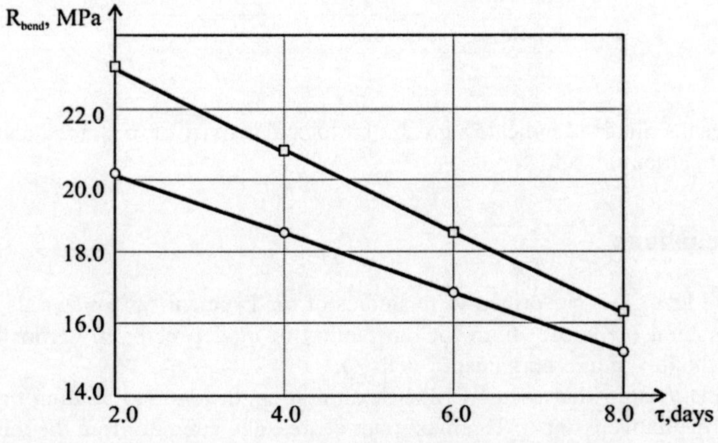

Fig. 2. Relationship between ultimate bending strength (R_{bend}) of RCMM samples and holding time (τ): 1 – in citric acid solution; 2 – in a mixture of acids (oxalic, malic, lactic, citric).

The obtained dependences show that strength characteristics decrease during the holding time, which indicates the destruction of samples. The nature of the linear dependences $R_{bend} = f(\tau)$ in the acid mixture from the second day is identical (although it has a different slope), and the rate of interaction (complexation) of the components of the model media with the samples in the main kinetic section is practically constant. When Ca2+ salts are washed out from the samples, they interact with the carboxylic acids of the medium with the formation of water-insoluble salts of calcium oxalate and citrate. As calcium salts go into the medium, fine-crystalline quartz sand is poured from the vertically standing samples and settles to the bottom of the test vessel of the kinetic installation. Thus, for interaction with medium, the surface areas of the tested samples are opened (updated) and the process, repeating many times, continues. Apparently, this determines the constant rate of interaction of calcium salts of samples with the model media components and the linearity of the kinetic dependences, the decrease in the strength level (R_{bend}).

The loss of strength in model media 1 and 2 was calculated using Eq. (3) (Table 5).

$$\Delta R_{\mathrm{bend}} = 100 - \frac{R_{\mathrm{bend}}(\text{in medium})}{R_{\mathrm{bend}}(\text{initial})} \cdot 100 \tag{3}$$

where ΔR_{bend} is the bending strength loss, %;

R_{bend} (in medium) is the ultimate bending strength after holding in the model medium, MPa;

R_{bend} (initial) is the ultimate bending strength before holding in the model medium, MPa;

Table 5. Change in bending strength after holding for 8 days in media 1 and 2.

Material, medium	Strength loss ΔR_{bend}, %
RCMM, 1	35.6
RCMM, 2	40.0

The results obtained indicate a greater biological activity of medium 2 containing a mixture of carboxylic acids.

4 Conclusions

1. For the first time, performed were studies of road cement-sand materials modified with natural (cellulose) fibre for biostability in model media of carboxylic acids – metabolites of microorganisms' activity.
2. The relationships between RCMM flexural strength loss and holding time in the tested media were linear. The maximum decrease in strength from the initial value was 40.0% after biodamage. It is recommended to consider these data when designing and constructing the regional and local highways, especially those connecting agricultural, livestock, and other complexes.
3. It was shown that the mixture of carboxylic acids, which simulates the culture fluid of molds of the Aspergillus niger genus, has the greatest ability to biodegrade road cement-sand materials modified with natural (cellulose) fibre. This is explained by

the synergy effect of the destructive action of carboxylic acids of various basicities and functionalities, in comparison with the action of tribasic citric acid, which prevails in the culture fluid of molds of the Penicillium genus.

References

1. Li, X., Zhang, S., Wu, H., Tian, D.: Study on the material properties of cement-sand-gravel. In: IOP Conference Series Earth and Environmental, Science, vol. 513, p. 012036 (2020). https://doi.org/10.1088/1755-1315/513/1/012036
2. Shabaev, S., Ivanov, S.: Selection of the optimal grain-size distribution of crushed stone sand and gravel sand mixtures for the pavement of truck haul roads of coal open pits. In: E3S Web of Conferences, vol. 174, p. 01036 (2020). https://doi.org/10.1051/e3sconf/202017401036
3. Xu, T., Jin, G.: Effects of unit cement content and aggregate maximum particle size on compressive strength of cemented sand and gravel materials. In: IOP Conference Series Earth and Environmental, Science, vol. 508, p. 012203 (2020). https://doi.org/10.1088/1755-1315/508/1/012203
4. Bao, L., Wei, X., Yu, L., Chen, A.: Analysis of stresses and processing method on the reflection cracks of semi-rigid base road resulting from the vehicle loads, Shenyang Jianzhu Daxue Xuebao (Ziran Kexue Ban). J. Shenyang Jianzhu Univ. (Nat. Sci.) **33**(3), 458–466 (2017). https://doi.org/10.11717/j.issn:2095-1922.2017.03.09
5. Pradena, M.: Analysis of the cracking process of lean concrete road bases. In: Conference: 14th SGEM GeoConference on Nano, Bio and Green Technologies for a Sustainable Future (2014). https://doi.org/10.5593/SGEM2014/B62/S27.059
6. Thomas, J., Ramaswamy, A.: Mechanical properties of steel fibre-reinforced concrete. J. Mater. Civ. Eng. **19**, 385–392 (2007). https://doi.org/10.1061/(asce)0899-1561(2007)19:5(385)
7. Matveika, M., Zarapin, V., Artimovich, V.: Modification of the steel fibres surface for disperse reinforcement of concrete. Vestn. Viteb. State Technol. Univer. **1**(32), 171–178 (2017)
8. Lukmanova, L.V., Mukhametrakhimov, R.Kh., Gilmanshin, I.R.: Investigation of mechanical properties of fiber-cement board reinforced with cellulosic fibers. In: IOP Conference Series: Materials Science and Engineering, p. 012113 (2019). https://doi.org/10.1088/1757-899X/570/1/012113
9. Khozin, V., Khokhryakov, O., Nizamov, R.: IOP Conference Series: Materials Science and Engineering (2020)
10. Kluev, S.V., Avilova, E.N.: Fine-grained fibre-reinforced concrete using polypropylene fibre for road surface. Vest. BSTU im. V.G. Shukhova **1**, 37–40 (2013)
11. Kiyanets, A.V.: Concrete with recycled polyethylene terephthalate fibre. Mag. Civ. Eng. **8**(84), 109–118 (2018)
12. Tang, C.S., Li, J., Wang, D.Y., Shi, B.: Investigation on the interfacial mechanical behavior of 3 wave-shaped fibre reinforced soil by pullout test. Geotext. Geomem. **44**(6), 872–883 (2016)
13. Volkov, I.V.: Problems of fibre-reinforced concrete in domestic construction. Stroit. Mater. **6**, 12–13 (2004)
14. Rabinovich, F.N.: Composites based on dispersion-reinforced concrete. In: Theory and Design Issues, Technology, Designs. ASV (2004)
15. Kluev, S.V.: Fibre concrete and products based on it. Mezhd. Nauch.-Issledov. Zhurn. **13**(34), 70–73 (2015)
16. Correia, A.A.S., Oliveira, P.J.V., Custodio, D.G.: Effect of polypropylene fibres on the compressive and tensile strength of a soft soil, artificially stabilised with binders. Geotext. Geomem. **43**(2), 97–106 (2015)

17. Zhan, Y., Bui, H.G., Ninic, J., Mohseni, S.A., Meschke, G.: Numerical modeling of steel fibre reinforced concrete on the meso and macro-scale institute for structural mechanics. Ruhr-Univ. Bochum Universitatsstraße **150**, 44801 (2014)
18. Khan, M., Ali, M.: Effectiveness of hair and wave polypropylene fibres for concrete roads. Constr. Build. Mater. **166**, 581–591 (2018). https://doi.org/10.1016/j.conbuildmat.2018.01.167
19. Lukmanova, L.V., Mukhametrakhimov, R.K., Gilmanshin, I.R.: Investigation of mechanical properties of fiber-cement board reinforced with cellulosic fibers. In: IOP Conference Series: Materials Science and Engineering, p. 012113 (2019). https://doi.org/10.1088/1757-899X/570/1/012113
20. Mukhametrakhimov, R.K., Lukmanova, L.V., Khuzin, A.F.: Investigation of the influence of organosilicon compounds on fiber cement panels' characteristics. In: IOP Conference Series: Earth and Environmental Science (2017). https://doi.org/10.1088/1755-1315/90/1/012132
21. Kosior-Kazberuka, M., Krassowskaa, J., Vidales Barrigueteb, A., Piña Ramirez, C.: Fracture parameters of basalt fibre reinforced concrete. Ana. Edificación **4**(3), 52–58 (2018)
22. Shen, D., Liu, X., Zeng, X., Zhao, X., Jiang, G.: Effect of polypropylene plastic fibres length on cracking resistance of high-performance concrete at early age. Constr. Build. Mater. **244**, 117874 (2020). https://doi.org/10.1016/j.conbuildmat.2019.117874
23. Wei, B., Cao, H., Song, S.: Surface modification and characterization of basalt fibres with hybrid sizings. Compos.: Part A **42**, 22–29 (2011)
24. Khan, M., Cao, M., Ali, M.: Cracking behaviour and constitutive modelling of hybrid fibre reinforced concrete. J. Build. Eng. **30**, 101272 (2020). https://doi.org/10.1016/j.jobe.2020.101272
25. Trainor, K.J., Foust, B.W., Landis, E.N.: Measurement of energy dissipation mechanisms in fracture of fibre-reinforced ultrahigh-strength cement-based composites. J. Eng. Mech. **139**(7), 771–779 (2013). https://doi.org/10.1061/(ASCE)EM.1943-7889.0000545
26. Trainor, K.J.: 3-D analysis of energy dissipation mechanisms in steel fibre reinforced reactive powder concrete. M.S. thesis, University of Maine, Orono, ME (2011)
27. Hang, Y., Hu, X., Li, Z.: Experimental study on reinforcement of biological enzyme for road soil Shenyang university. In: 6th International Conference on Energy and Environmental Protection ICEEP (2017)
28. Stroganov, V.F., Sagadeev, E.V.: Biodamage of construction materials. Stroit. Mater. **5**, 5–10 (2015)
29. Patent of the Russian Federation №. 170410, bull. №. 12, 24 April 2017
30. Yakovleva, G., Sagadeev, E., Stroganov, V., Kozlova, O., Okunev, R., Ilinskaya, O.: Metabolic activity of micromycetes affecting urban concrete constructions. Sci. World J. (2018). https://doi.org/10.1155/2018/8360287
31. Yakovleva, G.Y., et al.: Assessment of biodamage resistance of various concrete grades. Int. J. Pharm. Technol. **8**(4), 24291–24299 (2016)
32. Stroganov, V.F., Sagadeev, E.V.: Introduction to biodamage of building materials: monograph, p. 200. KSUAE, Kazan (2014)

Alternative Mineral Powders for Asphalt Concrete

Marina Vysotskaya[1], Evgenii Vdovin[2](✉), Dmitry Kuznetsov[1], and Artem Shiryaev[1]

[1] Belgorod State Technological University named after V.G. Shukhov, 308012 Belgorod, Russia
[2] Kazan State University of Architecture and Engineering, 420043 Kazan, Russia

Abstract. During the operation of asphalt concrete mix coatings in conditions of alternating temperatures, the formation of seasonal plastic deformations is often observed, expressed in the formation of ruts, waves and surges that violate the evenness of the upper layers of the road surface and temperature cracks that contribute to premature chipping and destruction of the coating. Such surface defects are primarily due to the specific properties of the organic binder used in the industry, expressed in the dependence of its rheological and physico-mechanical parameters on the ambient temperature. An alternative method that synthesizes both an increase in the working temperature range of the binder and the heat and crack resistance of asphalt concrete can be the use of alternative dispersed materials that serve as an active structuring filler in the technology of preparing asphalt concrete mixtures. In the present work, a series of mineral powders was considered: limestone, perlite, zeolite and expanded clay. The indicated powders differ from the classic limestone powder by their increased porosity and developed specific surface area. The study of the properties of asphalt binders prepared on the basis of various methods for the selection of bitumen confirmed that in the case of using porous mineral powders, it is most appropriate to take the indicator of water saturation, rather than residual porosity, as the criterion of optimal composition. In order to justify the prospects of using the studied powders for the preparation of organomineral composites, the quality level of asphalt binders was calculated. In accordance with the calculations, the quality level for water resistance was 1.29 for limestone, 1.21 for perlite, 1.23 for expanded clay, and 1.31 for zeolite. The results obtained allow us to judge the dependence of the quality level of the asphalt binder on the filler used, and indicate the feasibility of further research of dispersed mineral powders from perlite and zeolite.

Keywords: Mineral powder · Asphalt binder · Asphalt concrete · Road construction

1 Introduction

Road infrastructure is a vital system of any economic sector of the state. There is a direct relationship between the transport and operational condition of road surfaces and the dynamics of economic development, which requires a constant increase in the

N. Vatin (Ed.): STCCE 2021, LNCE 169, pp. 297–307, 2021.
https://doi.org/10.1007/978-3-030-80103-8_32

intensity and load capacity of traffic on roads. Therefore, it is obvious that it is necessary to strengthen the majority of road surfaces and eliminate existing and only emerging defects.

Thus, during the operation of asphalt concrete mix coatings in conditions of alternating temperatures, it is increasingly possible to observe the formation of seasonal plastic deformations, expressed in the formation of ruts, waves and surges that violate the evenness of the upper layers of the road surface and temperature cracks that contribute to premature chipping and destruction of the coating.

Such surface defects are primarily due to the specific properties of the organic binder used in the industry, expressed in the dependence of its rheological and physico-mechanical parameters on the ambient temperature.

In general, these problems are solved from two positions: an increase in the plasticity interval of bitumen due to the introduction of polymers and an increase in the heat and crack resistance of the composite by filling the bitumen with fibrous, dispersed mineral powders and other substances [1].

An alternative method that synthesizes both an increase in the working temperature range of the binder and the heat and crack resistance of asphalt concrete can be the use of porous dispersed materials that act as an active structuring filler in the technology of preparing asphalt concrete mixtures.

Obviously, therefore, recently there is a growing interest in the research and application of porous polyfractional and fine mineral materials for the preparation of organomineral road composites [1–3].

It is known that the composition, structure and properties of the material are directly related to each other. The disclosure of these relationships and the identification of patterns of changes in properties when changing the composition and structure serve as the basis for active regulation of the quality of building materials and the prevention of characteristic defects that occur during the operation of the coating. Therefore, the primary task in designing the composition of asphalt concrete is the rational selection of raw materials and the development of technical and technological measures that minimize the possibility of the appearance and development of such defects.

The analysis conducted by L.A. Gorelysheva [4] showed that the maximum contribution to the stability, quality and durability of asphalt concrete pavement is made by an asphalt binder (AB), whose structure, unlike asphalt concrete, is characterized by the absence of «synthesis of defects due to heterogeneity of the structure» [4]. The active component structuring bitumen is mineral powder, its specific surface area reaches up to 90–95% of the surface of all mineral grains that make up asphalt concrete. Thus, this indicator from the point of view of structuring systems is one of the fundamental ones.

For a comparative analysis and evaluation of the performance of various mineral powders in asphalt concrete, it is important to know what the dynamics of changes in the physical and mechanical properties of the binder in the composite are, since it is the synthesis of the two components of the mixture, bitumen and mineral powder, that determines the strength and deformative qualities, as well as the durability of the coating.

The aim of the research was to study the effect of porous fine powders on the properties of asphalt binder (ABB).

2 Materials and Methods

The study used BND 70/100 bitumen, Table 1, and fillers: perlite, expanded clay, zeolite, as well as traditional limestone.

Table 1. Indicators of BND 70/100 bitumen properties.

Indicators	Requirements GOST 33133	Actual results
Needle penetration depth 0.1 mm		
at 25 °C at 0 °C	71–100 >21	77 24
Extensibility, cm		
at 25 °C at 0 °C	>62 >3.7	100 3.7
Maximum tensile force, N, at temperature,		
at 25 °C at 0 °C	– –	1.2 118
Softening point, °C	>47	49.5
Brittleness temperature, °C	<−18	−22
Dynamic viscosity, Pa·s, at 60 °C, Condition 1	–	245.5
Change in dynamic viscosity after shear action, Pa·s, at 60 °C, Condition 2	–	12.97
Post-aging tests by the method of RTFOT		
Change in sample mass after aging, %	<0.6	0.2
Change in softening temperature, °C, after aging	<7	6.2
Dynamic viscosity after aging, Pa·s, at 60 °C, Condition 1	–	667.11
Change in dynamic viscosity after shear action, after aging, Pa·s, at 60 °C, Condition 2	–	24.72
Brittleness temperature after aging, °C	<−15	−19

The data on the mineral powder used is presented in the Table 2.

The studied fillers differ from the traditional raw materials by their increased porosity. So, in the mineral powder of limestone, this indicator was 28%, perlite and expanded clay was 39% and zeolite was 55%.

The specific surface area of the studied fillers was controlled using a device for measuring the specific surface area of PSH-2, the principle of operation of which is based on the passage of air through a compacted layer of mineral powder. In addition, the specific surface area of the fillers was measured using the SORBI-M device, designed to measure the specific surface area of dispersed and porous materials by comparing the volumes of adsorbate gas sorbed by the test sample and a standard sample of a material

Table 2. Indicators of mineral powder properties.

Indicators	Mineral powder			
	Limestone	Perlite	Expanded clay	Zeolite
Grain composition, % by weight				
smaller than 1.25 mm	100	100	100	100
<0.315	96	96	98	98
<0.071	85	80	84	80
Porosity, %	28	39	39	55
Swelling of AB samples, %	1.15	2.45	3.7	3.2
Water Resistance AB, %	0.91	0.85	0.83	0.92
Bituminous capacity index, g	68	78	96	109
Humidity, % by weight	0.40	0.48	2.56	2.95

with a known specific surface area. The specific surface area was measured using the 4-point BET method. The advantage of this method is that it is based on the measurement of the specific surface of the filler, both «external» and «internal» by means of liquid nitrogen adsorption and takes into account the presence of pores with a radius of less than 19.4 nm in the structure of the mineral material. Thus, the second method allows us to determine the «true» specific surface area of mineral powder, which is especially important when studying porous raw materials.

The determination of the Brensted acid adsorption centers on the surface of the fillers was carried out by titrimetric method according to the method [5], the method is based on the determination of the value of the exchange capacity with respect to calcium ions. To conduct the test, a 20 g weight of the material was placed in a conical flask, and 20 ml of a saturated solution of $Ca(OH)_2$ was poured into it. The resulting suspension was mixed on a shaking table for 2 h, after which the solution was analytically separated from the mineral material, an indicator was added to the resulting water extracts and titrated with a solution of hydrochloric acid. The difference in the volume of acid used for titration of the saturated solution of $Ca(OH)_2$ and the solution after interaction with the mineral material shows the degree of surface activity.

The results were processed according to the following formula, mg-eq/g:

$$K = \frac{(V_1 - V_2) \cdot 0.1 \cdot 1000}{m}, \tag{1}$$

where V_1 is the volume of acid used for titration of 20 ml of the initial solution $Ca(OH)_2$ («idle» experience), ml;

V_2 – the volume of acid used for titration of the solution after interaction with the mineral material, ml;

m – weight of the mineral material suspension, g.

The size distribution of the particles of the considered fillers was estimated using a laser particle analyzer Microsizer 201.

In the work, the objects of research were AB samples made according to the standard method. Previously, for each composition, the optimal content of bitumen was determined according to two different criteria: the value of water saturation of the samples and the value of residual porosity.

In order to justify the prospects of using the studied powders for the preparation of organomineral composites, the AB quality level was calculated – a relative characteristic of product quality based on comparing the values of the quality indicators of the evaluated products with the basic values of the corresponding indicators [6].

The regulatory documentation regulating the requirements for asphalt concrete mixes and mineral powder for their preparation contains requirements for ABB only for water resistance. Therefore, the determination of the quality level was carried out only for the ABB water resistance index on the studied fillers, according to the Eq. 2 [6].

$$Y_{ki} = \frac{P_i}{P_{i6}}, \tag{2}$$

where Y_{ki} – quality level for the i-indicator; P_i– the value of the i-quality indicator; P_{i6} – the basic value of the i indicator.

3 Results and Discussion

At the first stage of the study, the values of the specific surface area of the fillers were determined. From Fig. 1, it can be seen that the indicator determined on the PSH-2 device is approximately the same. However, measurements on the SORBI-M device indicate higher values of the specific surface area of the studied mineral powders.

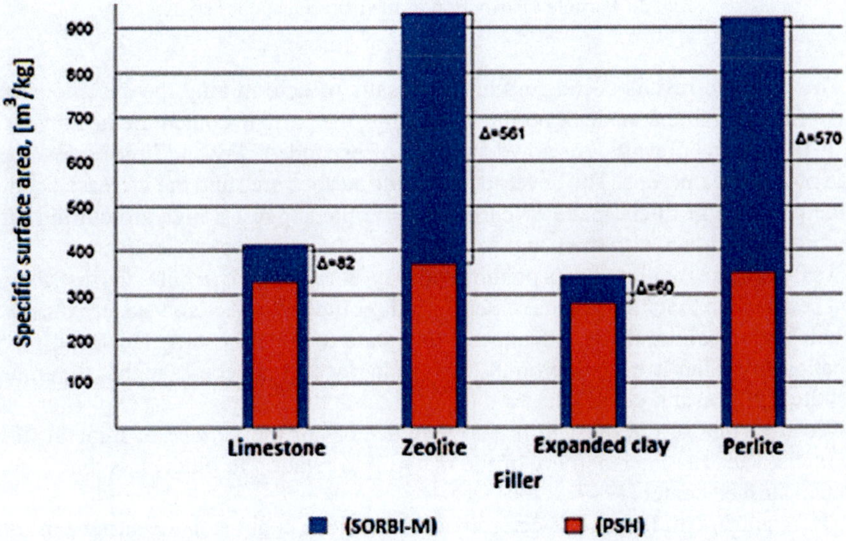

Fig. 1. Specific surface area of mineral powders by various methods.

Thus, the specific surface area of the zeolite filler is more than twice the value of the specific surface area of the standard limestone mineral powder and is 931 m^2/kg. The specific surface area of the perlite powder also increased significantly, both in relation to the value of its own surface determined on the PSH-2 device, and relative to limestone. The smallest change in the surface area was recorded in the study of dispersed limestone, which is characterized by minimal porosity in the series of fillers under study. It can be assumed that the adsorption and structuring capacity of the powders will significantly increase with this factor.

The next step in setting up the experiment was to estimate the particle size distribution, Fig. 2.

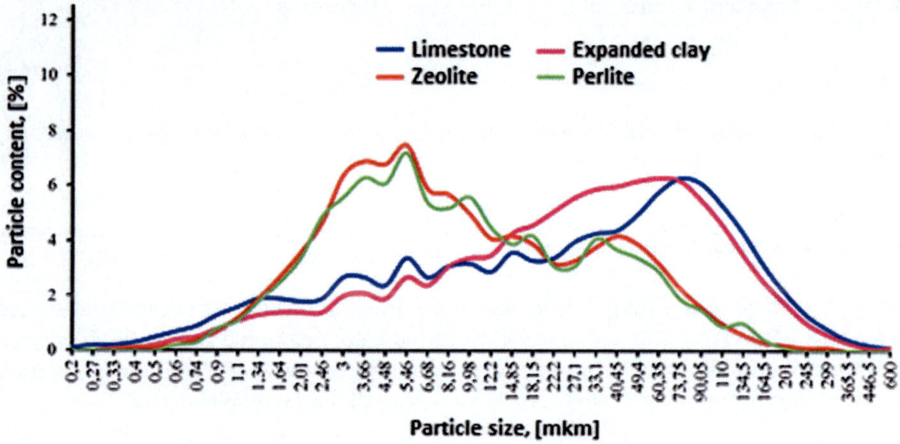

Fig. 2. Particle distribution of the materials under study.

The obtained results correlate with the results of determining the specific surface area of fillers: in fillers made of perlite and zeolite, the particle content in the range from 2.5 to 10 microns prevails, for powders made of expanded clay and limestone – in the range of 20–150 microns. The developed specific surface area and the increased content of fine particles in fillers made of perlite and zeolite suggest a high structuring effect when filling bitumen with them and forming a high-quality asphalt binder.

Analyzing the totality of data on the geometry, porosity and particle size distribution, it can be assumed that fillers made of perlite and zeolite will effectively adsorb the binder on their surface and transfer it from the bulk state to the film state. However, for the formation of strong irreversible bonds at the interface, it is necessary that the mineral substrate has a certain potential.

Recently, one of the most informative indicators of the surface of mineral fillers, which characterizes the reactivity of the material in relation to binders, is the presence of active surface centers [7–9].

The results of studies on the determination of active centers on the surface of fillers also suggested a high reactivity of the studied filler in contact with the binder, Fig. 3. It is worth noting the dynamics of changes in the content of active centers on the surface of porous fillers with an increase in the contact time of raw materials with a solution of calcium hydroxide. For example, in the study of limestone mineral powder, the time of the experiment does not affect the studied indicator, and for fillers made of zeolite and perlite, this indicator increased by more than 10%.

For a preliminary assessment of the performance of porous mineral powders and predicting the properties of road composites based on them, studies of the asphalt binder were conducted.

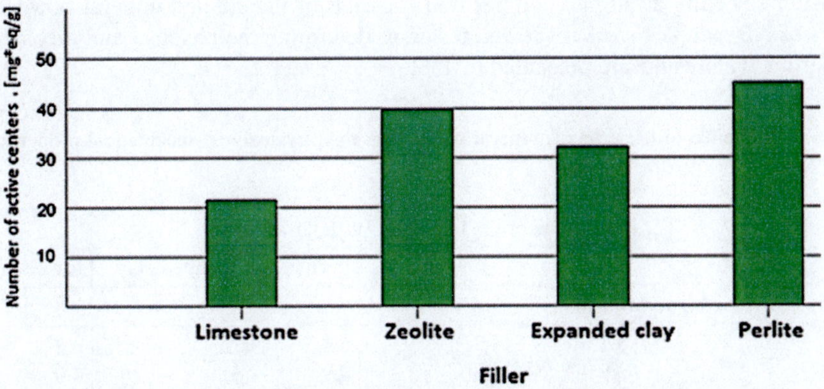

Fig. 3. Content of active centers on the surface of mineral powders.

In the course of the experiment, it was found that in the study of porous and dense powders, at the stage of selecting the composition of the asphalt binder according to different criteria, there are differences in the values of the optimal bitumen content. Moreover, the difference increases as the porosity of the filler increases. The AB compositions are presented in Table 3.

Table 3. Asphalt binder constituents.

Filler	Selection method			
	By residual porosity		By water saturation	
	The content of the components in the mixture, %			
	Mineral powder	Bitumen	Mineral powder	Bitumen
Limestone	87.5	12.5	88.5	11.5
Zeolite	64.1	35.9	66.7	33.3
Expanded clay	80.6	19.4	80.6	19.4
Perlite	68	32	71.9	28.1

It can be seen that the maximum content of bitumen in the samples prepared using perlite and zeolite is a logical consequence of the high porosity of the fillers. If the residual porosity index is chosen as the optimal composition criterion, the amount of bitumen increases by 12.2 and 7.2%, respectively, while for the series of asphalt binder prepared on limestone, the difference in the binder content is insignificant. Similar results were obtained in the study of expanded clay.

A more detailed study of the properties of asphalt binders prepared on the basis of various methods for the selection of bitumen confirmed that in the case of using porous mineral powders, it is most appropriate to take the indicator of water saturation, rather than residual porosity, as the criterion of optimal composition.

After selecting a rational composition for each of the studied mineral powders, a series of AB samples were made and tested to determine the physical and mechanical properties. The results are presented in Table 4.

Table 4. Influence of the type of mineral powder on the physical and mechanical properties of AB.

Indicators	The value of the indicator for the filler			
	Zeolite	Perlite	Expanded clay	Limestone
Compressive strength, MPa, at				
0 °C	5.1	6.4	6.6	6.2
20 °C	3.9	4.9	4.2	4.0
50 °C	3.7	3.6	3.6	3.4
Water Resistance	0.92	0.89	0.90	0.90
Water resistance after 15 days	0.80	0.73	0.68	0.82

As can be seen, the genesis of the fine filler used affects the strength characteristics of asphalt binders. The results obtained were a confirmation of the existing theories [7, 9, 10]. The high strength characteristics of AB correspond to fillers made of perlite and zeolite. Moreover, these fillers are characterized by the largest true specific surfaces.

In accordance with the calculations, the quality level for water resistance was 1.29 for limestone, 1.21 for perlite, 1.23 for expanded clay, and 1.31 for zeolite.

For greater informativeness and comparison of the quality level of mineral powders in the composition of composites, it is advisable to build a cyclogram, on the axes of which the values of the considered quality indicators for each composition of the asphalt binder will be deposited, Fig. 4.

As additional initial data for the construction of the cyclogram, we took the specific characteristics of the ABB, which affect the quality and efficiency of the final road construction composite-asphalt concrete. These characteristics include: the amount of change in the softening temperature of bitumen, the coefficients of heat resistance and crack resistance. The polygon corresponding to each composition is characterized by a certain area of the figure, which can be evaluated to judge the quality level of the composites [6].

The areas of the constructed figures presented in Fig. 3 were for AB prepared on zeolite – 17708 units^2, perlite – 18207 units^2, expanded clay – 11984 units^2 and limestone

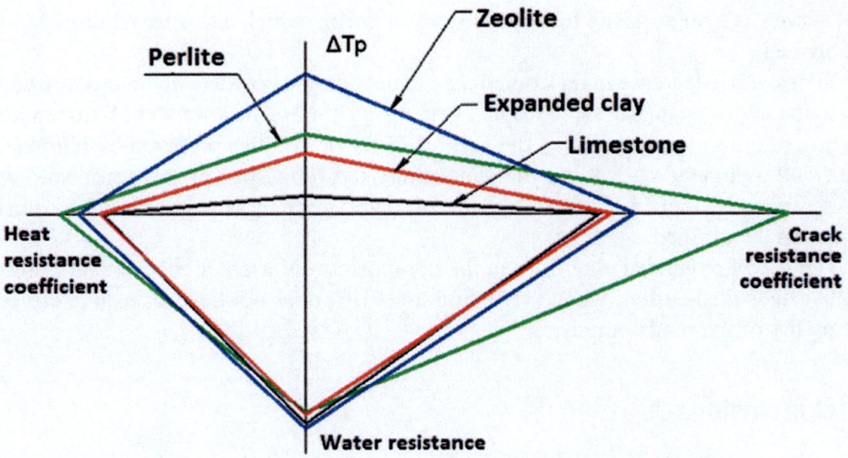

Fig. 4. Cyclogram of physical-mechanical parameters of AB.

– 10192 units^2. As can be seen, the smallest area of the three considered porous fillers has a polygon built according to the results of tests of asphalt binder on dispersed expanded clay.

At this stage of the study, the results obtained allow us to judge the dependence of the quality level of asphalt binder on the filler used, and indicate the feasibility of further research of dispersed mineral powders from perlite and zeolite.

There is a theory [7, 8], according to which the most important characteristics of AB are the optimal bitumen content and strength at this content. In this case, the strength indicators are related to the size and activity of the surface of the mineral powders used, as well as to the size of their grains.

However, there is another point of view [9], according to which only the total contact area of the mineral core particles with each other (according to which the bitumen most structured by the surface can be included in the work) is the main factor determining the strength and other properties of ABB. This area depends on the surface modulus of the mineral material used and on the porosity of the mineral core.

Comparison of the obtained results with the existing theories [7–16] allows us to predict that asphalt concrete coatings, using well-selected porous mineral powders, with a rational selection of asphalt concrete mixture, will be able to work in a wider temperature range, both at high summer temperatures and at low winter temperatures.

The validity of the hypothesis about the possibility of synthesizing not only effective ABB based on porous fillers, but also fine-grained asphalt concrete of type «B» was proved experimentally.

The studies carried out by the team of authors [10–12] previously established the possibility of reducing the content of dispersed perlite in the composition of asphalt concrete, which allowed not only to approach the control composition of the composite on limestone in terms of the content of bitumen, but also to achieve high indicators in a number of characteristics, such as strength at 0 and +50°C. The results obtained

[10] served as a prerequisite for further studies, during which a similar relationship was established.

Thus, with a decrease in the content of porous mineral powders in the mixture to 4%, due to the high specific surface area and structuring ability, the amount of bitumen in the mixture of type «B» approached the control series of samples prepared on a limestone filler, and amounted to 5%. At the same time, the properties of the composite were characterized by increased strength values at 0 and +50°C, as well as crack resistance indicators determined.

The use of expanded clay filler in the composition of asphalt concrete mixtures, as it was established earlier, when evaluating the ABB, does not have a positive effect on the quality of the road composite.

4 Conclusions

According to the results of the study, it can be concluded that the methods for studying the properties of dense powders do not provide objective and reliable information when considering porous fillers, which, with a rational approach, can effectively regulate the heat and crack resistance of road composites.

The study and generalization of questions about the quality of AB will allow you to make more informed decisions at the stage of selecting raw materials and designing the composition of asphalt concrete mix, as well as increase the efficiency of asphalt concrete pavement and investment of financial resources.

References

1. Borisenko, O.A.: Bituminous composition, modified by dropping cleavage of expanded clay for asphalt materials with high thermal stability and cracking. Stavropol 23 (2008)
2. Soldatov, A.A.: Increasing sdvigoustoychivosti and durability of road surfaces by using highly screenings of crushing claydite in asphalt mixtures. Voronezh 19 (2011)
3. Prokopets, V.S., Galdina, V., Undercut, G.A.: Electron microscopic study of the structure of the components on the porous asphalt filler. Ann. Sci. Session Assoc. Asphalt concr.: Sat. Stat. Rep. 68–74 (2012)
4. Gorelysheva, L.A.: Main obtain asphalt materials with desired properties. Proc. Roads Bridges **19**(1), 238–251 (2008)
5. Yadykina, V.: Managing the processes of formation and quality of construction composites-based state of the surface of dispersed materials, 374 p. Publishing House of the DIA, Moscow (2009)
6. Vereshchagin, V.P.: Assessing the level of quality asphalt binder with serobitumnogo. Sci. Technol. Road Sect. **3**, 17–18 (2002)
7. Rybev, I.A.: Asphalt concrete, p. 399. High School, Moscow (1969)
8. Korolev, I.V.: Ways to save bitumen in road construction, p. 149. Transport, Moscow (1986)
9. Zhelezko, T., Zhelezko, E.P.: Structure and properties of asfaltovyazhuschih. Math. Univ. Constr. **3**, 35–42 (1997)
10. Korotayevand, A.P.: Road composites based on particulate expanded perlite. Build. Mater. **5**, 42–44 (2009)

11. Gazaryan, S.O.: Stabilizing additives for crawler-mastic asphalt concrete based on high-fine porous mineral materials. Bull. Volgograd State Archit. Constr. Univ. Ser.: Constr. Archit. **1**(78), 149–156 (2020)
12. Ozeryanskaya, V.V.: Analysis of methods for processing and use of by-products of the technological process for producing asphalt concrete component. Saf. Man-Gener. Nat. Syst. **1**, 76–80 (2020)
13. Mavliev, L., Vdovin, E., Stroganov, V., Konovalov, N.: Road cement-mineral materials with granulometric and hydrophobic additives (2021)
14. Vdovin, E.A., Stroganov, V.F.: IOP Conference on Series: Materials Science and Engineering, p. 890 (2020)
15. Nizamov, R.K., Galeev, R.R., Abdrakhmanova, L.A., Khozin, V.G., Naumkina, N.I., Lygina, T.Z.: Stroit. Mater. 18 (2005)
16. Nizamov, R.K., Nagumanova, E.I., Abdrakhmanova, L.A., Khozin, V.G.: Stroit. Mater. 14 (2004)

Modification by Zeolite-Containing Additive the Road-Building Materials Based on Carbonate Crushed Stone-Sand Mixtures and Cements

Evgenii Vdovin(✉), Lenar Mavliev, and Nikita Konovalov

Kazan State University of Architecture and Engineering, 420043 Kazan, Russia

Abstract. Carbonate stone materials are one of the most widespread rocks. Due to insufficient strength and frost resistance, their usage in road construction is limited. This problem can be solved by treatment of carbonate stone materials and their mixtures with soils by binders and modifiers. A promising direction for expanding the application field of carbonate crushed stone-sand mixtures in road construction is treatment by cement modified by zeolite-containing additive. The zeolite-containing additive usage leads to increase the compressive strength, flexural tensile strength and frost resistance coefficient of strengthened materials. The usage of crushed stone-sand mixture treated by cement and modified zeolite-containing additive in upper layer of road pavement base and the soil strengthened by cement in lower layer leads to achieve the greatest economic effect.

Keywords: Crushed stone-sand mixture · Carbonate crushed stone · Soils · Portland cement · Zeolite-containing additive · Physical and mechanical properties · Economic efficiency

1 Introduction

One of the promising directions in road building materials development is the modification of local mineral raw materials by different additives [1–5]. This work is devoted to the study of zeolite-containing additives effect on physical and mechanical properties of carbonate crushed stone-sand mixtures treated by cement, and its economic efficiency assessment. In road construction industry, the transporting materials take a large proportion of financial costs. It is possible to abandon the imported stone materials usage by local mineral raw materials strengthened and treated by cement in road pavements construction [6–8]. In many countries there are crushed carbonate deposits, the demand of which is constantly decreasing due to low strength and frost resistance [9]. Today various ways of its usage have been proposed [10–12], including in road construction [13]. In this regard, imparting the required physical and mechanical properties to carbonate crushed stone is the actual task.

Many scientific works are devoted to increasing of strength and frost resistance of carbonate crushed stone-sand mixtures. The method of preparing mixtures treated by

N. Vatin (Ed.): STCCE 2021, LNCE 169, pp. 308–317, 2021.
https://doi.org/10.1007/978-3-030-80103-8_33

organic binder [14], sulfur-containing aqueous solution of calcium polysulfide [15], phosphogypsum [16] and others is known. These methods of increasing the physical and mechanical characteristics of crushed stone-sand mixtures are characterized by high energy consumption of impregnation technological process, as well as the fragility of obtained modified material. The particular interest is the treatment of carbonate stone material by cement mortar, due to the low cost and technology simplicity [17–19].

It should be noted that one of the most promising way to improve the cement quality is the various additives introduction into its composition, which actively affect the cement stone structure formation and properties [20]. Natural zeolite rocks occupy a special place among numerous additives for cements. [21–26]. Zeolites are referred to as pozzolanic additives due to SiO_2 and Al_2O_3 presence, which react with Portlandite during the Portland cement hydration, form hydrosilicates and calcium aluminates. It improves the cement stone microstructure. It is noted that the cement content increase in mixture increases the using zeolite efficiency [27]. In this regard, their usage as modifiers can be effective in cement-soil mixtures, because they usually have higher cement content. In cement-soil mixtures, the cement addition leads to increase the maximum density and optimal moisture content, however, the zeolite content increase acts in opposite way [28]. Only the optimal additives dosage including zeolite additives at processing and strengthening local mineral raw materials by cement, has the most effect on physical and mechanical characteristics of materials obtained.

The aim of the work is the modification by zeolite-containing additives the carbonate crushed stone-sand mixtures treated by cement, for the road building materials production.

2 Materials and Methods

Carbonate crushed stone was used as a starting material in the study, from which a crushed stone-sand mixture of optimal granulometric composition was prepared.

For the preparation of crushed stone-sand mixture, treated by an inorganic binder, Portland cement CEM I 42,5N was used. The mineralogical composition of Portland cement is represented by the following minerals, %: C_3S – 56.8 ± 3.0; C_2S – 19.0 ± 2.0; C_3A – 7.0 ± 0.5; C_4AF – 12.9 ± 1.0. The specific surface area was in the range of 320,0–360,0 m^2/kg. The content of Portland cement was 4.00; 6.00; and 8.00% of total mass.

The modification was carried out by a natural active mineral pozzolanic additive – a zeolite-containing rock. It is a ground rock, in the form of a light-colored powder. The rock mineral composition is represented by clinoptilolite – 14–28%, calcite – 18–23%, free silica – 24–30% and clay minerals – 24–30%. Density is 2.2–2.3 g/cm^3. The pH value is 6.8–7.2. Ion exchange capacity is not less than 1.23 mg-eq/g. The additive content was 1.25%, 2.50%, 3.75%, 5.00% of total weight.

The compressive strength of crushed stone-sand mixture treated by cement was determined on samples of 10.0 × 10.0 × 10.0 cm water-saturated for 2 days.

The bending tensile strength was determined on samples with a size of 10.0 × 10.0 × 40.0 cm water-saturated for 2 days.

Frost resistance was studied on cubes with an edge of 10.0 cm after 28 days of normal hardening storage. Freezing time was at least 2.5 h at temperature (−18 ± 2) °C, thawing time for 2.0 ± 0.5 h in water at a temperature of (20 ± 2) °C. The frost resistance coefficient was determined as the ratio of the sample strength after 15 freeze-thaw cycles to the sample strength before testing.

The cement and zeolite-containing additive effect on the strength and frost resistance of cement-crushed stone mixtures was studied using the method of experiment planning according to V. Kleiman contour-graphic scheme, which we used in our research earlier [29]. The content of cement and zeolite-containing additives (cement, zeolite) was changed in accordance with the plan of two-factor experiment (Fig. 1). Based on test results on contour-graphic models of response function, isolines of equal values of strength and frost resistance coefficient were constructed (Fig. 2, 3 and 4).

a)

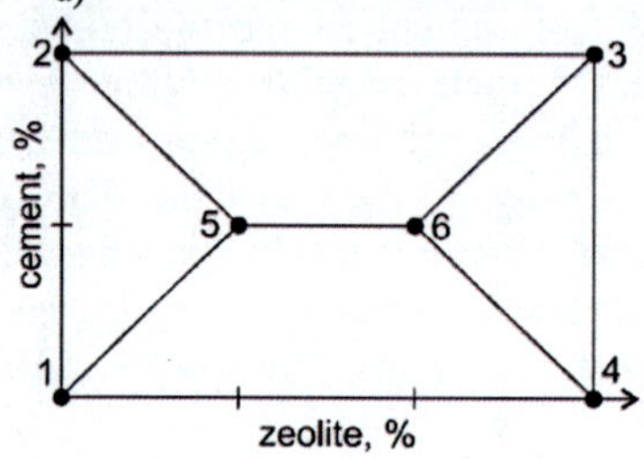

b)

№ experiment	Additives content, %		Test results		
	Cement	Zeolite	R_c	R_f	K_f
1	4.00	1.25	R_{c1}	R_{f1}	K_{f1}
2	8.00	1.25	R_{c2}	R_{f2}	K_{f2}
3	8.00	5.00	R_{c3}	R_{f3}	K_{f3}
4	4.00	5.00	R_{c4}	R_{f4}	K_{f4}
5	6.00	2.50	R_{c5}	R_{f5}	K_{f5}
6	6.00	3.75	R_{c6}	R_{f6}	K_{f6}

Fig. 1. Experiment planning: a) contour-graphic scheme of experiment planning; b) experimental design.

The design of pavement structures was carried out according to permissible elastic deflection, shear resistance, bending resistance, static load, frost resistance and drainage layer. The reliability factor for the IV category is 0.80, for the V category is 0.70. The elastic modulus for IV category is 150.0 MPa, for V category is 50.0 MPa.

The device cost of road pavements designed structures was determined by a local resource estimate. Economic efficiency was determined by comparing the estimated cost of cheapest pavement construction with the rest options, separately for IV and V roads categories.

3 Results and Discussion

Taking into account the work purpose and the experiment plan, the results of indicators changes of physical and mechanical characteristics of crushed stone-sand mixtures treated by Portland cement with a content of 4.00 were obtained; 6.00; 8.00% by mixture weight, which are the control material parameters without modifier (Table 1).

The cement content increase from 4.00% to 6.00% and 8.00% at processing a crushed stone-sand mixture leads to increase the compressive strength by 31.74–90.25%, the bending tensile strength by 38.04–101.09% and the coefficient of frost resistance by 60.0–92.0%.

Table 1. Physical and mechanical characteristics of crushed stone-sand mixtures treated by Portland cement.

Portland cement content, %	Characteristics		
	Compressive strength, MPa	Flexural tensile strength, MPa	Frost resistance coefficient
4.00	4.41	0.92	0.50
6.00	5.81	1.27	0.80
8.00	8.39	1.85	0.96

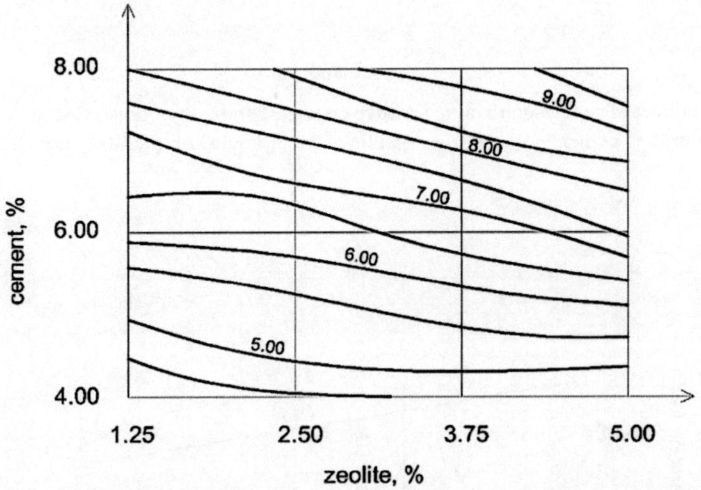

Fig. 2. The effect of zeolite-containing additive on compressive strength of crushed stone-sand mixture treated by Portland cement: ——— isolines the ultimate compressive strength (MPa).

When the Portland cement content is 4.00% and a zeolite-containing additive is 1.25%, the compressive strength does not increase (Table 1, Fig. 2). However, the further increase of the Portland cement and zeolite-containing additive content leads to increase of compressive strength. In comparison with the control compositions of the material with the Portland cement content of 6.00% and a zeolite-containing additive of 2.50%, 3.75%, there is an increase of compressive strength by 6.54% and 17.38%, respectively. With a Portland cement content of 8.00% and a zeolite-containing additive of 5.00%, the indicator increases by 22.17%.

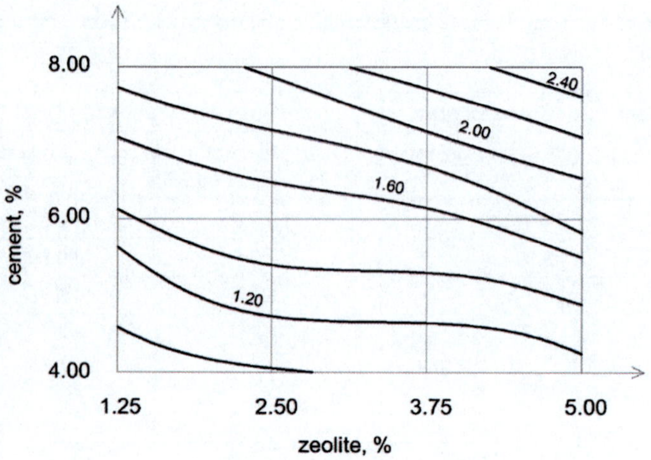

Fig. 3. The effect of zeolite-containing additive on tensile strength of crushed stone-sand mixture treated by Portland cement: ———— isolines the ultimate compressive strength (MPa).

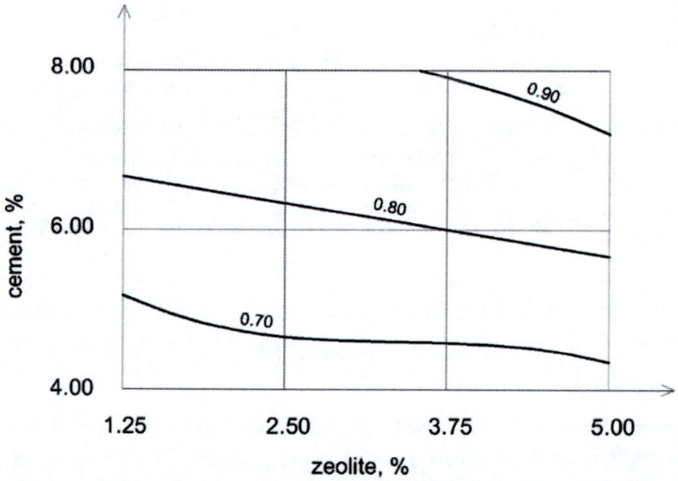

Fig. 4. The effect of zeolite-containing additive on frost resistance of crushed stone-sand mixture treated by Portland cement: ———— isolines the ultimate compressive strength (MPa).

It was found that with the Portland cement content of 4.00% and zeolite-containing additive of 1.25%, the tensile strength also does not change (Table 1, Fig. 3). With the Portland cement content of 6.00% and zeolite-containing additive of 2.50%, 3.75%, the strength increase is observed by 20.47% and 23.62%. With the Portland cement content of 8.00% and zeolite-containing additive of 5.00%, the indicator increases by 38.38%.

The frost resistance coefficient increases with the Portland cement content of 4.00% and zeolite-containing additives 1.25% is 28.00% in comparison with the control samples

(Table 1, Fig. 4). However, with the content of Portland cement increase from 4.00% up to 8.00% and zeolite-containing additive from 2.50% up to 5.00%, significant changes in frost resistance coefficient are not observed.

The economic efficiency of crushed stone-sand mixture treated by cement and modified zeolite-containing additive is assessed on four designed constructions option under equal conditions and loads (Fig. 5).

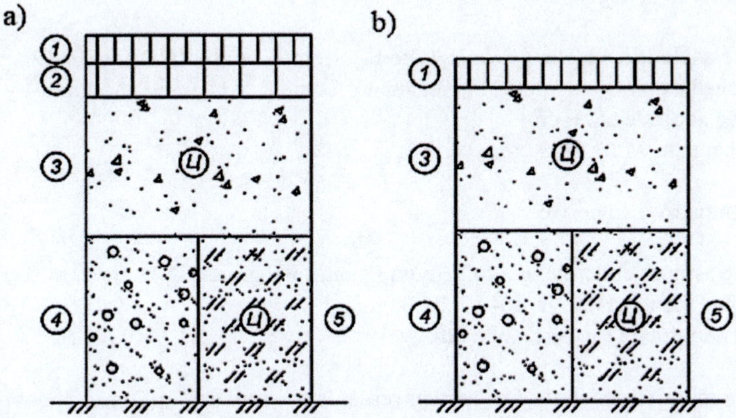

Fig. 5. Pavement constructions with different base layer options: a – IV category; b – V category.

I option of pavement construction (IV category) consists of:

1 – Dense asphalt concrete, grade II, type B, bitumen grade BND-50/70, 5.0 cm;
2 – Coarse porous asphalt concrete, grade II, bitumen grade BND-50/70, 6.0 cm;
3 – Crushed stone-sand mixture, treated by cement, M40 and modified by zeolite-containing additive, 24.0 cm;
4 – Sand and gravel mixture, 30.0 cm.

Option II (IV category):

1 – Dense asphalt concrete, grade II, type B, bitumen grade BND-50/70, 5.0 cm;
2 – Coarse porous asphalt concrete, grade II, bitumen grade BND-50/70, 6.0 cm;
3 – Crushed stone-sand mixture, treated by cement, M40 and modified by zeolite-containing additive, 23.0 cm;
5 – Soil-cement, M20, 30.0 cm.

III option (IV category):

1 – Dense asphalt concrete, grade II, type B, bitumen grade BND-50/70, 5.0 cm;
2 – Coarse porous asphalt concrete, grade II, bitumen grade BND-50/70, 6.0 cm;
3 – Crushed stone M400, fr. 40–70, 24.0 cm;
4 – Sand and gravel mixture, 30.0 cm.

IV option (V category):

1 – Dense asphalt concrete, grade II, type B, bitumen grade BND-50/70, 4.0 cm;
3 – Crushed stone-sand mixture, treated by cement, M40 and modified by zeolite-containing additive, 21.0 cm;
4 – Sand and gravel mixture, 29.0 cm.

V option (V category):

1 – Dense asphalt concrete, grade II, type B, bitumen grade BND-50/70, 4 cm;
3 – Crushed stone-sand mixture, treated by cement, M40 and modified by zeolite-containing additive, 20.0 cm;
5 – Soil-cement, M20, 29.0 cm.

VI option (V category):

1 – Dense asphalt concrete, grade II, type B, bitumen grade BND-50/70, 4.0 cm;
3 – Crushed stone M400, fr. 40–70, 21.0 cm;
4 – Sand and gravel mixture, 29.0 cm.

The comparison cost of developed pavements construction options is shown in Fig. 5 (Fig. 6).

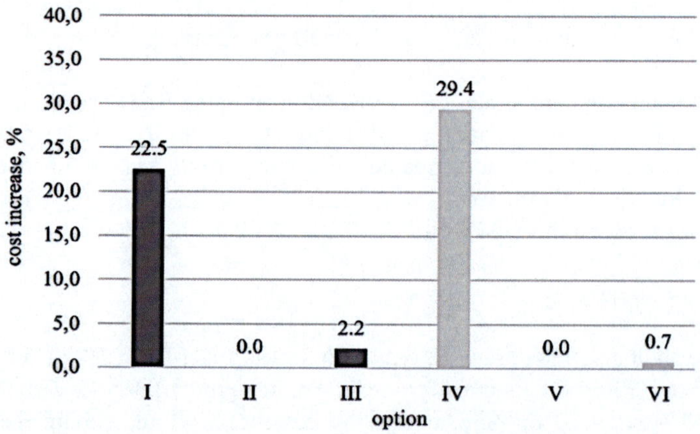

Fig. 6. Cost comparison of road pavements device (I-III option for the IV road category, IV-VI for the V road category).

The cost calculation of designed road pavements structures device showed that the greatest economic effect is achieved for IV road category with the II option of road pavement construction with usage of processed crushed stone-sand mixture and strengthened soil (cost reduction is 2.2–22.5%), for V category with the V option with similar materials (cost reduction is 0.7–29.4%).

In accordance with the scientific works results [20–23, 25–28], the zeolite-containing additive introduction improves the physical and mechanical characteristics of carbonate crushed stone-sand mixture treated by Portland cement due to silica (SiO_2) and aluminum oxide (Al_2O_3) active centers in zeolites.

4 Conclusions

1. The effect of zeolite-containing additive on physical and mechanical properties of crushed stone-sand mixture treated by Portland cement was investigated. The zeolite-containing modifier in content of 2.50 to 5.00% is effectively used and increases the strength when the Portland cement content was more than 6.00%. When cement content is lower (4.00%), the positive modification result was insignificant and the resulting material did not match the requirements for pavement materials.
2. It was established that the modification of crushed stone-sand mixture provided the maximum increase of compressive strength by 29.78% when the Portland cement content was 6.00% and zeolite-containing additive was 5.00%. Flexural tensile strength is maximally increased by 38.38% when the Portland cement content was 8.00% and zeolite-containing modifier was 5.00%.
3. It was found that the zeolite-containing additive introduction significantly affects the frost resistance coefficient. When the Portland cement content is 4.00%, the increase was up to 38.00%. When the Portland cement content is 6.00% and 8.00%, the frost resistance coefficient was not increased significant.
4. The greatest economic effect for IV and V road categories is achieved at option of road pavement construction with usage of treated crushed stone-sand mixture and strengthened soil. The cost reduction amounted to 22.50% and to 29.40%, respectively.
5. The improvement of physical and mechanical characteristics of carbonate crushed stone-sand mixture treated by Portland cement and modified zeolite-containing additive may be associated with the action of silica (SiO_2) and aluminum oxide (Al_2O_3) active centers in zeolites. In this regard, this is the urgent topic for further research.

References

1. Van Quan, H., Van Tuoi, N., Nguyen, C.V.: Effect of fly ash on the mechanical properties and drying shrinkage of the cement treated aggregate crushed stone. In: Ha-Minh, C., Dao, D., Benboudjema, F., Derrible, S., Huynh, D., Tang, A. (eds.) CIGOS 2019, Innovation for Sustainable Infrastructure. LNCE, vol. 54, pp. 421–426. Springer, Heidelberg (2020). https://doi.org/10.1007/978-981-15-0802-8_65
2. Rakhimova, N.R., Rakhimov, R.Z.: Reaction products, structure and properties of alkali-activated metakaolin cements incorporated with supplementary materials - a review (2019). https://doi.org/10.1016/j.jmrt.2018.07.006.
3. Mukhametrakhimov, R., Galautdinov, A., Lukmanova, L.: Influence of active mineral additives on the basic properties of the gypsum cementpozzolan binder for the manufacture of building products. In: MATEC Web of Conferences (2017). https://doi.org/10.1051/mateconf/201710603012

4. Khozin, V., Khokhryakov, O., Nizamov, R.: A «carbon footprint» of low water demand cements and cement-based concrete. In: IOP Conference Series: Materials Science and Engineering. Institute of Physics Publishing (2020). https://doi.org/10.1088/1757-899X/890/1/012105
5. Makarenko, S.V., L.B.M., K.O.V., K.V.G.: The effect of active pozzolanic fillers on the properties of fine-grained cement concrete. Izv. KGASU, pp. 39–46 (2020)
6. Mavliev, L., Vdovin, E.: Structure of road soil cement compositions modified by complex additive based on organosilicon compounds and electrolytes. In: E3S Web Conference, vol. 140 (2019). https://doi.org/10.1051/e3sconf/201914002016
7. Bulanov, P.E., Vdovin, E.A., Mavliev, L.F., Kuznetsov, D.A.: Development of road soil cement compositions modified with complex additive based on polycarboxylic ether. In: IOP Conference Series: Materials Science and Engineering (2018). https://doi.org/10.1088/1757-899X/327/3/032014
8. Vdovin, E., Mavliev, L., Stroganov, V.: Interaction of clay soil components with portland cement and complex additive based on octyltriethoxysilane and sodium hydroxide. In: IOP Conference Series: Materials Science and Engineering (2020). https://doi.org/10.1088/1757-899X/890/1/012031
9. Kameneva, E.E., Kamenev, E.V.: Manufacture of road stone and respective quality requirements in the pre-revolutionary Russia. Obogashchenie Rud. 2020 (2020). https://doi.org/10.17580/or.2020.05.08
10. Antonius, F., Susilo, A.J.: Analisis cara peningkatan daya dukung fondasi dangkal pada konstruksi gedung bertingkat. JMTS J. Mitra Tek. Sipil. **3** (2020). https://doi.org/10.24912/jmts.v3i3.8278
11. Kibriya, T., Tahir, L.: Sustainable construction – high performance concrete containing limestone dust as filler. World J. Eng. Technol. **05** (2017). https://doi.org/10.4236/wjet.2017.53034
12. El-Anwar, E.A.A.: Some carbonate rocks utilized as a building material rocks, Egypt. Carbon. Evapor. **34** (2019). https://doi.org/10.1007/s13146-018-0423-4
13. Kaźmierczak, U., Blachowski, J., Górniak-Zimroz, J., Wirth, H.: Quantitative and qualitative research on the waste from the mining of rock raw materials in Lower Silesia. Minerals **8** (2018). https://doi.org/10.3390/min8090375
14. Burenina, O.N., Savvinova, M.E., Andreeva, A.V.: The effectiveness of using bitumen emulsion in the technology of dedusting roads with a transitional type of coating. In: Materials Science Forum (2020). https://doi.org/10.4028/www.scientific.net/MSF.992.9
15. Massalimov, I.A., Chuykin, A.E., Massalimov, B.I., Urakaev, F.H., Uralbekov, B.M.: Improvement of operational properties of shell limestone building materials by polysulfide solution impregnation. Nanotechnol. Constr. A Sci. Internet-J. **9** (2017). https://doi.org/10.15828/2075-8545-2017-9-3-66-80.
16. Zhang, H., Cheng, Y., Yang, L., Song, W.: Modification of lime-fly ash-crushed stone with phosphogypsum for road base. Adv. Civ. Eng. **2020** (2020). https://doi.org/10.1155/2020/8820522
17. Strazzeri, V., Karrech, A., Elchalakani, M.: Mathematical models of homogenization for a rammed earth blend made of crushed limestone and cement in linear micro-poro-elasticity. In: MATEC Web Conference, vol. 303 (2019). https://doi.org/10.1051/matecconf/201930301003
18. Fedrigo, W., Núñez, W.P., Kleinert, T.R., Matuella, M.F., Ceratti, J.A.P.: Strength, shrinkage, erodibility and capillary flow characteristics of cement-treated recycled pavement materials. Int. J. Pavement Res. Technol. **10** (2017). https://doi.org/10.1016/j.ijprt.2017.06.001
19. Prado, K., Albuquerque, F.S., de Oliveira Almeida, G.B.: Influence of granulometry, cement content, and moisture in the mechanical performance of the cemented crushed stone. Transportes. **28** (2020). https://doi.org/10.14295/transportes.v28i1.1861

20. Skhvitaridze, R., et al.: Scientific principles and practice in the use of natural mineral and industrial resources in the Georgian cement industry. Cem. Int. **16** (2018)
21. Mariri, M., Ziaie Moayed, R., Kordnaeij, A.: Stress-strain behavior of loess soil stabilized with cement, zeolite, and recycled polyester fiber. J. Mater. Civ. Eng. **31**, 04019291 (2019). https://doi.org/10.1061/(asce)mt.1943-5533.0002952
22. Rakhimova, N.R., Rakhimov, R.Z., Morozov, V.P., Eskin, A.A.: Calcined low-grade clays as sources for zeolite containing material. Period. Polytech. Civ. Eng. **65**, 204–214 (2021). https://doi.org/10.3311/PPci.14084
23. Linek, M., Zebrowski, W., Wolka, P.: The application of natural zeolite to cement concrete intended for communications surfaces. In: E3S Web of Conferences (2018). https://doi.org/10.1051/e3sconf/20184500047
24. Pushkareva, K.K., Gonchar, O.A., Kaverin, K.O.: The role of the crystallo-chemical factor in the evaluation and improvement of the nanomodification efficiency of mortar and concrete. In: IOP Conference Series: Materials Science and Engineering (2019). https://doi.org/10.1088/1757-899X/708/1/012102
25. Girskas, G., Skripkiūnas, G.: The effect of synthetic zeolite on hardened cement paste microstructure and freeze-thaw durability of concrete. Constr. Build. Mater. **142**, 117–127 (2017). https://doi.org/10.1016/j.conbuildmat.2017.03.056
26. Małek, M., Jackowski, M., Zyciński, W., Wachowski, M.: Characterization of new filler additions affecting the mechanical strength of concrete. Mater. Tehnol. **53**, 399–403 (2019). https://doi.org/10.17222/mit.2018.155
27. Mola-Abasi, H., Khajeh, A., Naderi Semsani, S.: Variables controlling tensile strength of stabilized sand with cement and zeolite. J. Adhes. Sci. Technol. **32**, 947–962 (2018). https://doi.org/10.1080/01694243.2017.1388052
28. Ahmadi Chenarboni, H., Hamid Lajevardi, S., MolaAbasi, H., Zeighami, E.: The effect of zeolite and cement stabilization on the mechanical behavior of expansive soils. Constr. Build. Mater. **272** (2021). https://doi.org/10.1016/j.conbuildmat.2020.121630.
29. Vdovin, E.A., Stroganov, V.F.: Properties of cement-bound mixes depending on technological factors. Mag. Civ. Eng. **93**, 147–155 (2020). https://doi.org/10.18720/MCE.93.12

Coupling Agents Based on Single-Walled CNTs for Polyvinylchloride Wood-Polymer Composites

Ayaz Khantimirov(✉), Lyaylya Abdrakhmanova, and Vadim Khozinv

Kazan State University of Architecture and Engineering, 1, Zelenaya Street, Kazan 420043, Russia

Abstract. In this work the modification of wood-filled composites based on polyvinylchloride with single-walled carbon nanotubes was carried out. To increase the degree of interaction between the organic filler and the polymer the studies on the application of small doses of single-walled carbon nanotubes as coupling agents in the composition of wood-polymer composites were carried out. The influence of nanomodifier on technological and operational properties of rigid wood-polymer composites was estimated, namely, an increase in tensile strength by 13%, the melt flow rate limit by 10 times, and thermal stability by 84 min.

Keywords: Polyvinylchloride · Wood-polymer composites · Coupling agent · Carbon nanotubes · Profile mouldings

1 Introduction

Wood-polymer composites (WPC) are currently one of the most demanded composite materials, which gained popularity due to the combination of the positive characteristics of both wood and polymer resins. For profiled moldings as an organic filler of WPC the most interesting is wood flour. These products, unlike natural wood, have a low value of water absorption, increased biopersistence, weathering resistance and isotropic physical and mechanical properties [1, 2]. For WPC as polymeric matrices such thermoplastics as polyethylene (PE), polypropylene (PP) and polyvinylchloride (PVC) are most often used. Among them, the most advantageous in terms of performance properties is PVC [3–6].

Increasing the content of wood flour in PVC-based WPC leads to deterioration of their properties, which is caused by low adhesion between the hydrophilic wood particles and the hydrophobic polymer. This problem is usually solved by introducing effective coupling agents into composites. The most common coupling agents nowadays are nanoscale additives, especially those with a carbon base. Such additives include graphene structures, fullerenes, nanoscale technical carbon, carbon nanotubes (single- and multiwalled), etc.

Various polymer composites modified with nanoscale technical carbon have been obtained in works [7–15]. For example, composites based on high-density polyethylene and technical carbon with a particle size of 20–35 nm are known [7]. Significant

N. Vatin (Ed.): STCCE 2021, LNCE 169, pp. 318–324, 2021.
https://doi.org/10.1007/978-3-030-80103-8_34

improvements in tensile strength, tensile strain, electrical conductivity, and melt flow rate have been observed in the range of 12–18% W.

The influence of carbon nanotubes on the properties of PVC composites is also well studied [16–23]. Nanomodified composites had improved electrical, physical and mechanical, rheological properties, structure and durability.

However, due to the high specific surface area, the aforementioned coupling agents are characterized by a tendency to aggregation, which prevents their uniform distribution over the composite volume. Therefore, the efficiency of a particular additive depends not only on the structure of the carbon skeleton, but also on the carrier medium in which these nanomodifiers can be distributed [24–27], as well as on the presence of functionalization of their surface. For example, in [24], carbon nanotubes were first mechanically grinded in water with the addition of a water-soluble polymer and dispersion by ultrasound. Then, the resulting suspension was applied to the surface of polyolefin granules and subjected to drying. The resulting composites have high electrical conductivity, thermal conductivity and stiffness. Tensile modulus of elasticity increased by 50%, tensile strength - by 30%.

The effect of superconcentrate of single-walled carbon nanotubes (SWCNTs) in fatty carboxylic acids [26] on the properties of thermosetting polymers, in particular those based on epoxy resins, was studied. The properties of the modified samples were distinguished by increased values of strength and Young's modulus.

In [27] dispersions of SWCNTs in water and in fatty carboxylic acid esters in the amount from 0.001 to 0.007 m.p. were introduced into PVC composites. The composites nanomodified by aqueous dispersions of SWCNTs showed an increased thermal stability; the samples with SWCNTs in fatty acids were characterized by a decrease in melt viscosity. The strength properties remained at the level of the original composition.

In this work, a superconcentrate of SWCNTs in fatty carboxylic acids esters was used as a coupling agent in PVC-based WPCs, which had a positive effect on the properties of various polymer composites [24, 27]. This choice is also due to the fact that fatty carboxylic acids in the production of BPC on the basis of PVC are often used as plasticizers, which should also have a positive effect on the recyclability of compositions.

The aim of the work is to study the effect of single-walled carbon nanotubes distributed in fatty carboxylic acid esters on the technological and operational properties of wood-polymer composites based on PVC.

2 Materials and Methods

In our research we used suspension PVC of C-7059-M brand (GOST 14332-78), a complex stabilizer - two-base lead stearate, a stabilizer-lubricant - calcium stearate (TU 6-09-4104-87), an acrylic impact strength modifier KaneAce FM-50, wood flour M180 (GOST 16361-87). Often, to facilitate the recyclability of wood-polymer composites based on PVC a small amount of molecular plasticizers is introduced into the mixture, which does not affect the strength properties. Dioctyl phthalate in an amount of 10 m.p. was used as a plasticizer (GOST 8728-88).

TUBALL™ MATRIX 202 single-walled carbon nanotubes from OCSiAl LLC (Novosibirsk, Russia) were used as a nanomodifier. They represent a superconcentrate of single-walled carbon nanotubes with 10% by mass in derivatives of esters of fatty carboxylic acids. The average outer diameter of the tubes is 1.8 nm, the length of the tubes is not less than 5 μm, the specific surface area is 500 m^2/g.

To study the influence of SWCNTs on the properties of WPCs, compositions are presented in Table 1.

Table 1. Components of experimental compositions

Component	Concentration, m.p									
PVC	100	100	100	100	100	100	100	100	100	100
Two-base lead stearate	5	5	5	5	5	5	5	5	5	5
Calcium stearate	3	3	3	3	3	3	3	3	3	3
Impact strength modifier FM-50	7	7	7	7	7	7	7	7	7	7
Wood flour	50	50	50	50	50	50	50	50	50	50
Dioctyl phthalate	10	10	10	10	10	10	10	10	10	10
SWCNT	0	0.01	0.02	0.03	0.04	0.05	0.07	0.1	0.3	0.5

The introduction of the modifier was carried out by preparing a premix with 0.1 and 0.5 m.p.h. carbon nanotubes in PVC by mixing in a RetschPM 100 CM laboratory planetary ball mill. The resulting premixes were mixed to the required concentrations of SWCNTs with the initial PVC composition. The compositions were mixed in a laboratory mixer LDU-3 MPR with a propeller nozzle for 4–5 min at 700 rpm.

Film and extruded samples in the form of bundles were obtained for research of PVC compositions. Film samples were prepared by thermoplasticizing method on laboratory rollers LB 200 100/100E at temperature of rollers 140…150 °C for 5 min.

Laboratory twin-screw extruder LabTechScientific LTE 16–40 with round die was used to obtain samples in the form of bundles. All samples were molded at the same extruder zone temperature profile with a screw rotation speed of 14 rpm. The composition was kept in the extruder barrel for 8–10 min. The bundles had a transverse diameter of 3 mm or more, depending on the swelling coefficient of the samples.

Tensile strength was determined using a PM-250 tensile machine at a tensile speed of 100 mm/min on five prepared samples 150 × 15 mm in size.

The average density of the samples was determined by measuring and weighing.

The melt flow rate i (MFR) was determined on a Franка plastometer at 190 °C and a load of 21.6 g. The thermostability time τ was determined by the method "кongo red".

Water absorption W was evaluated by estimating the change in mass of the sample immersed for 24 h in distilled water. Samples were weighed on analytical scales ADV-200M.

3 Results

In the experimental studies different amounts of superconcentrated SWCNT in esters of fatty carboxylic acids were introduced into WPCs.

To determine the optimal concentrations of SWCNTs, at which, first of all, a significant improvement of the technological properties of composites is observed, such parameters as the melt flow rate and thermal stability of film samples were determined. For compositions with optimal concentrations of SWCNTs extruded samples were made, on which the performance characteristics were determined.

The test results are presented in Table 2.

Table 2. Characteristics of nanomodified wood-polymer composites

Parameters	Concentration of SWCNTs, m.p		
	0	0,01	0,1
Tensile strength, MPa	30	29	34
Melt flow rate, g/10 min	0.4	1.5	4.0
Thermal stability, min	136	210	220
Swelling factor, %	1.08	1.10	1.05
Water absorption, %	4.1	3.5	2.5
Average density, g/cm^3	1.015	1.03	1.15

According to the complex of properties the best indicators are observed for WPC with the content of SWCNT in the amount of 0.1 m.p. The growth of strength and density by 10–15% is observed. There is a significant increase in thermal stability and melt flow rate. Nanomodified composites also have lower water absorption and swelling coefficient, which is especially valuable for extrusion of WPC products.

4 Discussion

From the presented data it is possible to draw conclusions that at use of a superconcentrate of nanotubes, obviously, at small concentration (0,01 m.p.) their uniform distribution in a matrix of a composite is not reached, therefore reduction of strength characteristics is observed. This fact is illustrated by the data of optical microphotographs of the composite sections with 0.01 m.p. SWCNTs (Fig. 1), showing the separation of SWCNTs in the form of individual inclusions of different sizes. When the concentration is increased (up to 0.1 m.p.), it is evident that the nanotubes are evenly distributed throughout the matrix volume. This leads to an ordering of the structure in the amorphous phase of the polymer, which causes an increase in strength, average density. The swelling of the extrudate jet decreases with increasing concentration of SWCNTs, which indicates a decrease in the contribution of highly elastic deformations. These effects also explain

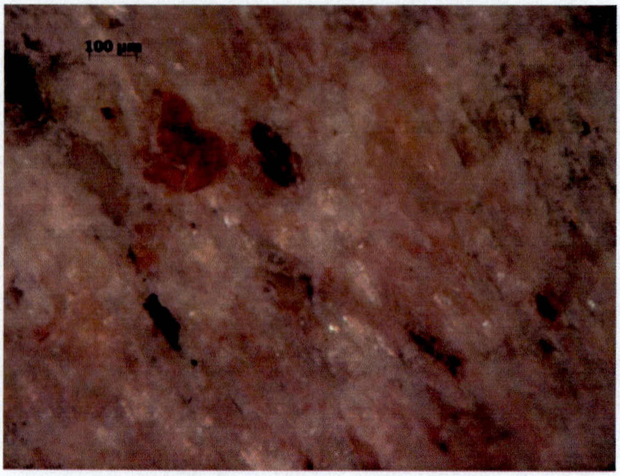

Fig. 1. Microphotograph of a composite section with 0.01 m.p. SWCNTs

the decrease in water absorption with increasing concentration of SWCNTs, which is especially important for wood-filled polymer composites.

As a rule, an increase in melt viscosity is observed with increasing wood flour content in WPC. MFR for wood-filled compositions, equal to 0.4 g per 10 min, is almost five times lower than the fluidity of unfilled PVC compositions. At the introduction of carbon nanotubes their interaction with macromolecules of polymer or filler has weak van der Waals nature, nanotubes weakly influence the mechanical properties, but with the increase of SWCNT concentration in WPC composition MFR grows (at the content of 0.1 m.p. SWCNT fluidity index increases by one order). This fact indicates a decrease in melt viscosity, which positively affects the quality of the extrudate.

The thermostabilizing effect of nanotubes in thermoplastics is due to their interference with the transfer of products of polymer degradation and chemical interaction of CNTs with macroradicals formed in the process of polymer degradation with the formation of stable radicals [28, 29]. In PVC compositions to a greater extent realized the possibility of sorption by carbon nanotubes of hydrogen chloride, released during polymer degradation, and organic acids that are released when heating the wood flour and are catalysts of PVC decomposition, so their negative effect is negligible. The radical molecule penetrates inside the nanotube under the capillary effect and is retained inside due to sorption forces [30].

5 Conclusion

1. The introduction of superconcentrates of SWCNTs in esters of fatty carboxylic acids as a coupling agent in the compositions of wood-polymer compositions based on PVC was carried out. The use of stable highly concentrated SWCNTs dispersions in plasticizers for PVC allowed to exclude the operation of wood flour pretreatment with aqueous SWCNTs dispersions with its subsequent drying.

2. Introduction of SWCNTs into extruded samples in an amount of 0.1 mass part improves technological and operational properties of WPC composition, namely, it increases strength by 13% and density by 13%, decreases water absorption value by 1.5%, increases thermal stability time by 84 min. The main advantage of such nanomodification is easier recyclability of the wood-filled composition, which is confirmed by a significant increase in MFR and this effect is due to the fact that carbon nanotubes, on which molecules of plasticizer are adsorbed, are the medium that provides the occurrence of layer-by-layer flow on the interface PVC-wood flour.

References

1. Klyosov, A.A.: Wood-plastic composites (2007). https://doi.org/10.1002/9780470165935
2. Wilkes, C.E., Summers, J.W., Daniels, C.A.: PVC Handbook (2007)
3. Matseevich, T., Askadskiy, A.: Mechanical properties of decking on the base of polyethylene, polypropylene and polyvinylchloride. Stroit. Stvo Nauk. i Obraz. **3**, 49–59 (2017). https://doi.org/10.22227/2305-5502.2017.3.4
4. Fayzullin, I.Z., Volfson, S.I., Musin, I.N., Fayzullin, A.Z., Nikiforov, A.A.: Influence of the type of wood flour and nanoadditives on the structure and mechanical properties of polypropylene-based wood-polymer composites. In: AIP Conference Proceedings, pp. 309–306 (2016). https://doi.org/10.1063/1.4967155
5. Nizamov, R., Abdrakhmanova, L.: Influence of fillers on polyvinyl chloride materials thermal resistance (2016). https://doi.org/10.4028/www.scientific.net/MSF.871.84
6. Galeev, R., Nizamov, R., Abdrakhmanova, L., Khozin, V.: Resource-saving polymer compositions for construction purposes. In: IOP Conference Series: Materials Science and Engineering (2020). https://doi.org/10.1088/1757-899X/890/1/012111
7. Igumenova, T.I., Kleimenova, N.L., Akatov, E.S., Popov, G.V.: Application fullerene containing technical carbon for modification of polyethylene properties. Trans. TSTU **8719**, 1071–1076 (2011)
8. Baccaro, S., Cataldo, F., Cecilia, A., Cemmi, A., Padella, F., Santini, A.: Interaction between reinforce carbon black and polymeric matrix for industrial applications. In: Nuclear Instruments and Methods in Physics Research, pp. 191–194 (2003). https://doi.org/10.1016/S0168-583X(03)00638-4
9. Hatthapanit, K., Sae-Oui, P., Sombatsompop, N., Sirisinha, C.: Enhancement of rubber-carbon black interaction by amine-based modifiers and their effect on viscoelastic and mechanical properties. J. Appl. Polym. Sci. **126**, 315–321 (2012). https://doi.org/10.1002/app.36969
10. Mazlina, M.K.: Influence of fillers surface characteristics on bound rubber properties of filled natural rubber compounds. In: Advanced Materials Research, pp. 412–416 (2014). https://doi.org/10.4028/www.scientific.net/AMR.845.412
11. Borukaev, T.A., Gaev, D.S.: PACS: 61.82.Pv + 72.80.Le Physico-mechanical properties of composites based on high-density polyethylene and Prikl. Fiz., pp. 76–81 (2017)
12. Saatchi, M.M., Shojaei, A.: Effect of carbon-based nanoparticles on the cure characteristics and network structure of styrene-butadiene rubber vulcanizate. Polym. Int. **61**, 664–672 (2012). https://doi.org/10.1002/pi.4132
13. Bulanov, P.E., Vdovin, E.A., Mavliev, L.F., Kuznetsov, D.A.: Development of road soil cement compositions modified with complex additive based on polycarboxylic ether. In: IOP Conference Series: Materials Science and Engineering (2018)

14. Mavliev, L.F., Vdovin, E.A.: Structure of road soil cement compositions modified by complex additive based on organosilicon compounds and electrolytes. In: E3S Web of Conferences 2019 (2019)
15. Vdovin, E.A., Stroganov, V.F.: Optimization of complex frost-resisting additives in line with operational requirements towards freezing temperature in road dressing constructions. In: IOP Conference Series: Materials Science and Engineering, p. 832 (2020)
16. Silvano, L.T., Vittorazzo, J., Atilio, L., Araujo, G.R.: Effect of preparation method on the electrical and mechanical properties of PVC/carbon nanotubes nanocomposites. Mater. Res. **21** (2018). https://doi.org/10.1590/1980-5373-mr-2017-1148
17. Shieh, Y.T., Hsieh, K.C., Cheng, C.C.: Carbon nanotubes stabilize poly (vinyl chloride) against thermal degradation. Polym. Degrad. Stab. **144**, 221–230 (2017). https://doi.org/10.1016/j.polymdegradstab.2017.08.017
18. Aygubova, A.C., Kozlov, G.V., Magomedov, G.M., Zaikov, G.E.: Structural analysis of water sorption for nanocomposites of poly(vinyl chloride)/carbon nanotubes. Sci. J. Volgogr. State Univ. **4**, 40–46 (2015)
19. Pachina, A.N., Kassin A.S ., Tikhonov, N.N., Komissarov, D.M.: Methods of increasing the atmospheric resistance of polyvinylchloride. Uspekhi Him. Him. Tekhnol. 93–95 (2017)
20. Zaporotskova, I.V, Krutoyarov, A.A., Polikarpova, N.P.: Theoretical studies of polymer nanocomposites based on polyethylene, polypropylene and PVC, reinforced with carbon nanotubes. Perspekt. Mater. 5–12 (2015)
21. Ge, S., et al.: Potential use of different kinds of carbon in production of decayed wood plastic composite. Arab. J. Chem. **11**, 838–843 (2018). https://doi.org/10.1016/j.arabjc.2017.12.026
22. Vanyorek, L., et al.: Nanotubes as polymer composite reinforcing additive materials – a comparative study. Arab. J. Chem. (2020). https://doi.org/10.1016/j.arabjc.2019.01.001
23. Skórczewska, K., Tomaszewska, J., Piszczek, K.: Influence of MWCNT on the processing properties and structure of PVC composites. In: Macromolecular Symposia (2018). https://doi.org/10.1002/masy.201600177
24. Amirov, R.R., Nekljudov, S.A., Amirova, L.M.: Method of obtaining compositions based on carbon nanotubes and polyolefins (2011)
25. Gataullin, A.R., Bogdanova, S.A., Rakhmatullina, A.P., Galyametdinov, Y.G.: Dispersion of carbon nanotubes in solutions of oxyethylated isononylphenols. Russ. J. Appl. Chem. 33–34 (2017). https://doi.org/10.1134/S1070427217110118
26. Gusev, K., Solovyev, V.: Effect of carbon nanotubes on the mechanical properties of the polymeric composites. Vestn. Pskov. Gos. Univ. 150–153 (2017)
27. Khozin, V.G., Nizamov, R.K., Abdrakhmanova, L.A.: Modification of construction polymers (Polyvinylchloride and Epoxy) by single-wall carbon nanotubes. Stroit. Mater. 55–61 (2017)
28. Marosfoi, B.B., Szabó, A., Marosi, G., Tabuani, D., Camino, G., Pagliari, S.: Thermal and spectroscopic characterization of polypropylene-carbon nanotube composites. J. Therm. Anal. Calorimet. 669–673 (2006). https://doi.org/10.1007/s10973-006-7891-9
29. Ivanova, S.R., Minsker, K.S., Nagumanova, E.I., Nizamov, R.K., Kazar'yans, S.A.: The stabilizing properties of synthetic zeolites in plasticized polyvinyl chloride compounds. Plast. Massy Sint. Svojstva Pererab. Primen. 39–43 (2005)
30. Eletskiĭ, A.: V: Sorption properties of carbon nanostructures. Uspekhi Fiz. Nauk. **174**, 1191–1231 (2004)

Nanomodified Polymer-Bitumen Binders

Damir Ayupov(✉), Dmitry Makarov, and Rauf Kazakulov

Kazan State University of Architecture and Engineering, 420043 Kazan, Russia

Abstract. A modification of petroleum bitumen for road building was carried out by devulcanizing rubber crumb and adding modified dispersion, consisting of single-walled carbon nanotubes (SWCNT), distributed in industrial oil (I-20A). To solve the problem of distribution of nanotubes in industrial oil I-20A it is necessary to «break» their aggregates using ultrasound [26–28]. The SWCNTs distribution was observed using a HoribaLA-950 laser analyzer and a KFK-3 photoelectric photometer for optical density evaluating. Softening point, penetration index (PI), penetration at 25 °C (P25), ductility at 25 °C (D25) and elasticity at 25 °C (E25) were respectively determined to GOST 32054-2013, GOST 33134-2014, GOST 33136-2014, GOST 33138-2014, GOST EN 13398-2013 (Russian standards). With the introduction of nanotubes, the effect of the dispersed phase on the properties decreases and the presence of nanotubes in dispersion, when introduced after devulcanization, does not give a plasticizing effect. But it simultaneously increases both ductility and elasticity, which is rarely observed. The introduction of nanotubes before devulcanization plasticizes the binder by reducing the thermal distillation of the oil during devulcanization or by the plasticizing effect of nanotubes.

Keywords: Bitumen · Bitumenpolymer binder · Nanomodified · Nano · Bitumen modifications

1 Introduction

It is generally accepted that physical or chemical modification is one of the most effective ways to obtain or improve the properties of almost any composite building material [1–4].

Modifying petroleum bitumen with polymers allows, as it is known from [5–7], eliminating a number of disadvantages associated with binders. The most important of them are a narrow temperature range of plasticity, insufficient resistance to atmospheric aging, and low elasticity [8–10]. However, the high cost of the polymer leads to a strong increase in the cost of the final product.

Earlier [11], a bitumen-polymer composition was developed, obtained by devulcanization of rubber crumb in a bitumen environment, which features a high range of properties and is able to solve the problem of high cost of bitumen-polymer binders (BPB) and environmental pollution. The modification technology was as follows: rubber crumb and a devulcanizing agent (DA) were introduced into the heated petroleum bitumen with constant stirring. The composition, however, was characterized by low elongation. These disadvantages are common to all BPBs [12, 13], which nevertheless

N. Vatin (Ed.): STCCE 2021, LNCE 169, pp. 325–333, 2021.
https://doi.org/10.1007/978-3-030-80103-8_35

limits their application. These disadvantages can be eliminated by introducing plasticizers, but their use leads to a decrease in the softening point and an increase in the tendency to aging due to thermal distillation [14–16]. Moreover, the paradoxical effect of reducing the viscosity of binders (and couplers) when introducing carbon nanomaterials, e.g., in bitumen, is known [17–19].

2 Materials and Methods

To achieve a uniform distribution of nanotubes in a bitumen-polymer binder, it is also necessary to use a plasticizer, but the total amount of it in the binder is much lower. We used I-20A industrial oil as a plasticizer.

It is known from literature [20–22] that when single-walled carbon nanotubes (SWCNT) are added to bitumen, the complex of the obtained properties is higher than when it is modified with multi-walled carbon nanotubes (MWCNT). SWCNTs are characterized by such factors as high strength of sp 2 C–C bonds, high packing density of atoms in graphenes, absence or low density of structural defects [23–25]. We used SWCNT by «Tubal», Novosibirsk. The characteristics of the used SWCNTs are presented in Table 1.

Table 1. Characteristics of Tuball TM (Batch number 73-21052015).

Parameters	Unit	Values	Assessment method
Carbon content	wt. %	92 ± 1	TGA, EDA
CNT content	wt. %	76 ± 1	TEM, TGA
Metallic impurities	wt. %	8 ± 1	EDA, TGA
Number of walls in CNT	units	1–2	TEM
Length	microns	>5	AFM
Outer average diameter of CNT	nm	1.4 ± 0.15	Raman spectroscopy, TEM
G/D ratio	units	161	Raman spectroscopy, 488 nm
Total specific surface	m^2/g	450	Adsorption of N_2 at 77K

Initially, CNTs were introduced into industrial oil, and then the binder was modified with the resulting dispersion. This dispersion, as well as the model system in the form of pure I-20A oil without SWCNTs, were introduced in two ways: either simultaneously with rubber crumb (RC), or after the devulcanization process into the finished bitumen-polymer binder. The optimal concentration of nanotubes in bitumen was considered $5 \cdot 10^{-5}$ ppm, according to [20]. The content of the plasticizer was calculated in such a way that the optimal concentration of nanotubes in the bitumen was achieved at a concentration of plasticizer of 2%, which corresponds to its usual content. The maximum plasticizer concentration was initially limited to four percent.

To solve the problem of distribution of nanotubes in industrial oil I-20A, it is necessary to «break» their aggregates using ultrasound [26–28]. The SWCNTs distribution was observed using a HoribaLA-950 laser analyzer and a KFK-3 photoelectric photometer for optical density evaluating. Softening point, penetration index (PI), penetration at 25 °C (P_{25}), ductility at 25 °C (D_{25}) and elasticity at 25 °C (E_{25}) were respectively determined to GOST 32054-2013, GOST 33134–2014, GOST 33136-2014, GOST 33138-2014, GOST EN 13398-2013 (Russian standards).

3 Results and Discussions

For clarity, the particle size distribution of the dispersion was determined up to (Fig. 1) and after (Fig. 2) sonication.

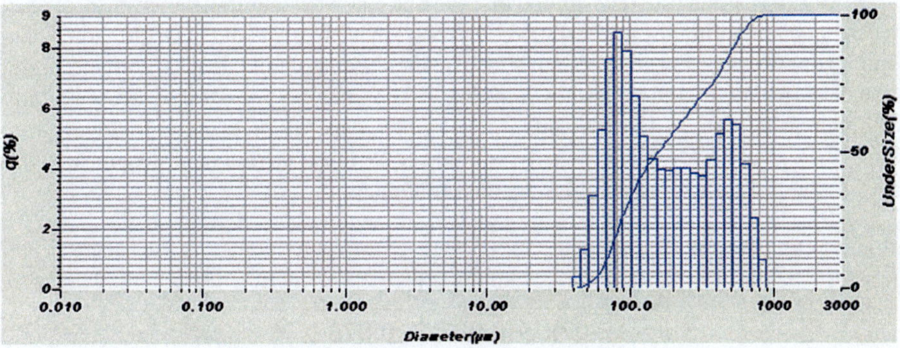

Fig. 1. Particle size before sonication.

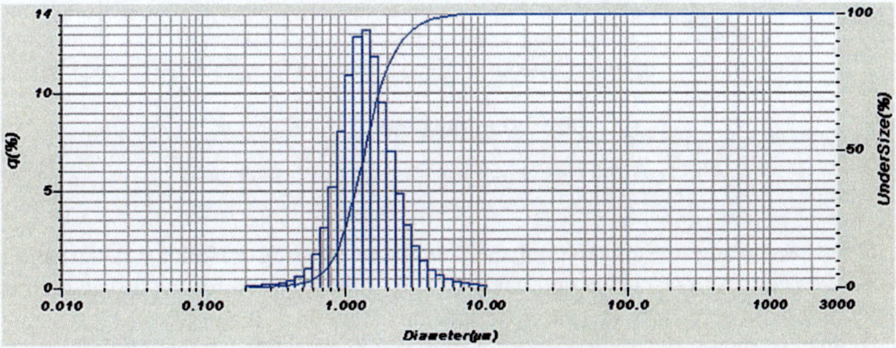

Fig. 2. Particle size after 10 min of sonication.

With dispersion, the histogram shifts to the left, which can be seen when comparing Fig. 1 and 2, i.e., the particle size after sonication became significantly smaller. The optimal dispersion time t was 10 min, since further ultrasonic exposure was not accompanied by a change in the histogram. Furthermore, the absolute particle size, according to the laser analyzer, does not reach the «nano» level, but it must be remembered that the device is designed to determine the size of spherical particles. It should also be noted that the Gaussian distribution in Fig. 2 indicates the presence of one dominant size in the initial SWCNT product, i.e., its purity.

Photoelectric photometer KFK-3 (photometer) is designed to measure the transmittance and optical density of transparent solutions and transparent solid samples [29, 30].

The principle of operation of the photometer is based on comparing the luminous flux F0, passed through the standard solution, in relation to which the measurement is carried out, and the luminous flux F, passed through the investigated solution. The photodetector converts light fluxes F0 and F into electrical signals U0, U and UT (UT signal when the receiver is not illuminated) which are processed by a microcomputer photometer and presented on the digital board as the transmittance (T) and optical density (D).

It was experimentally determined that the optimal wavelength for measuring optical density and transmittance is $\lambda = 430$ nm. Dispersion was carried out in an ultrasonic bath. Optical density was measured every 30 s. The increase in optical density with increasing time of ultrasonic treatment is shown in Fig. 3. The transmission ratio decreases rapidly with increasing optical density, and, as it is known, depends on it in the following way: T = 10-D. When the optical density is equal to 2, the transmittance becomes so low (0.01), that it ceases to be recorded by the photometer.

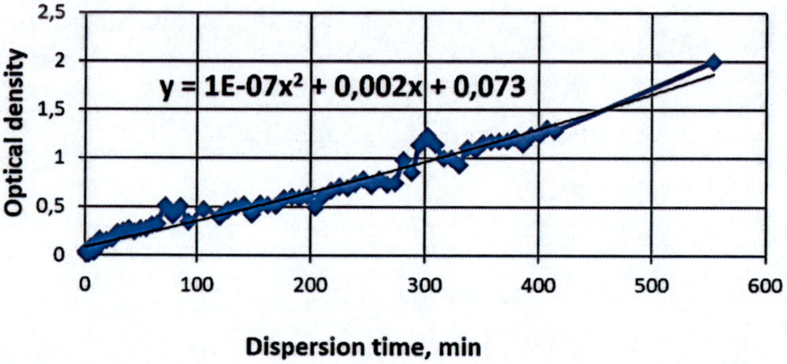

Fig. 3. A graph of optical density versus dispersion time.

Thus, according to the results of the data obtained using a Horiba LA-950 and a KFK-3 spectrophotometer, the optimal dispersion time required for the distribution of nanotubes in oil was accepted as $t = 10$ min.

Figure 4 shows the dependence of the softening point of the binder on the concentration of the modifying dispersion.

Comparing curves 3 and 4 it can be seen that the introduction of nanotubes induces the plasticization of the binder probably due to a decrease in the thermal distillation of the oil during devulcanization. A comparison of curves 1 and 2 shows that the presence of nanotubes in the dispersion, when introduced after devulcanization, has no effect. Curve 3 lies above curve 1, which indicates the volatilization of oil during devulcanization; however, curves 2 and 4 are close to each other, which again indicates that, in the presence of nanotubes, the oil either does not thermally distill during the devulcanization process, or thermal distillation is compensated by the plasticizing effect of the tubes. This effect is confirmed by penetration data (Fig. 5). In this case, the introduction of SWCNTs, regardless of the method of preparation of the binder, increases its hardness, which can be seen from a comparison of curves 1 and 2, as well as curves 3 and 4 in Fig. 5.

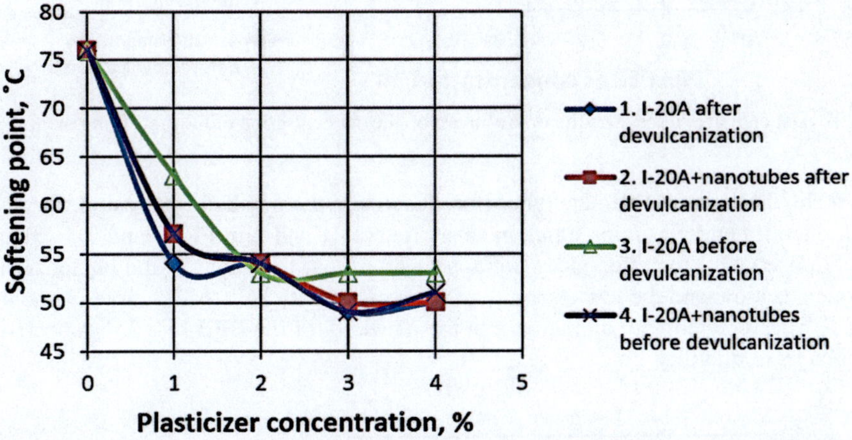

Fig. 4. Dependence of the softening point of the binder on the concentration of the dispersion.

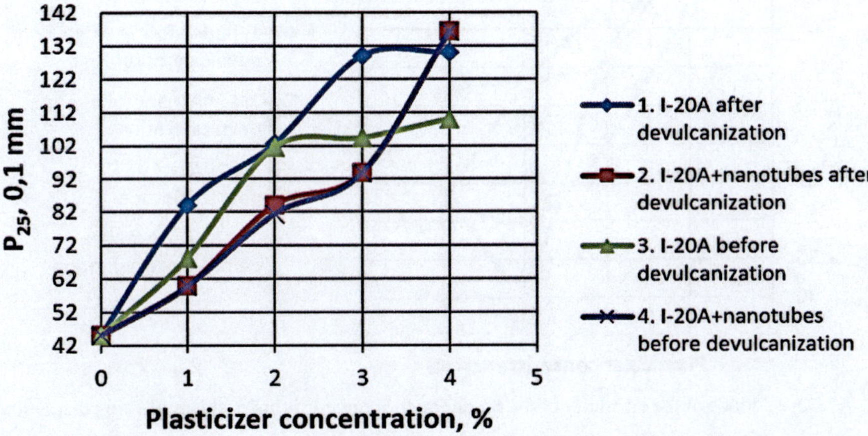

Fig. 5. Dependence of the binder penetration on the concentration of the modifying dispersion.

Figure 6 shows the dependence of the ductility of the binder on the concentration of the modifying dispersion. Noteworthy is the extremum with 3% of variance. It is obvious that the tubes significantly increase the extensibility of the binder. It is known [31–33] that the more bitumen deviates from the Newtonian flow, the less it is extensible. Put in other words, carbon nanotubes do not structure the binder, but, rather, bring the nature of its flow closer to the Newtonian one.

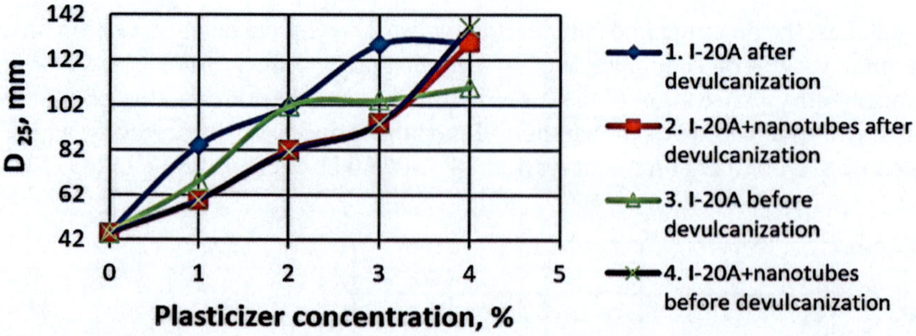

Fig. 6. Dependence of the ductility of the binder on the concentration of the modifying dispersion.

It should be noted that nanotubes introduced into the finished composition increase both ductility and elasticity simultaneously (curves 1 and 2 in Fig. 6 and 7), which is rarely observed. This effect does not appear after heat treatment of the binder during devulcanization, and the low elasticity of composites with SWCNTs (curves 3 and 4 in Fig. 7) is associated with an increase in the plasticity of the BPB [34, 35], reflected in the results of ductility.

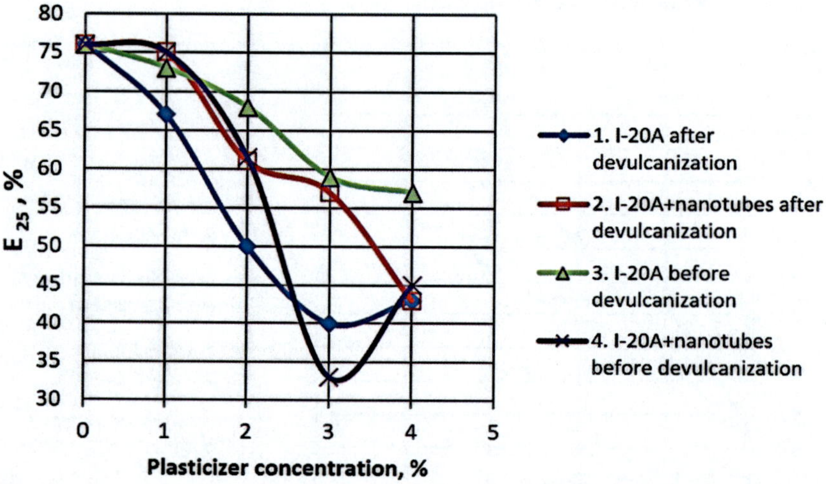

Fig. 7. Dependence of the elasticity of the binder on the concentration of the modifying dispersion.

Curves 2 and 4, characterizing binders with nanotubes, lie lower than curves 1 and 3 (Fig. 8), i.e., with the introduction of nanotubes, the degree of colloid of the binder decreases, that is the effect of the dispersed phase on the properties.

It should be noted that all developed binders have disadvantages typical for plasticized bitumen. However, the introduction of carbon nanotubes through a plasticizer makes it possible to evenly distribute them in the volume of the binder, and the above results allow us to see their effect on the properties of the composition.

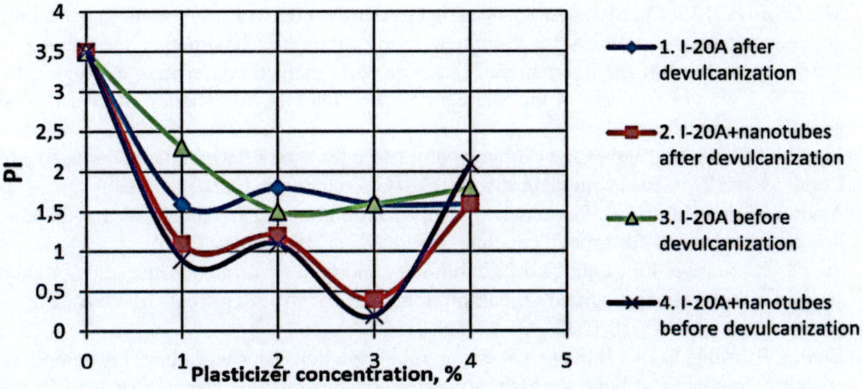

Fig. 8. Dependence of the bitumen penetration index on the SWCNT concentration.

4 Conclusion

1. The introduction of nanotubes before devulcanization plasticizes the binder by reducing the thermal distillation of the oil during devulcanization or by the plasticizing effect of nanotubes.
2. The presence of nanotubes in a dispersion, when introduced after devulcanization, does not give a plasticizing effect, but it simultaneously increases both ductility and elasticity, which is rarely observed.
3. With the introduction of nanotubes, the effect of the dispersed phase on the properties decreases.

References

1. Nizamov, R.K., Nagumanova, E.I., Abdrakhmanova, L.A., Khozin, V.G.: Stroit. Mater. **14** (2004)
2. Khozin, V., Khokhryakov, O., Nizamov, R.: IOP Conference Series: Materials Science and Engineering (2020)
3. Vdovin, E.A., Stroganov, V.F.: Properties of cement-bound mixes depending on technological factors. Mag. Civ. Eng **93**(1), 147–155 (2020)
4. Vdovin, E.A., Stroganov, V.F.: Modification of cement-bound mixtures with sodium formate additives for the construction of pavement bases at low air temperatures. In: IOP Conference Series: Materials Science and Engineering, vol. 786, no. 1 (2020)
5. Volfson, S.I., et al.: Modification of bitumen as a way to improve their performance. Technol. Univ. Bull. **19**(17), 29–33 (2016)
6. Shabaev, S.N., Vakhyanov, E.M., Ivanov, S.A.: The analysis of structure bituminous knitting, modified by a rubber crumb. Int. J. Appl. Eng. Res. **10**(24), 45345–45349 (2015)
7. Oner, J., Sengoz, B.: Effect of polymers on rheological properties of waxy bitumens. Rev. Constr. **17**(2), 279–295 (2018). https://doi.org/10.7764/RDLC.17.2.279
8. Kiselev, V.P., Shevchenko, V.A., Vasilovskaya, G.V., Ivanova, L.A., Voronchikhin, V.D.: Resistance to aging of oil road bitumen modified with low carboxylated polybutadienes. Bull. High. Educ. Inst. Constr. **3**(675), 78–84 (2015)

9. Mardirosova, I.V., Chernykh, D.S.: Investigation of the effect of a rubber-polymer modifier on the structural properties of road bitumen during aging. In: Construction and architecture-2015: Proceedings of the international scientific and practical conference. Rostov-on-Don: RSUCE, Union of Builders of the Southern Federal District, Association of Don Builders, pp. 175–177 (2015)
10. Lin, P., et al.: On the rejuvenator dosage optimization for aged SBS modified bitumen. Constr. Build. Mater. **271** (2021). https://doi.org/10.1016/j.conbuildmat.2020.121913
11. Ayupov, D.A., Murafa, A.V., Makarov, D.B., Khakimullin, Yu.N., Khozin, V.G.: Nanomodified bitumen binders for asphalt concrete. Build. Mater. **10**, 34–35 (2010)
12. Farina, A., Kutay, M.E., Lanotte, M.: Laboratory and field performance investigation of pre-swollen crumb rubber modified asphalt mixtures. Int. J. Pave. Res. Technol. **14**(4), 513–518 (2020). https://doi.org/10.1007/s42947-020-0191-0
13. Huang, J., Wang, Q.-A.: Influence of crumb rubber particle sizes on rutting, low temperature cracking, fracture, and bond strength properties of asphalt binder. Mater. Struct. **54**(2), 1–15 (2021). https://doi.org/10.1617/s11527-021-01647-4
14. Iliopolov, S.K.: Modern ways of increasing the durability of asphalt concrete surfacing. Bull. Kharkiv Natl. Automob. Highway Univ.
15. Olabemiwo, O.M., Lateef, A., Agunbiade, F.O., Akanji, S.B., Bakare, H.O.: The effects on oxidative aging, physical and flow properties of agbabu natural bitumen modified with silver nanoparticles. Heliyon **6**(6), (2020). https://doi.org/10.1016/j.heliyon.2020.e04164
16. Çubuk, M., Gürü, M., Çubuk, M.K., Arslan, D.: Improvement of properties of bitumen by organic-based magnesium additive. J. Fac. Eng. Archit. Gazi Univ. **28**(2), 257–264 (2013)
17. Urkhanova, L.A., Shestakov, N.I., Buyantuev, S.L., Semenov, A.P., Smirnyagina, N.N.: Improving the properties of bitumen and asphalt concrete by introducing a carbon nanomodifier. In: The Collection: Science-Intensive Technologies and Innovations Anniversary International Scientific and Practical Conference Dedicated to the 60th Anniversary of V.G. Shukhov BSTU (XXI scientific readings), pp. 391–398 (2014)
18. Zhang, X., et al.: VOCs reduction in bitumen binder with optimally designed $Ca(OH)_2$ – incorporated zeolite. Constr. Build. Mater. **279** (2021). https://doi.org/10.1016/j.conbuildmat.2021.122485
19. Bindu, C.S., Joseph, M.S., Sibinesh, P.S., George, S., Sivan, S.: Performance evaluation of warm mix asphalt using natural rubber modified bitumen and cashew nut shell liquid. Int. J. Pave. Res. Technol. **13**(4), 442–453 (2020). https://doi.org/10.1007/s42947-020-0241-7
20. Shekhovtsova, S.Yu.: Effective asphalt concrete based on nanomodified polymer-bitumen binder. V.G. Shukhov BSTU
21. Vysotskaya, M.: Polymer-bitumen binder with the addition of single-walled carbon nanotubes (2013). https://doi.org/10.4028/www.scientific.net/AMR.699.530
22. Babagoli, R.: Laboratory investigation of the performance of binders and asphalt mixtures modified by carbon nano tube, poly phosphoric acid, and styrene butadiene rubber. Constr. Build. Mater. **275** (2021). https://doi.org/10.1016/j.conbuildmat.2020.122178
23. Irfan, A., Chaudhry, A.R., Muhammad, S., Al-Sehemi, A.G.: Exploring the potential of boron-doped nanographene as efficient charge transport and nonlinear optical material: a first-principles study. J. Mol. Graph. Modell. **75**, 209–219 (2017). https://doi.org/10.1016/j.jmgm.2017.05.017
24. Ul. Haq, M.F., Anwar, W., Ahmad, N., Khitab, A., Jamal, M., Hllssan, S.: Carbon nanotubes and their use for asphalt binder modification: a review. Emerg. Mater. Res. **9**(2) (2020). https://doi.org/10.1680/jemmr.18.00115
25. Loise, V., et al.: The effect of multiwalled carbon nanotubes on the rheological behaviour of bitumen. Coll. Surf. A: Physicochem. Eng. Aspects **566**, 113–119 (2019). https://doi.org/10.1016/j.colsurfa.2019.01.021

26. Abdullah, H.B., Irmawati, R., Ismail, I., Yusof, N.A.: Utilization of waste engine oil for carbon nanotube aerogel production using floating catalyst chemical vapor deposition. J. Clean. Prod. **261** (2020). https://doi.org/10.1016/j.jclepro.2020.121188
27. Kohli, H.P., Gupta, S., Chakraborty, M.: Stability and performance study of emulsion nanofluid membrane: a combined approach of adsorption and extraction of ethylparaben. Coll. Surf. A: Physicochem. Eng. Aspects **579** (2019). https://doi.org/10.1016/j.colsurfa.2019.123675
28. Zhu, L., Dong, X., Xu, M., Yang, F., Guiver, M.D., Dong, Y.: Fabrication of mullite ceramic-supported carbon nanotube composite membranes with enhanced performance in direct separation of high-temperature emulsified oil droplets. J. Mem. Sci. **582**, 140–150 (2019). https://doi.org/10.1016/j.memsci.2019.04.001
29. Aben, E., Markenbayev, Z., Khairullaev, N., Myrzakhmetov, S., Aben, K.: Study of change in the leaching solution activity after treatment with a cavitator. Min. Miner. Deposits **13**(4), 114–120 (2019). https://doi.org/10.33271/mining13.04.114
30. Ilyin, V.K., Morozova, Y.A., Usanova, N.A., Gotovskiy, M.Y., Roik, O.A., Matiushin, A.O.: Ultra-weak electromagnetic signals: effects of storing and playback on example of saccharomyces. Int. J. High Dilut. Res. **17**(2), 40 (2018)
31. Gun, R.B.: Petroleum bitumen, p. 432. Chemistry, Moscow (1973)
32. Lemarchand, C.A., Bailey, N.P., Todd, B.D., Daivis, P.J., Hansen, J.S.: Non-newtonian behavior and molecular structure of cooed bitumen under shear flow: a non-equilibrium molecular dynamics study. J. Chem. Phys. **142**(24) (2015). https://doi.org/10.1063/1.4922831
33. Mousavi, M., Oldham, D., Fini, E.H.: Using fundamental material properties to predict the moisture susceptibility of the asphalt binder: polarizability and a moisture-induced shear-thinning index. ACS Appl. Bio Mater. **3**(11), 7399–7407 (2020). https://doi.org/10.1021/acsabm.0c00374
34. Murafa, A.V., Makarov, D.B., Nuriev, M.A., Khozin, V.G.: Nanomodified bitumen emulsions for construction purposes **2**(14), 245–249 (2010)
35. Khafizov, E.R., Fomin, A.Yu.: Application of modified bitumen for road construction in the Republic of Tatarstan. Izvestija KSUAE **4**(30), 303–307 (2014)

The Effect of Additives of Mechanically Activated Mineral Fillers on the Properties of Composite Gypsum Binders

Marat Khaliullin(✉) and Alsu Gilmanshina

Kazan State University of Architecture and Engineering, Kazan 420043, Russia
khaliullin@kgasu.ru

Absract. Problem statement. The purpose of this research is to study the influence of the method of grinding quartz and carbonate fillers in a ball mill and in an electro-mass classifier with intensive mechanical activation of the surface of the filler particles on the efficiency of their introduction into the composite gypsum binders. Results. The effect of mechanochemical activation of the surface of quartz filler particles during grinding in the electro-mass classifier to the optimal specific surface of 500 m^2/kg with the introduction of 5% by weight of a ground additive in the composite gypsum binders, an increase in strength parameters by 15% is achieved compared to compositions without adding filler additives. With the introduction of additives in the amount of up to 15% by weight, the strength and water resistance are maintained at the level of control samples. Compared to the introduction of a similar amount of quartz filler, ground in a ball mill, a maximum increase in strength indicators by 10% is achieved. When grinding carbonate fillers in an electro-mass classifier compared to grinding in a ball mill, there is a slight increase in the strength of the binder. Conclusions. The significance of the results for the construction industry lies in the fact that it is effectually to use an electro-mass classifier in the preparation of quartz filler additives introduced into the composition of composite gypsum binders.

Keywords: Composite gypsum binder · Mineral filler · Mechanochemical activation · Gypsum-cement-pozzolanic composition

1 Introduction

One of the ways to improve the production of binders is the introduction of ground mineral additives into their composition. Currently, there is considerable experience in the use of ground mineral additives of natural, artificial and industrial origin in the production of most types of binders. The introduction of ground mineral additives is one of the effective technological methods, which allows ensuring the control of the structure and properties of binders with an increase in their economic, environmental and technical efficiency [1–7]. Economic efficiency with the introduction of ground mineral additives into the composition of binders is achieved by increasing the production volume of the resulting binders with a simultaneous decrease in their cost while reducing

N. Vatin (Ed.): STCCE 2021, LNCE 169, pp. 334–342, 2021.
https://doi.org/10.1007/978-3-030-80103-8_36

energy and labor costs for calcination and grinding of the calcinating component [8]. A significant environmental effect is achieved due to the reduction of emissions of by-products formed during the calcination of the initial components for the production of binders (for example, carbon dioxide in the production of Portland cement), as a result of replacing a part of the calcinating component of the binders with ground mineral additives. A decrease in the ecological impact on the environment, the volume of use of natural raw materials occurs in cases of utilization of ground industrial waste when used as additives in the production of binders.

Russian and foreign scientists have carried out a significant amount of research in the field of obtaining and studying the physical and mechanical properties of gypsum binders with finely dispersed mineral additives. They are active, forming compounds with binding properties (slags, ash, zeolite-containing rocks, glass cullet, perlite, flask, glauconitic sand, volcanic tuff, expanded clay dust, microsilica, etc.), and inert fillers that do not form such compounds (quartz sand, limestone, dolomite waste from magnetic separation of iron ores, etc.) [9–12]. The research results show that both active and inert mineral additives take part in the structure formation of binder systems [13–15].

Thus, in [16] electron microscopic research method has shown that mineral fillers are able to influence the formation of gypsum crystals in the process of hardening gypsum binders. Differences in the shape, orientation, size and growth density of gypsum crystals formed on substrates, which are freshly cleaved single crystals of various minerals (dolomite, calcite, gypsum, quartz) have been revealed. It was found that the growth density of gypsum crystals increased depending on the type of the substrate mineral in the series: quartz – dolomite – calcite – gypsum, i.e. as the crystal-chemical similarity of the filler to gypsum increased. Thus, optimally selected mineral fillers, acting as substrates for the crystallization of newly formed gypsum during hardening, contribute to an increase in the density and strength of the forming gypsum stone.

Modern technologies for the production of mineral composite binders with the increased physical and mechanical properties widely use methods associated with mechanochemical activation, based on change in the crystal structure of the components of binder systems, in particular, due to increase in the concentration of surface defects of binder particles or mineral additives under mechanical action, for example, during grinding [17–19].

Based on previous A.V. Volzhensky principles of obtaining gypsum-cement-puzzolanic binders A.V. Ferronskaya, V.F. Korovyakov and other scientists made further development in obtaining waterproof high-strength composite gypsum binders. Composite gypsum binders are obtained on the basis of gypsum binders with the introduction of a complex of modifying components. They are Portland cement or lime, pozzolanically active finely dispersed silica-containing materials (microsilica, glass cullet, ceramsite dust, zeolite-containing rocks, slags, ash, etc.) plasticizers and other functional additives, and they may also contain mineral fillers subjected to mechanochemical treatment in a grinding unit [20].

Earlier studies conducted by the authors of this article [16] showed that in terms of increasing the strength of composite gypsum binders with the introduction of an optimal amount of mineral filler additive, as well as the maximum possible reduction in the consumption of the calcinating component of the binder without reducing the strength characteristics, carbonate fillers are more effective than ground quartz sand, while for carbonate fillers, ground limestone is more effective than ground dolomite. It was found that carbonate minerals such as calcite and dolomite were more preferable as substrates for gypsum crystallization, in comparison to quartz; calcite as a mineral substrate was more effective than dolomite, since it had a greater crystal-chemical similarity to gypsum, which caused the formation of denser contacts between gypsum and the calcite mineral substrate.

The purpose of this research is to study the influence of the method of grinding quartz and carbonate fillers in a ball mill and during processing in an electro-mass classifier with intensive mechanical activation of the surface of filler particles on the efficiency of their introduction into the composition of composite gypsum binders.

2 Materials and Methods

Composite gypsum binder was obtained on the basis of G-5BII gypsum in accordance with GOST 125 (State Standard of the Russian Federation) with pozzolanic component, thermally activated according to the optimal mode and ground to a specific surface area of 500 m^2/kg clay from the Saray-Chekurchinskoye deposit [21]; the powdery plasticizer and mineral filler were added. G-5BII gypsum was produced by the Volgograd gypsum plant of Volma Corporation when mixed with 500-D0-N Portland cement produced by JSC "Mordovcement" in accordance with GOST 10178 (State Standard of the Russian Federation). The chemical, mineral and grain size composition of the pozzolanic component is presented in the initial state, respectively, in Tables 1, 2, 3.

Table 1. Chemical composition of clay from the Saray-Chekurchinskoye deposit (content in % by weight)

SiO_2	TiO_2	Al_2O_3	Fe_2O_3	MnO	CaO	MgO	Na_2O	K_2O	P_2O_5	SO_3/S	Ignition loss
68.52	0.86	13.42	6.18	0.10	1.33	1.66	1.20	1.82	0.09	<0.05	4.62

Table 2. Mineralogical composition of clay from the Saray-Chekurchinsky deposit (content in % by mass)

Quartz	Mica	Orthoclase	Plagioclases	Mixed-layered clay mineral*	Chloride
28	10	7	8	40	4

*mixed-layered mineral contains up to 40% non-swelling layers.

The selection of the gypsum-cement-pozzolanic composition was carried out according to the method developed at Moscow State University of Civil Engineering named after V.V. Kuibyshev [20]. It is based on the ability of pozzolanic additives to absorb calcium hydroxide from its saturated solution in the study of special preparations, which

Table 3. Grain size composition of clay from the Saray-Chekurchinskoye deposit (content in% by mass)

Less than 0.005 mm (clayish)	0.005–0.05 mm (pulverous)	0.05–1.0 mm (areneceous)
49.5	37.1	13.4

are aqueous suspensions of semi-aqueous gypsum, Portland cement and pozzolanic additives.

Melflux 2651 F additive based on polycarboxylate (manufactured by BASF Constraction Polymers) was used as a plasticizer in an amount of 0.8% by weight of the binder [21].

Ground quartz sand of the Vasilievsky deposit, carbonate rocks such as limestone of the Aldermysh deposit and dolomite of the Bimsky deposit, the chemical compositions of which are presented, respectively, in Tables 4, 5 were used as mineral fillers introduced into the composition of the composite gypsum binders.

Table 4. Chemical composition of quartz sand of the Vasilievsky deposit

Test №	SiO_2	Al_2O_3	Fe_2O_3	CaO	MgO	R_2O	Ignition loss
1	95.52	3.13	0.4	0.10	0.05	0.04	0.55

According to the test results based on GOST 8735 (State Standard of the Russian Federation), the initial quartz sand has a grain size modulus of 1.01; the content of dust-like and clay particles is 2.8%; the content of organic impurities is normal.

Table 5. Chemical composition of carbonate rocks

Test №	Rock name	Deposit	CaO	MgO	SO_3	Ignition loss	Insoluble residue
1	Limestone	Aldermysh	49.74	1.86	1.23	42.15	3.46
2	Dolomite	Bimsky	30.01	20.09	0.44	46.49	0.47

Pre-dried quartz sand and carbonate rocks were ground to specific surfaces of 90, 300, 500 m^2/kg.

Quartz sand was ground in its original form and washed in a 1KSP-12 hydroclassifier, which removed dust-like and clay particles.

40-ML laboratory ball mill and an MS-EMK-050M electro-mass classifier produced by LLC "Poliprom" were used to grind the fillers.

The 40-ML laboratory ball mill has a rotation speed of 60 rpm. The mass of grinding bodies loaded into the mill was 12 kg. The ratio between the mass of the crashed material and the grinding media (steel balls) is 1:8.

The electro-mass classifier MS-EMK-050M operates on the principle of a centrifugal mill, where grinding operations, particle size classification, disaggregation, and mechanical activation of the particle surface are combined. The creation of rigid conditions for the formation of gas and dust flows in a closed space under the influence of electrostatic forces contributes to a more intensive mechanical activation of the surface of the crushed particles, in comparison with the grinding process in ball mills [22].

Tests of the composite gypsum binder were carried out in accordance with GOST 23789 (State Standard of the Russian Federation). Samples of the composite gypsum binder for testing to determine the strength, average density, water absorption by weight, softening coefficient were kept in a normal hardening chamber for 28 days. After that they were dried at 55 °C until constant weight was reached. The softening coefficient was determined as the ratio of compression strength of water-saturated artificial stone samples based on a composite gypsum binder to the tensile strength of artificial stone samples based on a composite gypsum binder dried to constant weight.

3 Results and Discussions

Figure 1 shows the research results of dispersion effect of quartz filler crushed in a ball mill (when the composition of the composite gypsum binder is 5% by weight) and limestone filler (when the composition of the composite gypsum binder is 10% by weight) on the compression strength of the artificial stone.

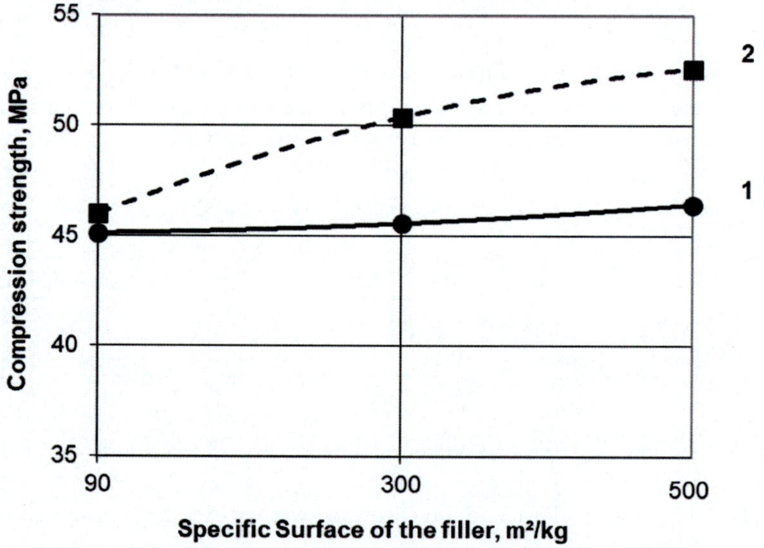

Fig. 1. Dispersion effect of mineral fillers, crushed in a ball mill, on the compression strength of the artificial stone based on a composite gypsum binder 2 – limestone filler introduced in an amount of 10% by weight (- ■ -)

The presented results show that the greatest strength is achieved with the introduction of fillers ground to a specific surface area of 500 m^2/kg. Quartz filler was obtained by grinding sand after washing it.

At the next stage of work, carbonate and quartz fillers were ground to a specific surface area of 500 m^2/kg in a ball mill and in the electro-mass classifier and added to the composition of a composite gypsum binder in an amount of up to 25% by weight.

Figure 2 shows the results of studies of the effect of introducing quartz filler ground to a specific surface area of 500 m^2/kg in various grinding units on the compression strength of the artificial stone based on a composite gypsum binder.

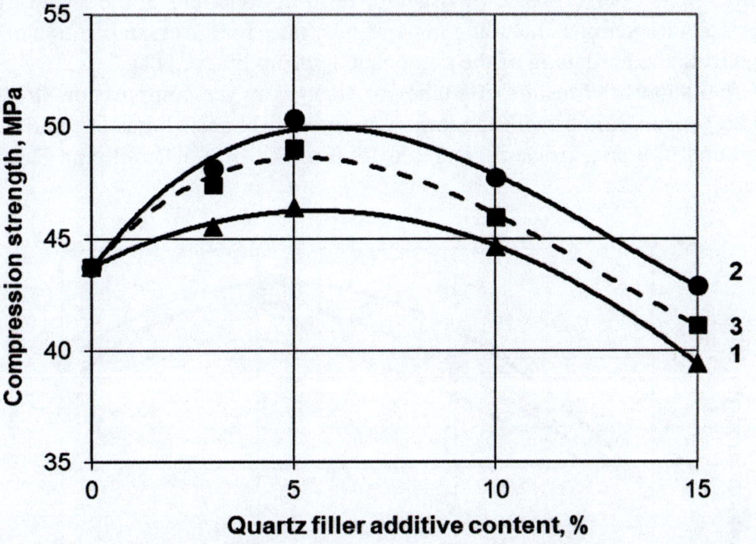

Fig. 2. Influence of the type of grinding unit and the content of quartz filler (with a specific surface area of 500 m^2/kg) on the compression strength of the artificial stone based on a composite gypsum binder. 1 – quartz filler made of washed sand, crushed in a ball mill (-▲-); 2 – quartz filler made of washed sand, crushed in the electro-mass classifier (-•-); 3 – quartz filler made of unwashed sand (the content of dust-like and clay particles is 2.8% by weight), crushed in an electro-mass classifier

The analysis of the presented data shows that the most effective is the introduction of the washed quartz filler additive, subjected to grinding and additional mechanochemical activation in an electro-mass classifier, in an amount of 5% by weight of the binder. At the same time, the compression strength of the artificial stone based on a composite gypsum binder increases by 12–15% compared to control samples without the introduction of the filler. With the introduction a quartz filler ground up to 15% by weight in the electro-mass classifier, the compression strength of the artificial stone remains at the level of control samples. The softening coefficient of gypsum stone slightly decreases from 0.72 to 0.70–0.69, remaining at the level of indicators that meet the requirements for gypsum binders with increased water resistance (Кр > 0.65) [14]. With the introduction of up to 15% by weight of a quartz filler, crushed and additionally mechanochemically activated in the electro-mass classifier, into the composition of a composite gypsum binder, in comparison to the introduction of a similar amount of quartz filler ground in a ball mill, a maximum increase in strength parameters by 10% is achieved. The presence of dust-like and clay particles in the amount of 2.8% by weight in the composition of the quartz filler leads to a slight decrease in the compression strength of the artificial stone based on a composite gypsum binder.

An increase in the strength characteristics of the artificial stone based on a composite gypsum binder with the introduction of crushed and additionally mechanochemically activated in an electro-mass classifier quartz filler in comparison to its grinding in a ball mill can be associated with higher values of the surface energy of filler particles and more efficient performance of structure-forming functions at the level of physical and chemical interactions, including mineral substrates for the crystallization of gypsum formed during the hardening of the composite gypsum binder [11].

Figure 3 shows the results of studies of changes in the compression strength of a composite gypsum binder with the introduction of carbonate (limestone and dolomite) fillers ground to a specific surface area of 500 m^2/kg in a ball mill and electro-mass classifier.

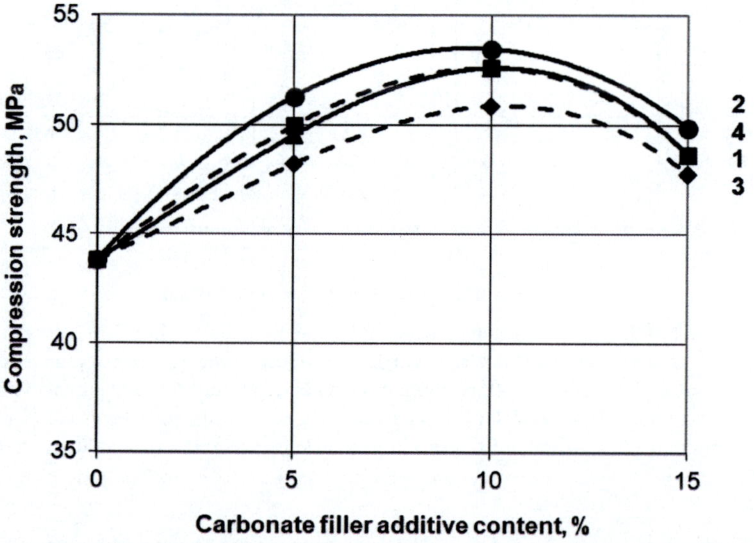

Fig. 3. Influence of the type of grinding unit, the amount and type of addition of carbonate fillers (specific surface area of 500 m^2/kg) on the compression strength of the artificial stone based on a composite gypsum binder. 1 - limestone filler, crushed in a ball mill (-●-); 2 - limestone filler, crushed in an electro-mass classifier (-▲-); 3 - dolomite filler, crushed in a ball mill (-♦-); 4 - dolomite filler, crushed in an electro-mass classifier (-■-)

The presented results show that when using the electro-mass classifier for grinding carbonate fillers introduced into the composition of a composite gypsum binder in the amount of up to 25% by weight, a slight increase in strength characteristics compared to the introduction of a carbonate filler ground in a ball mill is observed.

4 Conclusion

1. The introduction into the composition of the composite gypsum binder quartz filler subjected to grinding and additional mechanochemical activation in the electro-mass classifier to an optimal specific surface area of 500 m^2/kg is more effective compared to the introduction of a similar amount of quartz filler ground in the ball mill. With the optimal content of ground quartz filler in the electro-mass classifier in the amount of 5% of the binder weight, the compression strength of the artificial stone based on a composite gypsum binder increases by 15% compared to control samples without the addition of a filler, with the introduction of up to 15% by weight of quartz filler. The consumption of the calcinating part of the binder decreases without reducing the strength and water resistance compared to the control samples. In this case, compared to the introduction of a similar amount of quartz filler ground in a ball mill, a maximum increase in strength indicators by 10% is achieved.
2. The use of the electro-mass classifier for grinding carbonate fillers is less effective, since compared to the introduction of a carbonate filler ground in the ball mill. There is a slight increase in strength indicators.
3. Thus, it is reasonable to use the electro-mass classifier when preparing the quartz filler additive to the composition of a composite gypsum binder for grinding it to the required specific surface area and additional mechanochemical activation.

References

1. Rakhimova, N.R., et al.: Marl-based geopolymers incorporated with limestone: a feasibility study. J. Non-Crystal. Solids. **492**, 1 (2018). https://doi.org/10.1016/j.jnoncrysol.2018.04.015
2. Zhang, S., Lu, D., Xu, Z.: Effect of dolomite powders on the hydration and strength properties of cement mortars. In: XIV International Congress on the Chemistry of Cement (2015). https://doi.org/10.14062/j.issn.0454-5648.2016.08.08
3. Nocun-Wczelik, W., Szybilski, M., Zugaj, E.: Hydration of Portland cement with Dolomite. In: Proceedings of XIV International Congress on the Chemistry of cement (2015)
4. Vdovin, E.A., Stroganov, V.F.: Properties of cement-bound mixes depending on technological factors. Mag. Civ. Eng. **93**, 147 (2020)
5. Bulanov, P.E., Vdovin, E.A., Mavliev, L.F., Kuznetsov, D.A.: Development of road soil cement compositions modified with complex additive based on polycarboxylic ether. In: IOP Conference Series: Materials Science and Engineering (2018)
6. Galeev, R., Nizamov, R., Abdrakhmanova, L., Khozin, V.: Resource-saving polymer compositions for construction purposes. In: IOP Conference Series: Materials Science and Engineering (2020)
7. Galeev, R., Nizamov, R., Abdrakhmanova, L.: E3S Web Conference (2020)
8. Khozin, V., Khokhryakov, O., Nizamov, R.: A «carbon footprint» of low water demand cements and cement-based concrete. In: IOP Conference Series: Materials Science and Engineering (2020)
9. Mukhametrakhimov, R.K., Galautdinov, A.R., Lukmanova, L.V.: Influence of active mineral additives on the basic properties of the gypsum cementpozzolan binder for the manufacture of building products. In: MATEC Web of Conferences, vol. 106 (2017). https://doi.org/10.1051/matecconf/201710603012

10. Tokarev, Y., et al.: Modern building materials, structures and techniques. Proc. Eng. 12. **172**, 1161 (2017)
11. Morozova, N.N., Kuznetsova, G.V., Maysuradze, N.V., Akhtariev, R.R., Abdrashitova, L.R., Nizamutdinova, E.R.: Research in the activity of a pozzolanic component and superplasticizer for gypsum cement pozzolanic binder of white colour (GCPB). Stroitelnye Mater. **8**, 26 (2018)
12. Novichenkova, T.B., Petropavlovskaya, V.B., Zavadko, M., Buryanov, A.F., Pustovgar, A.P., Petropavlovskii, K.S.: The use of dusty wastes of basalt production as a filler for gypsum compositions). Stroitelnye Mater. **8**, 9 (2018)
13. Mukhametrakhimov, R.K., Galautdinov, A.R., Gilmanshin, I.R.: Modified gypsum-cement-pozzolanic composites reinforced with polypropylene fibers. In: IOP Conference Series: Materials Science and Engineering (2019). https://doi.org/10.1088/1757-899X/570/1/012068
14. Batova, M.D., et al.: Complex mineral additives for the modification of calcium sulphate based materials. Stroitelnye Mater. **1–2**, 13 (2021). https://doi.org/10.31659/0585-430X-2021-788-1-2-13-21
15. Ruzina, N.S., Yakovlev, G.I., Gordin, A.F., Pervushin, G.N., Semenova, Y., Begunova, E.V.: Modification of binders based on calcium sulfate with complex additives. Stroitelnye Mater. **7**, 18 (2020). https://doi.org/10.31659/0585-430X-2020-782-7-18-22
16. Khaliullin, M.I., Rakhimov, R.Z.: Researches of the mechanism of influence of mineral and chemical additives on the process of anhydrite binding hardening. In: Proceedings of a XII International Congress of the Chemistry of Cement (2011)
17. Kuzmina, V.P.: Mechanical Chemistry in Concrete. ASV, Moscow (2013)
18. Mukhametrakhimov, R.K., Galautdinov, A.R.: Mechanically activated gypsum-cement-pozzolanic binder based on modified low-grade raw materials. Izvestiya KGASU **1**(43), 187 (2018)
19. Vdovin, E.A., Stroganov, V.F., Konovalov, N.V.: Modification of road soil cement with activated fillers (2021)
20. Ferronskaya, A.V., et al.: Gypsum in Low-Rise Building. ASV, Moscow (2008)
21. Khaliullin, M.I., Rakhimov, R.Z., Faizrakhmanov, I.I.: The influence of thermally activated clay additives on the properties of composite gypsum binder. ZKG Int. **5**, 58 (2017)
22. Zyryanov, V.V., Zyryanov, D.V., Sadykov, V.A.: Fabrication of films by charge aerosol deposition method. Rossiiskie Nanotehnol. **5–6**, 118 (2008)

About Wear Courses and Concrete Pavements Abrasion of Highways

Viktor Ushakov(✉), Mikhail Goryachev, Grigory Diakov, and Sergey Yarkin

Moscow Automobile and Road Construction State Technical University (MADI), Moscow 125319, Russia

Abstract. Article deals with issues of rut formation on concrete pavements and asphalt concrete courses of road wear as a result of studded tires impact. Article presents study results to establish effect of concrete composition, crushed stone strength and size of its fractions, as well as concrete surface texturing technology on concrete pavements abrasion. Tests were carried out on universal complex CATRM-2 "Carousel" located at the training ground of MADI Educational and Scientific Center (TRC MADI). Road surface abrasion investigations in real operating conditions on M-4 "Don" highway at the section km 52 - km 71 (Moscow region) made it possible to establish rut formation dynamics in individual traffic lanes. On left and middle lanes of concrete pavement, after 7 years of operation, a thin asphalt concrete wear course 3.5 cm thick was made using membrane technology. On the right lane, wear course was not satisfactory and concrete pavement was operated since 2010 without any repair work.

1 Introduction

Due to the increase in traffic load on roads, change in cars dynamic characteristics and studded rubber use, one of the most common types of deformation has become rut appearance in many countries of the world. Operating highways experience shows that rut formation leads to significant deterioration in road safety, reduction in highways overhaul life, increase in maintenance and repair cost [1, 3, 11, 17, 20]. In contrast to non-rigid road pavements, rut formation on concrete pavements is not a consequence of plastic and permanent deformations, but a consequence of wear and tear of concrete pavements surface layer, especially in winter under studded rubber influence [4, 8, 10, 15, 16, 18]. Study of concrete composition effect, crushed stone strength and size of its fractions on concrete pavements abrasion, as well as effective technologies development for eliminating rut on road surfaces is an urgent task.

2 Materials and Methods

To determine studded tires impact effect on concrete pavement, special tests were carried out on "Carousel" installation at training ground of MADI Educational and Scientific Center (TRC MADI). Universal complex CATRM-2 "Carousel" is intended for comparative tests of road surfaces and automobile tires in conditions as close as possible

N. Vatin (Ed.): STCCE 2021, LNCE 169, pp. 343–350, 2021.
https://doi.org/10.1007/978-3-030-80103-8_37

to operational ones. Concrete pavements were arranged on passenger module track of universal ring complex in segments form - concrete slabs 100 x 120 cm in size and 10 cm thick. Concrete slabs were subjected to repeated abrasion by spiked rubber.

As a result of work, studies were carried out to establish effect of concrete composition, crushed stone strength and size of its fractions, as well as concrete surface texturing technology on abrasion of concrete pavements.

In preparation of concrete mixes for pavements construction, the following materials were used:

1. Cement PC 500-D0-N in accordance with GOST 10178–85.
2. Cement CEM I 42.5N in accordance with GOST 31108–2016.
3. Microsilica MKU-95 (bulk density 300–400 kg/m^3).
4. Crushed stone of various types and fractions (GOST 8267–93):

   - crushed stone from porphyrite fraction 5–15 mm;
   - crushed stone from gabbro-diabase fractions 5–10 mm and 10–20 mm;
   - crushed stone from gabbro-diabase after electrohydrodynamic crushing of 4–8 mm fraction;
   - crushed stone from granite fraction 5–10 mm and 10–20 mm.

5. Natural coarse sand of I class (washed) (GOST 8736–2014).
6. Chemical additives:

   - air entraining agent Micro Air 125;
   - «curing-TAL» set retarder.

Table 1. Crushed stone type and size of its fractions in concrete mixtures preparation.

Concrete composition number	Crushed stone type	Crushed stone fraction size, mm
№1	Gabbro-diabase	4–8
№2	Gabbro-diabase	5–10 and 10–20
№3	Granite	5–10 and 10–20
№4	Gabbro-diabase	5–10 and 10–20
№5	Granite	5–10
№6	Porphyrite	5–15
№7	Porphyrite	5–10
№8	Porphyrite	5–15
№9	Granite	5–10

Table 1 shows concrete compositions numbers with various types and sizes of crushed stone fractions used for manufacture of concrete pavement slabs.

On concrete slabs of compositions No. 3 and No. 9, freshly laid concrete surface was treated with a brush with plastic bristle to give texture. Such technological operation is often performed in concrete pavements construction in order to increase pavement adhesion coefficient to car wheel. Concrete pavement surface of composition No. 8 was cleaned from mortar part with metal brush using technology of obtaining "exposed" crushed stone – "Washbeton", to create a rough surface.

Every 25 thousand passes of wheel with spiked rubber, one track was taken to measure rut depth. Maximum number of wheel passes with studded tires was 250 thousand.

Specialists of Department of Roads Construction and Operation of MADI carried out research on road surfaces abrasion not only at ring test bench, but also in real operating conditions on M-4 Don highway at section from km 52 to km 71 (Moscow region) [4, 17, 18].

This road section was completed with reconstruction in 2009, during which carriageways were widened due to dividing strip, one lane in each direction with a pavement arrangement similar to the main carriageway.

In period from 2010 to 2011, on this road section, work was carried out to pavement overhaul of the second (central) and third (right) lane.

Replacement of worn out construction of road pavement with concrete pavement by means of its destructurization and laying a new pavement was made.

Pavement construction after reconstruction and repair completion of site is shown in Fig. 1.

In 2017, based on annual monitoring and a detailed survey results of section M-4 "Don" highway, km 52 - km 71, it was concluded that it was necessary to restore transverse evenness of concrete pavement. In this regard, recommendations were prepared on repair measures appointment.

In 2017 and 2018, concrete pavement was milled on the left and middle lane to depth of 35 mm and width of 7.3 m, with a wear course made of high-strength SMA-12 asphalt concrete, designed according to Superpave volumetric design methodology with thickness of 3.5 cm with priming base with polymer-bitumen (latex) emulsion with flow rate 0.8–1.2 l/m^2. SpreyJet spreader was used to lay wear course.

Transverse profile of section km 52 - km 71 of M-4 "Don" highway after rut elimination is shown in Fig. 2.

Rut depth measurement was carried out in accordance with GOST 32825–2014 "Public automobile roads. Road surfaces. Methods for measuring geometrical dimensions of damage" in June 2018, May 2019 and July 2020 using a mobile laboratory DVS-4IK, which ensures accuracy of rut depth measuring up to 1 mm and is calibrated in prescribed manner.

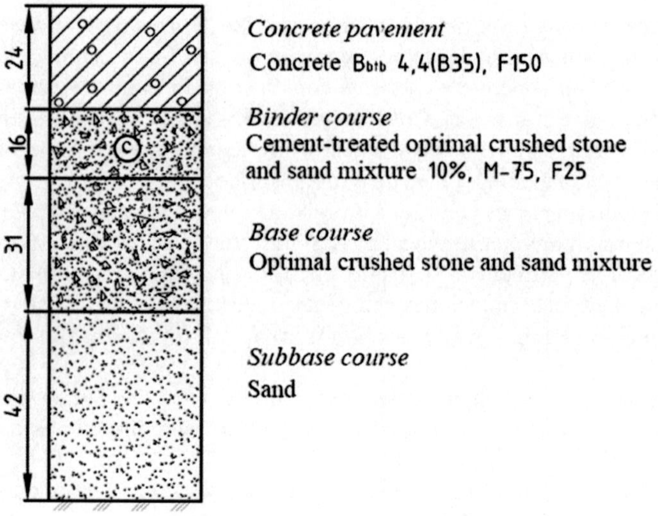

Fig. 1. Road pavement construction of M-4 “Don” highway on section from km 52 to km 71.

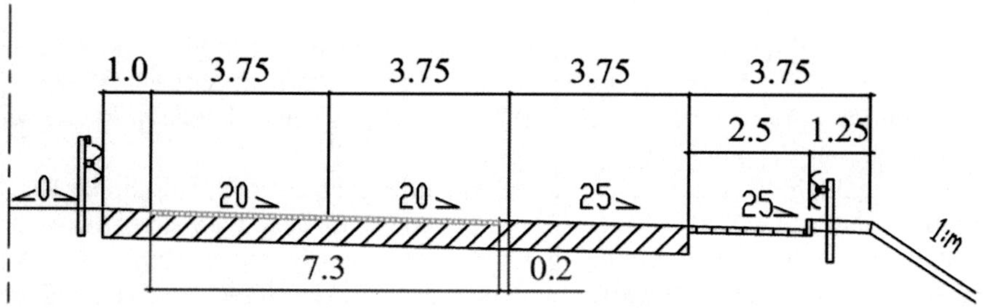

Fig. 2. Cross-section of M-4 “Don” highway at section from km 52 to km 71 after rut elimination.

3 Results

Performed studies of concrete coatings abrasion on circular universal stand CATRM-2 “Carousel” made it possible to establish that the highest wear resistance was demonstrated by concrete coatings of compositions No. 8 and No. 6, made using porphyrite of 5–15mm fraction as a coarse filler. After 250 thousand wheel passes with studded rubber average rut depth on concrete pavements was 4.67 and 5.03 mm, respectively. In addition, it should be noted that application of “Washbeton” technology, “exposure” of large aggregate on surface of road pavement, made it possible to slow down concrete abrasion under studded rubber action, when mortar part was removed from coating surface. Low wear resistance was shown by concrete coatings made of compositions No. 1 and No. 5, using crushed stone of fine fractions from gabbro-diabase fractions 4–8 mm and granite fractions 5–10 mm. Average rut depth on concrete pavements under consideration was 10.50 and 8.17 mm, respectively.

During the first 25 thousand wheel passes with spiked rubber, all concrete pavements, with exception of concrete pavement of composition No. 8, showed the highest abrasion. During this period, abrasion of mortar part formed on concrete pavement surface during its construction occurs. The greatest abrasion after the first 25 thousand wheel passes was shown by coatings made of concrete of compositions No. 3 and No. 9, made with crushed granite use. On these coatings, freshly laid concrete surface was treated with plastic bristle brush. Later on, wear on these concrete pavements slowed down.

Studies of road surface abrasion in real operating conditions on M-4 "Don" highway at section km 52 - km 71 (Moscow region) made it possible to establish dynamics of rut formation in individual traffic lanes. On the left and middle lanes of concrete pavement, a thin asphalt concrete wear course with thickness of 3.5 cm was arranged. On the right-hand lane, wear course was not satisfactory and concrete pavement was operated since 2010 without any repair work.

Based on measurements carried out in 2018, 2019 and 2020, average values of depth of the right and left rut were calculated for each lane. Obtained data in millimeters are presented in Table 2.

Table 2. Average values of rut depth for period 2018–2020, mm.

	−3R	−3L	−2R	−2L	−1R	−1L	+ 1L	+ 1R	+ 2L	+ 2R	+ 3L	+ 3R
2018	5,7	4,5	3,4	3,7	3,9	1,9	2,3	4,6	4,7	5,3	4,8	4,7
2019	6,1	5,2	5,3	6,3	5,3	3,4	4,8	6,5	8,0	7,5	5,4	5,0
2020	6,4	6,0	10,1	8,8	5,8	7,9	6,9	9,7	11,6	10,8	5,8	5,4

Graphical result that displays change in rut depth for each roll-off lane and traffic lane with detailed increment by year in accordance with Table 2 is shown in Fig. 3.

4 Discussion

Top course of road surface is exposed to the maximum impact of traffic loads and climatic factors, and special requirements must be imposed on it [6, 7, 9, 13, 14, 19].

In many European countries, two-layer technology for concrete pavements construction is used [5]. The main advantage of such coatings is ability to create a thin (5–6 cm) top coating course with special technical characteristics. Such a layer can be called a "wear course". Architecture of the layer allows you to give surface of road surface improved properties. At the same time, costs for coating construction as a whole are significantly reduced, due to the use of a cheaper and in many regions available materials for creating bottom concrete layer.

In the USA and a number of other countries, concrete coatings are arranged in one layer with a certain margin of thickness [2, 12]. In this case, work technology is simplified, productivity increases, and logistics of supplying materials and concrete mix is simplified. As top layer of coating wears out, it is milled with diamond discs

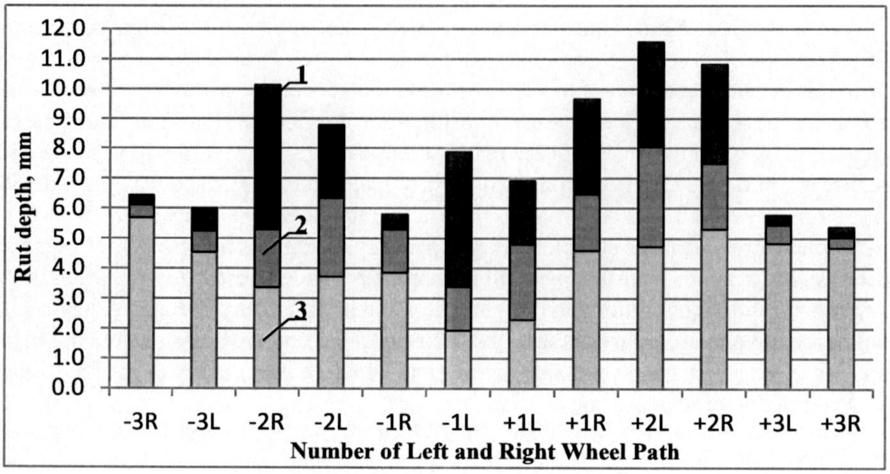

Fig. 3. Changing rut depth along wheel path: 1 - 2020; 2 - 2019; Oct 2018

to thickness of 20–25 mm. Milled surface of road surface has an ideal flatness, high adhesion coefficient and low noise level when driving vehicles.

Testing results on circular universal stand CATRM-2 "Carousel" MADI showed that crushed stone quality has a primary effect on wear of concrete pavements under studded rubber action. Use of durable filler (porphyrites) and its optimal fractions in concrete preparation allows reducing concrete coatings abrasion under studded rubber action by 1.5–2 times.

Based on survey and measurement results of rut on M-4 "Don" highway in section km 52 - km 71, depth values of the right and left rut were obtained for all traffic lanes in each direction, as well as cracks location and length. The maximum average values of rut depth of asphalt concrete wear course and its greatest increase are observed in the central lane in each direction of travel, which are presented in Table 3. Unevenness of increase in rut depth on the left and central traffic lanes with an asphalt concrete wear course can be explained by the different traffic intensity of vehicles.

Table 3. Rut depth average values of asphalt concrete wear course and its increment.

Year	−2 lane		−1 lane		1 lane		2 lane	
	Depth, mm	Increase, mm	Depth, mm	Increase, mm	Depth, mm	Increase, mm	Depth, mm	Increase, mm
2018	3,5	-	2,9	-	3,4	-	5,0	-
2019	5,8	+2,3	4,3	+1,4	5,6	+2,2	7,8	+2,7
2020	9,5	+3,6	6,8	+2,5	8,3	+2,7	11,2	+3,4

Rut depth obtained values of asphalt concrete wear course for direction from Moscow have greater values than in direction to Moscow. This may be due to start of coatings operation: section from Moscow was commissioned in December 2017, to Moscow - in June 2018.

After 3 years of operation, no areas with rut depth exceeding maximum permissible value according to GOST R 50597–2017 (more than 20 mm) have been identified. There is no need to carry out rut elimination work.

After 10 years of operation, rut depth of the right lanes of concrete pavement in the forward and backward directions has minimum values and does not exceed maximum permissible values. With a steady rate of increase in depth of concrete pavement rut, its maximum permissible values will be reached in 28–30 years.

5 Conclusion

1. As a result of studies performed, it was found that crushed stone strength affects concrete pavements wear under the studded rubber action. Use of durable filler (porphyrites) and its optimal fractions in concrete preparation allows to reduce concrete coatings abrasion under studded rubber action by 1.5–2 times. Construction of two-layer concrete pavements will significantly reduce expensive crushed stone consumption, using it only in the top layer of pavement.
2. Size of coarse aggregate fractions in concrete affects concrete pavements wear. With an increase in crushed stone fractions size, wear resistance of concrete increases. Removing a weak mortar part from coating surface using the "Washbeton" technology, allows to slow down concrete coating wear under studded rubber action.
3. Construction of a wear course made of crushed stone-mastic asphalt concrete with a thickness of 3.5 cm with priming of base with latex emulsion with consumption should be attributed to effective repair technologies and recommended for wide use as wear courses of concrete coatings of highways, which will significantly increase their service life and high transport and operational performance.

References

1. Borodin, R.K.: World and domestic experience review in assessing studded tires impact on road surface. Roads Bridges **1**(39), 14 (2018)
2. Cotter, A., Muench, S.T.: Studded Tire Wear on Portland Concrete Pavement in the Washington State Department of Transportation Route Network. Washington State Transportation Center (TRAC), Research Report (2010)
3. Dmitrieva, Y.: Rutting causes in road pavements structures on roads with asphalt pavement. Young Sci. **10**(352), 41–43 (2021)
4. Dyakov, G.G., Ponarin, G.A., Ushakov, V.V., Vasiliev, Yu.E.: Wear resistance of road cement concrete. Scientific and practical conference dedicated to the 85th anniversary of the Honored Scientist of the Russian Federation, Academician of the RAASN, Doctor of Technical Sciences Bazhenov Yuri Mikhailovich, pp. 174–180 (2015)
5. Eckert, A.: Construction and maintenance of concrete roads in PPP projects framework in Germany. World Roads **120**, 48–51 (2020)

6. Ekkel, S.V.: Some issues of construction and repair of cement concrete coverings of roads and aerodromes **6**, 78–86 (2017)
7. Nosov, V.P., Dobrov, E.M., Chistyakov, I.V., Borisiuk, N.V., Fotiadi, A.A.: Mathematical modelling of cracking process in concrete pavement highways **12**, 13158–13164 (2017)
8. Ushakov, V., Jelagin, D., Dyakov, G., Dmitriev, S., Goryachev, M.: Field evaluation of rutting in concrete pavements. IOP Conf. Ser.: Mater. Sci. Eng. **832** (2019). International Conference on Digital Solutions for Automotive Industry, Roadway Maintenance and Traffic Control (DS ART 2019) Cholpon-Ata, Kyrgyzstan
9. Nosov, V.P., Ushakov, V.V., Fotiadi, A.A., Stepush, A.P.: Improving the regulatory framework for the design of rigid pavements. IOP Conf. Ser.: Mater. Sci. Eng. **832**(1) (2020)
10. Park, S., Pour-Ghaz, M.: Another look at the abrasion resistance of concrete. Adv. Civil Eng. Mater. **8**(1), 423–434 (2019)
11. Permyakov, V.B., et al.: Negative factors analysis and defects accumulation in asphalt layers of road pavements during life cycle. In: Proceedings of Kazan State Architectural and Construction University, vol. 2, no. 14, pp. 261–265 (2010)
12. Ponomarev, A.: New technologies and materials introduction in highways construction. Highways **3**, 86–88 (2021)
13. Sheinin, A.M., Ekkel, S.V.: Modern requirements for road concrete. ALITinform: Cement Concrete Dry mixes **3**, 58–66 (2011)
14. Stepushin, A.P., Chistyakov, I.V., Goryachev, M.G., Sadovnikova, Y.S., Kyaing, A.K.: Refined assessment of strength of concrete pavement probabilistic method **832**(1) (2020)
15. Ushakov, V.: Modern trends in highways development with concrete coatings printed. Road Power **86**, 37–39 (2018)
16. Ushakov, V., Dyakov, G.: Abrasion investigation of highways concrete coatings. Sci. Technol. Road Ind. **67**, 30–32 (2014)
17. Ushakov V.V., Dyakov G.G.: Investigation of rutting on cement concrete pavements of highways. Collection of scientific works of MADI, pp. 138–150 (2013)
18. Ushakov, V., Jelagin, D., Dyakov, G., Dmitriev, S., Goryachev, M.: Field evaluation of rutting in concrete pavements **832**(1) (2020)
19. Yakobson, M.Ya., Kuznetsova, A.A., Vvedenskaya, A.S., Bychkov, A.V.: Relevance and prospects of cement concrete application in road construction. Syst. Technol. **1**(18), 132–140 (2016)
20. Yoshitake, I., Ueno, S., Ushio, Y., Arano, H., Fukumoto, S.: Abrasion and skid resistance of recyclable fly ash concrete pavement made with limestone aggregate. Constr. Build. Mater. **112**, 440–446 (2016)

Modification of Harsh Cement Pavement Concretes with Bitumen Emulsion

Aidar Garipov, Dmitry Makarov, Vadim Khozin, and Sergey Stepanov(✉)

Kazan State University of Architecture and Engineering, 420043 Kazan, Russia

Abstract. To this date, the total length of highways in Russia is more than 1.5 million km. And the problem of building high-performance roads is relevant due to the climatic features of Russia. Road surfaces must be frost-proof and chemical-resistant and must withstand prolonged cyclic compression and bending loads. These impacts lead to stresses in the concrete matrix, which cause cracks development. The primary objective in the development of road concrete is to increase the bending strength of concrete thus raising the service life of the coating from 20 to 50 years. This can be facilitated by new types of cement concrete, for example, roller compacted concrete pavement with a reduced modulus of elasticity in the road base courses. Roller compacted concrete pavement with the lowest possible content of water and adhesive has the maximum degree of material density, while ensuring the formation of a dense structure. Such concretes can be obtained by using damping bitumen emulsion. This paper examines the theoretical aspects of the structure formation of concrete with bitumen emulsion injected in it. It studies the main operational properties of road concretes modified with high-dispersion anion-active bitumen emulsion. It was found that modifying harsh concrete with anion-active bitumen emulsion boosts the performance properties of roller compacted concrete pavement, namely: compressive strength (by 32%), bending strength (by 38%), and cold-resistance (by 2 times). In addition, this paper offers a determination of the optimum modifying additive: 5% of the anion-active bitumen emulsion of cement mass for harsh ready-mix concretes.

Keywords: Road construction · Concrete · Pavement concrete · Bitumen emulsion · Anion-active emulsifier · Ready-mix concrete · Deflection strength · Compressive strength · Cold-resistance

1 Introduction

Over the last twenty years, the number of heavy-duty vehicles and traffic intensity on the roads has been increasing abroad. This leads to higher requirements for reliability and durability of road pavement, which determines the relevance of research in this area [1–8]. Asphalt concrete surfaces do not always meet these requirements and, in addition, they often need repair. One of the solutions to this problem is the construction of cement concrete coatings offering several advantages over asphalt concrete, the main of which are greater durability and resistance to sustained loading. The disadvantages of the monolithic cement concrete coatings construction with the concrete pouring in

N. Vatin (Ed.): STCCE 2021, LNCE 169, pp. 351–363, 2021.
https://doi.org/10.1007/978-3-030-80103-8_38

sliding formwork include a longer construction period due to the development of concrete strength when hardening; high initial cost of the coating compared to asphalt concrete; and low maintainability.

One of the most promising and effective areas of harsh road surface application is the use of roller-compacted concrete pavement as the base for asphalt pavement in the reconstruction and modernization of main-line and local highways.

The existing experience of such structures application in Russia and abroad has shown a number of advantages compared to the traditional road pavement, namely:

- low shrinkage when exposed to temperature;
- increased smoothness and hardness of the road base course, which reduces the number of upper bituminous concrete layers;
- long service life and, as a result, low road maintenance costs;
- simplified construction technology with the use of available tools and mechanisms.

Moreover, one of the main advantages is increased crack resistance of the bituminous concrete surface on such a base. These road structures are required for the construction of roads for vehicles with heavy axle and mechanical loads, for example, in the construction of large main-line roads, and reinforced roads with increased resistance to harsh climatic conditions [9].

However, the base itself made of roller compacted concrete pavement has a number of disadvantages that limit its use, such as low impact resistance, deformability, crack resistance, and cold-resisting quality.

New types of roller compacted concrete pavement manufactured by using damping bitumen, bitumen and polymer emulsions can meet the challenge. Roller compacted concrete pavement modified with bitumen emulsion will have a reduced modulus of elasticity in the road base course, low water absorption index, high deflection and compressive strength, cold-resistance, deformability and, as a result, increased durability.

The intention of the present study is to consider the possibility of bituminization of harsh ready-mix concretes and the way it affects the physical and mechanical properties of pavement concrete.

2 Materials and Methods

As a coarse aggregate, the study used gravel with characteristics M800-1000 and $F_1$150, as well as crushed stone made of this gravel.

The following additives were used as concrete modifiers:

- air-entraining additive Master Air based on surfactants, in a dosage of 0.1 to 0.6%;
- plasticizer additive Master Glenium based on polycarboxylate ester, in a dosage of 0.2 to 0.9%.

The following materials were selected for production of cement concrete modified with bitumen emulsion.

Table 1. Physical and mechanical properties of Portland cement

Normal density of cement paste, %	Setting time, hour: min		Compressive strength, MPa			Product size	Soundness
	Beginning	Ending	steaming	3 days	28 days		
24.6–26.4	3:15	4:15	43.3	36.3	56.5	97.8	OK

Table 2. Chemical composition of cement

Content of sulfur oxide (VI) SO_3	Fraction of alkalic oxides R_2O expressed as Na_2O	Content of chloride ion Cl^-
3.00	0.79	0.019

Table 3. Chemical and mineralogical composition of hard-burned brick

Tricalcium silicate C_3S	Dicalcium silicate C_2S	Tricalcium aluminate C_3A	Tetracalcium alumoferrite C_4AF	Magnesium oxide MgO
62.67	14.18	6.54	13.60	1.27

1. Portland Cement CEM I 42.5 N (Tables 1, 2 and 3).
2. Anionic slow-breaking bitumen road emulsion based on flotation asphaltic bitumen [10] (Table 4).
   As for its technical properties, ABE meets the requirements of regulatory documentation and belongs to the ASBRE brand (anionic slow-breaking bitumen road emulsion).

Table 4. Characteristics of ASBRE bitumen emulsion

Characteristics	Anion-active bitumen emulsion (hereinafter referred to as ABE)	ASBRE standard
1. Bitumen fraction, %	55	55–65
2. Max. relative viscosity of the emulsion at 20 °C, sec	20	<40
3. Max. uniformity in sieve No. 0.14, %	0.3	<0.4
4. Max. stability after 7-day storage, %	0.35	<0.5
5. Transport stability	Stable	No breaking of the emulsion into water and bitumen
6. Min. adhesion to aggregate, %	75	>50

3. Crushed stone of 5–10 M400 fraction (Table 5).
   Regarding its technical properties, the crushed stone meets the requirements of the regulatory documentation.
4. Sand and gravel mix (SGM) (Table 6).

Table 5. Technical properties of crushed stone

Item No	Indicator name	Actual indicator
1	Stone type	Carbonate rock
2	Bulk density, kg/m^3	1 300
3	Total breakdown product on sieves, %	
	25 mm	0.9
	20 mm	11.8
	10 mm	65
	5 mm	83
	1.5 mm	95
	0 mm	100
4	Content of silt and clay particles, %	≤ 3
5	Abrasion capacity grade	I-IV (И-IV)
6	Compressive strength grade	M400
7	Weak grain content, %	≥ 10
8	Cold-resisting quality	F50
9	Flatness, %	2–3
10	Clay content in clunches, %	≤ 0.25

Table 6. Technical properties of the SGM

Item No.	Indicator name	Actual indicator
1	Humidity, %	4.7
2	Content of gravel exceeding 5 mm, %	20.7
3	Grain content of sand fractions up to 5 mm, %	72.3
4	Modulus of SGM sand size	2.6
5	Content of silt and clay particles, %	1.2
6	Including clay in clunches	0.3
7	Content of organic impurities	Within limits
8	Bulk weight	1,750
9	Granularity grade	1,000
10	Cold-resisting quality grade	F100
11	Specific activity of radionuclides, Bq/kg	19
12	Filtration coefficient (at optimum water content and maximum density)	Min. 5 m/day

Regarding its technical properties, SGM meets the requirements of the regulatory documentation.

Concrete compression and bending under tension tests were carried out according to the standard regulatory methods of the Russian Federation.

3 Results and Discussion

The study of scientific and technical literature has revealed many ways to combine organic and cementing binders and methods of their modification. [11, 12] The most effective of which is the adding of ready-made bitumen emulsion into the ready-mix concrete at the stage of its preparation [13]. Slow-breaking emulsions are used to prepare organo-mineral mixtures.

However, though seemingly straightforward, the preparation of combined mixtures includes a separate issue: a process of distributing the bitumen emulsion in cement concrete and creating its uniform structure. Therefore, the right choice of the emulsion type is crucial for creating high-quality concrete.

Thermodynamic processes described by the laws of physical chemistry on surface phenomena in dispersed systems take place from the beginning of the preparation of a ready-mix concrete modified with bitumen emulsion to the finally-formed cement concrete. The processes of structure formation are spontaneous and occur towards the stable equilibrium state. Regarding the second law of thermodynamics, all spontaneous processes are exothermic and go with the release of energy, i.e. the total system energy decreases. The following issues were considered theoretically based on these speculations:

1) Wetting of solid components with water;
2) Stabilization of bitumen emulsion in cement concrete;
3) Concrete mix compaction;
4) Bituminous film formation;
5) Bitumen adhesion to mineral materials.

A theoretical analysis has shown that the distribution of the bitumen emulsion is carried out by adhesion to the cement particles. Cement grains adsorb Ca++ ions in water, and their surface acquires a positive potential (Fig. 1).

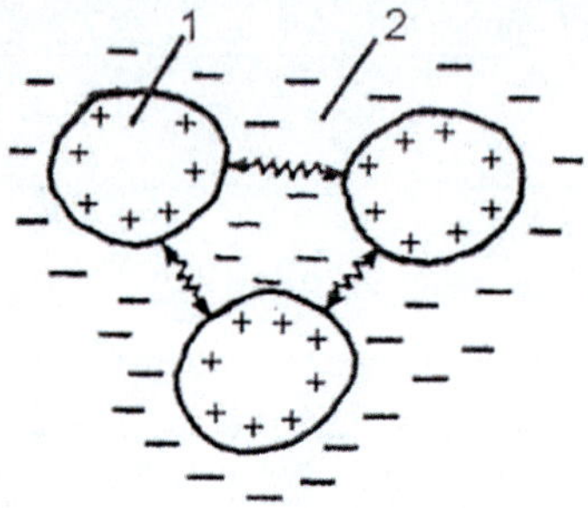

Fig. 1. Structure of cement grains sealed with water

1 – cement particle;
2 – water.

The higher the charge density and the smaller the grains are the greater the effect of repulsive forces is [14]. Bitumen emulsions stabilized by anion-active surface-active agents are dissociated in alkaline solutions this fact leads to the formation of negatively charged surface-active ions (anions).

Bitumen emulsion stabilization is based on the ideas of B.V. Deryagin [15] on boundary layers and P.A. Rebinder [16] on selective wetting by solid-body fluids. Bitumen emulsion stabilization by cement grains capable of hydrolytic dissociation is explained by bitumen emulsion stabilization within the scope of the dispersion medium and is carried out by the action of a double-electric layer resulting from the saponification reaction (Fig. 2).

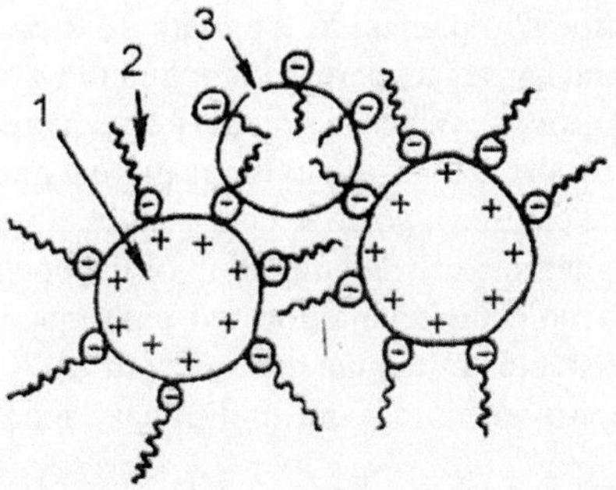

Fig. 2. Structure of cement grains sealed with water.

1 – cement particle;
2 – anion-active SAAs;
3 – bitumen grain stabilized with an anion-active SAA.

Anion-active surface-active agents (SAAs) give the bitumen grains a negative charge and attract them to the positively charged cement grains. It provides good conditions for their stabilization. When exposed to external forces, bituminous grains act as balls that ensure a mixture flow, which leads to increased mixture mobility and cohesion. Thus, modifying cement concrete with anion-active bitumen emulsion can be expected to result in increased mixture plasticization. However, in the case of a high concentration of bitumen emulsion, there is a high probability of slowing down the cement hydration since the bitumen screening layer will prevent the interaction of water and cement.

In addition, the aqueous solution of the emulsifier includes anion-active SAAs in the form of globules that are able to adsorb on the surface of cement grains and participate in the formation of spatial coagulation structures (Fig. 2). As shown in the figure, being adsorbed on the cement surface, the SAA is oriented in such a way that hydrophilous polar groups, including anions, chemisorptively bind to the hydrating surface of the cement grains, while hydrophobic hydrocarbon radicals are turned to water. SAA also creates a "palisade" of oriented molecules on the surface of the cement grains.

It was established that anion-active SAAs consisting of higher fatty acids reflect X-rays in a thin layer like properly formed crystals while not being crystals. This is due to the fact that the SAA molecules in a thin film on the surface of solid grains are oriented, remaining parallel to each other. Moreover, the orientation is not limited to one layer; each subsequent layer is under the influence of the previous one, but with distance from the solid surface of the cement grains, the interaction weakens. As a result, we can see a multilayer formation of molecules, turned to each other by their polar groups. Such a layered structure of the shell creates slip planes at the weaker bonds and thereby facilitates the mutual movement of solid particles, plasticizing the ready-mix concrete. An external force increases the orientation of molecules consequently increasing the flexibilizing effect of the ready-mix concrete.

Hydrocarbon chains (radicals) of anion-active SAAs, on the contrary, have a weak molecular field. The coagulation bonds which occur in the water medium between the methyl groups are weaker than those between the polar ends of the molecules, and, with forces applied, they become slip planes. In the ready-mix concrete, the stronger force field of cement minerals and aggregate particles applied to water is replaced by a weaker field of the SAA methyl groups. When the ready-mix concrete is exposed to external forces, the presence of SAA shells contributes to the mutual movement of solid particles. On the contrary, at rest one can see a positive coagulation effect caused by hydrocarbon chains, preventing sedimentation and delamination of the concrete mixture.

Since the degree of the molecule orientation and, consequently, the plasticizing effect depends on external mechanical action, the efficiency of SAAs and ready-mix concrete is increased by more intensive mixing and preliminary mechanical activation of the cement.

Concrete has a heterogeneous structure, where certain areas may differ significantly in terms of their properties (Fig. 3).

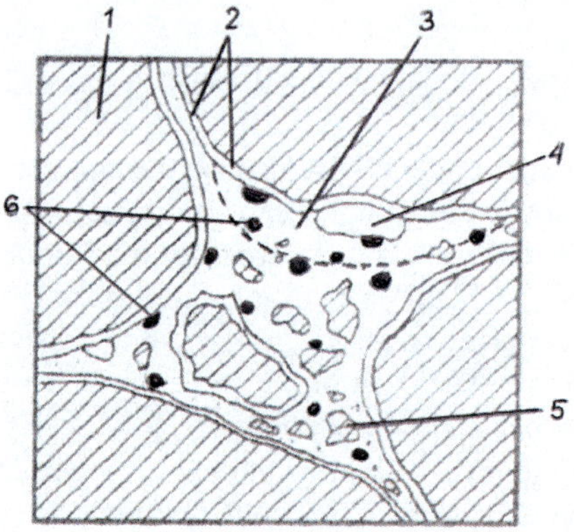

Fig. 3. Unit cell of concrete structure

1 – aggregate grains;
2 – contact area;
3 – area of structural weakening due to the formation of sedimentation holes;
4 – air bubbles;
5 – compacted structure area;
6 – bituminous grains formed after the emulsion decomposition.

Figure 3 shows a theoretical model of the distribution of bitumen extracted from the emulsion in the unit cell of the concrete structure. Concrete structure and properties can vary depending on the distribution of bituminous particles, anion-active SAA in the contact area, internal holes, and air bubbles.

Earlier, the authors had studied the most important operational characteristics of bitumen emulsion obtained on the basis of flotation asphaltic bitumen. They proved the applicability of the developed anion-active bitumen emulsion for the modification of cement concrete [10].

This paper examines the main operational properties of road cement concretes modified with high-dispersion anion-active bitumen emulsion. Ready-mix concretes were bituminized by introducing anion-active bituminous emulsions (ABE) (from 2.5% to 7.5% of the cement weight), which allowed increasing the performance properties of the hardened concrete. The ABE concentration was chosen in accordance with other studies [17–19].

The dependence of the compressive strength of cement concrete on the concentration of the bitumen emulsion was traced after 28 days of hardening under normal humidity conditions. As shown by Fig. 4, there is a sharp increase in concrete strength by 32% (from 11.5 MPa to 15.2 MPa) at 5% concentration of anion-active bitumen emulsion in modified concrete, followed by a decrease at 7.5% concentration of ABE. This stems from the fact that concrete modified with bitumen emulsion tends to gain strength more quickly over time due to the accumulation of clinker reserve and decreased water-solid ratio. With an increase of the bitumen emulsion concentration in concrete, the mechanical strength of concrete under compression decreases. This may be due to the fact that bitumen slows down cement hydration. With further aging of the modified concrete, the compressive strength will increase in relation to unmodified concrete. These results are consistent with other studies [20, 21].

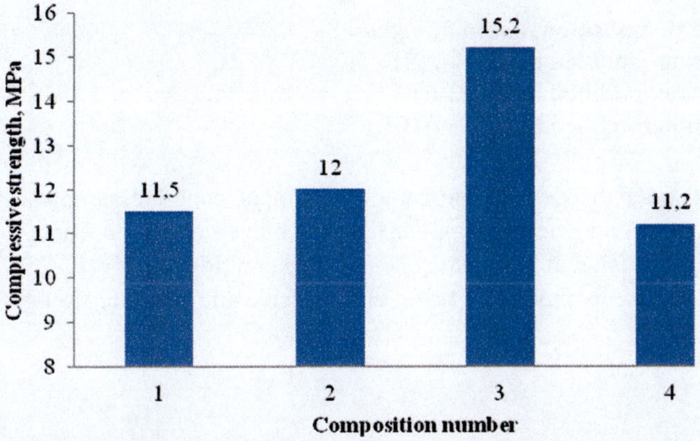

Fig. 4. Compressive strength of cement concrete samples modified with bitumen emulsion.

1 – Reference composition, not modified;
2 – Composition modified by 2.5% ABE;
3 – Composition modified by 5.0% ABE;
4 – Composition modified by 7.5% ABE.

The dependence of the bending strength of cement concrete completely associates with the indicators of compressive strength. With an increase of the bitumen emulsion concentration, the bending strength increases from 3.14 MPa to 4.35 MPa at 5% concentration of anio-active bitumen emulsion in modified concrete, followed by a decrease at 7.5% concentration of ABE to 3.9 MPa. As shown in Fig. 5, the bending strength of concrete increases by 38%.

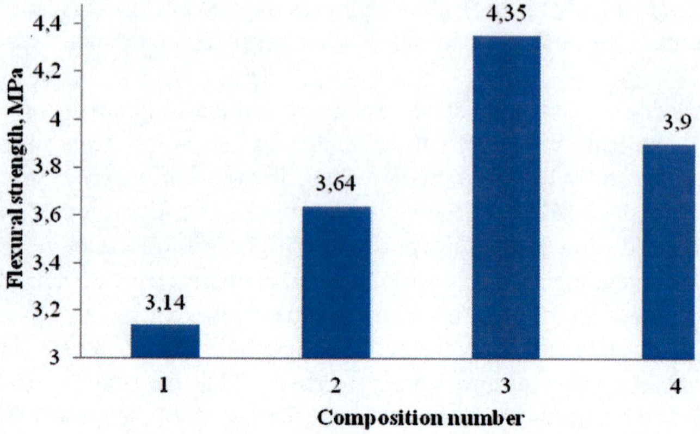

Fig. 5. Bending strength of cement concrete samples modified with bitumen emulsion

1 – Reference composition, not modified;
2 – Composition modified by 2.5% ABE;
3 – Composition modified by 5.0% ABE;
4 – Composition modified by 7.5% ABE.

Let us consider the density indicators of cement concrete samples. As shown in Fig. 6, the highest concrete packing density is achieved at 5% concentration of anion-active bitumen emulsion in modified concrete. The creation of a more uniform and dense structure of concrete explains its greater compressive and bending strength.

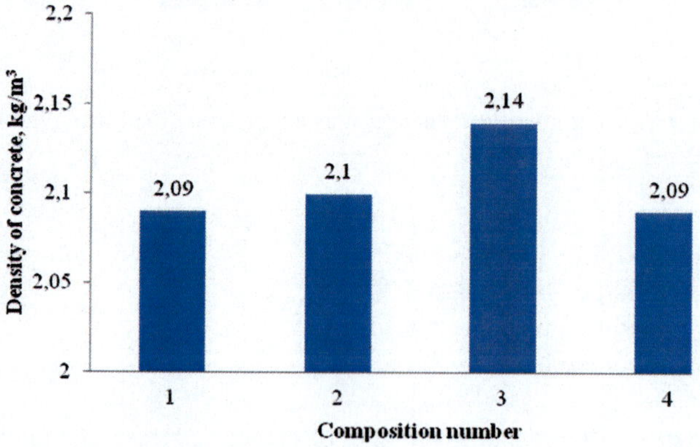

Fig. 6. Density of cement concrete samples modified with bitumen emulsion

1 – Reference composition, not modified;
2 – Composition modified by 2.5% ABE;
3 – Composition modified by 5.0% ABE;
4 – Composition modified by 7.5% ABE.

The cold resistance of cement concrete can be increased by the introduction of bitumen emulsion into the ready-mix concrete. This is due to the colmatation of the concrete holes with bitumen emulsion particles and a decrease in the water-solid ratio resulting from the increased mobility of the ready-mix concrete, which leads to a decrease in void volume and water absorption of concrete.

Table 7. Cold-resisting quality grade of cement concrete.

No.	Composition	Brand	Concrete grade
1	Reference composition, not modified;	F 50	B 7.5
2	Composition modified by 2.5% ABE;	F 75	B 12.5
3	Composition, modified 5.0% ABE;	F 100	B 15
4	Composition modified by 7.5% ABE	F 100	B 15

Table 7 shows that the cold resistance index increases by 2 times from 50F freezing and thawing cycles to F75 at 2.5% and to 100F at 5% concentration of anion-active bitumen emulsion, respectively. A further increase of the bitumen emulsion concentration in the concrete structure does not lead to an increase in cold resistance due to a drop in strength properties and a lower density of concrete.

4 Conclusions

1. A model of the distribution of bitumen extracted from the emulsion in the unit cell of the concrete structure is theoretically justified.
2. It has been proven that it is possible to bituminize cement concrete by introducing a high-dispersion anion-active bitumen emulsion into the ready-mix concrete.
3. The introduction of bitumen emulsion (from 2.5 to 7.5% from cement weight) improves the physical and operational characteristics of concrete: compressive strength increases by 32%, bending strength – by 38%, cold-resisting quality – by 2 times.
4. The optimal concentration of the modifying additive has been established. The introduction of 5% concentration of anion-active bituminous emulsion in the modified concrete creates a more uniform and dense concrete structure with the best physical and operational properties.

References

1. Yao, C., Guo, Y., Shen, A., Cui, W., He, Z.: Recycling of fine-asphalt-pavement solid waste for low-shrinkage rapid hardening Portland cement concrete pavement. Constr. Build. Mater. **289**, 123–132 (2021). https://doi.org/10.1016/j.conbuildmat.2021.123132
2. Rocha Segundo, I., Silva, L., Palha, C., Freitas, E., Silva, H.: Surface rehabilitation of Portland cement concrete (PCC) pavements using single or double surface dressings with soft bitumen, conventional or modified emulsions. Constr. Build. Mater. 281 (2021). https://doi.org/10.1016/j.conbuildmat.2021.122611
3. Fotiadi, A.A., Gnezdilova, S.A., Strekha, I.S.: Remote method for predicting damage to cement concrete pavements. In: Klyuev, S.V., Lesovik, V.S., Vatin, N.I. (eds.) BUILDINTECH BIT 2020. LNCE, vol. 95, pp. 333–339. Springer, Cham (2021). https://doi.org/10.1007/978-3-030-54652-6_50
4. Grosek, J., Zavrel, T., Stryk, J.: Mitigation possibilities of concrete pavement degradation. IOP Conf. Ser. Mater. Sci. Eng. **1039** (2021). https://doi.org/10.1088/1757-899X/1039/1/012018
5. Yang, M., Paudel, S.R., Gao, Z.J.: Snow-proof roadways using steel fiber-reinforced fly ash geopolymer mortar-concrete. J. Mater. Civ. Eng. 33 (2021). https://doi.org/10.1061/(ASCE)MT.1943-5533.0003537
6. Pavlov, V., Krainov, D., Akhmerova, G.: Influence of electric heating on concrete strength of individual sections of monolithic reinforced concrete multi-span slabs. Vestnik grazhdanskikh inzhenerov **77**, 111–113 (2019). https://doi.org/10.23968/1999-5571-2019-16-5-111-113
7. Wathiq Hammodat, W.: Investigate road performance using polymer modified concrete. Mater. Today Proc. **42**, 2089–2094 (2021). https://doi.org/10.1016/j.matpr.2020.12.290
8. Mukhametrakhimov, R., Aliullova, I.: Improvement of the quality control system for expansion joints with rubber compensators during the repair of bridge constructions. Izvestiya KGASU 53 (2020)
9. Pizhov, A.: Technology of road concrete with disperse bitumen. Vestnik volgogradskogo gosudarstvennogo arkhitekturno-stroitel'nogo universiteta. Seriya: stroitel'stvo i arkhitektura **38**, 51–57 (2010)
10. Khozin, V.G., Makarov, D.B.: Bitumen emulsion. Patent RF, no. 2185878 (2002)
11. Khozin, V., Khokhryakov, O., Nizamov, R.: A «carbon footprint» of low water demand cements and cement-based concrete. IOP Conf. Ser. Mater. Sci. Eng. **890** (2020). https://doi.org/10.1088/1757-899X/890/1/012105
12. Vdovin, E., Mavliev, L., Stroganov, V.: Interaction of clay soil components with portland cement and complex additive based on octyltriethoxysilane and sodium hydroxide. IOP Conf. Ser. Mater. Sci. Eng. 890 (2020). https://doi.org/10.1088/1757-899X/890/1/012031
13. Bushneva, E.: Cement mortars and concretes with modified bitumen emulsion additives. Dissertation of the candidate of technical sciences. Moscow (2011)
14. Bazhenov, Y.: Concrete Technology. ASV Publishing House, Moscow (2011)
15. Deryagin, B., Sherbakov, A.: Influence of surface forces on phase equilibria of polymolecular layers and contact angle. Kolloidnyy zhurnal. **1**, 65–69 (1961)
16. Rebinder, P.: Surface phenomena in dispersed systems. Physicochemical mechanics (The science), Moscow (1979)
17. Dareyni, M., Mohammadzadeh Moghaddam, A., Delarami, A.: Effect of cationic asphalt emulsion as an admixture on transport properties of roller-compacted concrete. Constr. Build. Mater. **163**, 724–733 (2018). https://doi.org/10.1016/j.conbuildmat.2017.12.156
18. Tian, Y., Lu, D., Ma, R., Zhang, J., Li, W., Yan, X.: Effects of cement contents on the performance of cement asphalt emulsion mixtures with rapidly developed early-age strength. Constr. Build. Mater. 244 (2020). https://doi.org/10.1016/j.conbuildmat.2020.118365

19. Dareyni, M., Mohammadzadeh Moghaddam, A.: Fresh and mechanical properties of roller compacted concrete containing Cationic Asphalt Emulsion admixture. Constr. Build. Mater. **198**, 226–236 (2019). https://doi.org/10.1016/j.conbuildmat.2018.11.186
20. Fang, X., Garcia, A., Winnefeld, F., Partl, M.N., Lura, P.: Impact of rapid-hardening cements on mechanical properties of cement bitumen emulsion asphalt. Mater. Struct. **49**(1–2), 487–498 (2015). https://doi.org/10.1617/s11527-014-0512-3
21. Miljković, M., Radenberg, M., Fang, X., Lura, P.: Influence of emulsifier content on cement hydration and mechanical performance of bitumen emulsion mortar. Mater. Struct. **50**(3), 1–14 (2017). https://doi.org/10.1617/s11527-017-1052-4

Numerical Study of the Influence of the Inlet Geometric Parameters on the Jet Characteristics

Juliya Kareeva(✉), Raushan Zakieva, and Kseniya Bliznjakova

Kazan State University of Architecture and Engineering, 420043 Kazan, Russia

Abstract. Calculation of air distribution in premises is one of the main stages when designing ventilation and air conditioning systems of a building. The microclimate quality and the efficiency of ventilation and air conditioning systems depend on the accuracy and correctness of the calculations. The purpose of this study is to assess the influence of the distance between the inlet and the 45-degree turn on the jet characteristics. The basic geometric and kinematic characteristics of the jet flowing from the opening located at various distances from the 45-degree turn were obtained. The change in the deflection of the outflow angle and the kinematic coefficient of the inlet opening were obtained.

Keywords: Jet flow · Numerical method · Computational fluid dynamics · Fluent · Air distribution · Flow characteristics

1 Introduction

Calculation of air distribution in the premises is one of the main stages when designing ventilation and air conditioning systems of a building. The microclimate quality and the efficiency of ventilation and air conditioning systems depend on the accuracy and correctness of the calculations. It should take into account the need for rational use of resources and compliance of the supplied air parameters with regulatory requirements. By changing the feed angle, one can provide the required microclimate parameters without additional equipment, and thereby optimize the costs of ventilation and air conditioning.

This is especially important in rooms with increased heat generation, where one has to use air exchange schemes with the supply of jets directly to the working area.

Many works consider an urgent issue of the influence of the geometric and kinematic parameters of the supply jet on the air distribution in rooms for various purposes. The results of a numerical experiment using the CFD software package are presented in [1–6]. The choice of a numerical model for various cases of air distribution is considered in [2–7]. The detailed studies of the air distribution scheme "from top to bottom" in a flat and spatial setting are presented in [8–13]. In the articles [2, 3, 13], the velocity fields in an air jet supplied by recirculation air diffusers were obtained numerically. The authors [14–16] carried out a series of studies on vortex zones and the local resistance coefficient (LRC) of exhaust openings.

N. Vatin (Ed.): STCCE 2021, LNCE 169, pp. 364–371, 2021.
https://doi.org/10.1007/978-3-030-80103-8_39

The works [17, 18] study the influence of the geometric characteristics of the inlet located at different distances from the 90-degree turn of an air duct. The geometric and kinematic characteristics of jet as a function of distance were obtained. However, due to the limited room space, the air in the working area must be supplied at other angles rather than 90°.

For geometrically complex air distributors, the kinematic coefficient is not constant along the length of the main jet section [19]. So, it becomes necessary to take into account the unevenness of velocity profiles in jets, formed during an outflow from nozzles with different geometry and direction [20].

The purpose of this work is to determine the influence of the distance between the inlet and a 45-degree turn on the jet characteristics.

To achieve this purpose, the following tasks were set:

- To study numerically the jet outflow from the inlet located at different distances from the 45-degree turn.
- To determine the patterns of change in geometric and kinematic characteristics as a function of the distance from the air duct turn.

2 Materials and Methods

The study was carried out numerically using the Fluent software package. The standard k-e turbulence model was adopted as in [17, 18]. Standard near-wall functions were used to model the near-wall boundary layer.

Figure 1 shows the geometry of the studied area. The jet is fed from an inlet opening with a width of $b_O = 0.2$ at a constant velocity $u_O = 3$ m/s. The inlet is located at a distance $l = 0...2$ m after the air duct turns at the angle $\alpha = 45°$. The considered area has the following dimensions: $b = 8.0$ m, $h = 4.0$ m, $h_l = 0.7$ m.

We adopted the following boundary conditions:

- AB (feed opening) – Velocity inlet: velocity is uniform and directed along the normal to the boundary: $u_O =$ const; $k = 0$; $\varepsilon = 0$;
- FG, GH, HI, IK (free flow boundaries) – Pressure Outlet: excess pressure $\Delta p = 0$; velocity is directed along the normal to the boundary - u - $= u_n$, $dk/dn = 0$, $d\varepsilon/dn =$ 0;
- AF, FE, BC and CD (impermeable walls) Wall: $u = 0$, $(du_n)/dn = 0$; d/dn are derivatives along the normal to the boundary.

As a result of the solution, the basic geometric and kinematic characteristics of the jet flowing from the opening located at different distances between $\bar{l}$ and the 45-degree turn were obtained.

The results are presented in dimensionless form:

- $\overline{u_x} = u_x/u_0$; $\overline{u_y} = u_y/u_0$; $\overline{u_{av}} = u_{av}/u_0$ are the longitudinal and transverse components and mean velocity, respectively;
- $\overline{P_{st}} = P_{st}/P_D$ is the static pressure;
- $\bar{x} = x/b_0$; $\bar{y} = y/b_0$ are the geometric parameters.

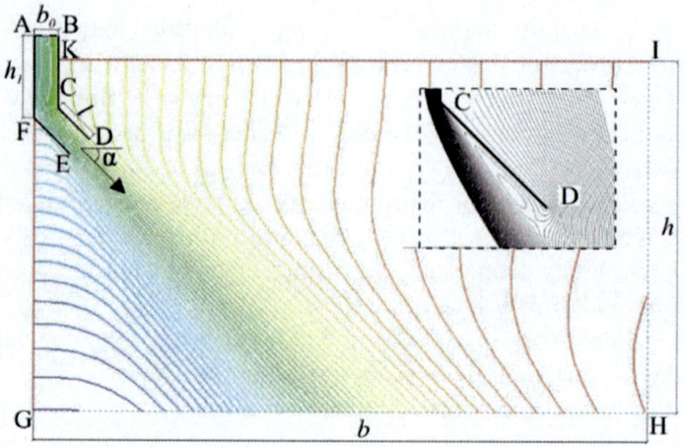

Fig. 1. Geometry of the considered area and streamlines

3 Results and Discussion

Figure 1 shows the streamlines of jet flowing out from the inlet located after a 45-degree turn. In the previously considered case [17, 18], a vortex region was shown, which had a significant influence on the jet characteristics. The jet flattened only at a distance of $\bar{l} > 3.0$ from the turn. In case of a 45-degree turn, a vortex zone is also formed along the upper wall CD, due to which the jet breaks off and the outflow angle changes. In this case, the vortex region affects the jet only at a small distance from the turn $\bar{l} < 1.0$.

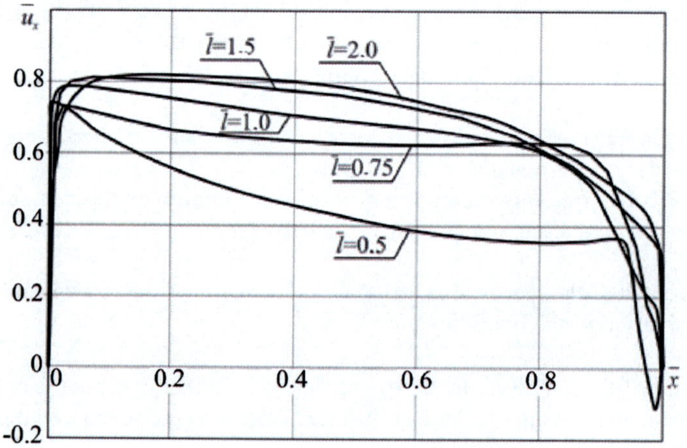

Fig. 2. Longitudinal velocity component

Figures 2 and 3 show graphs of changes in dimensionless longitudinal and transverse velocities, respectively. Since the jet is directed at an angle of 45 degrees, the longitudinal and transverse velocities are comparable at all lengths. At the length of $\bar{l} = 0.5$ velocities

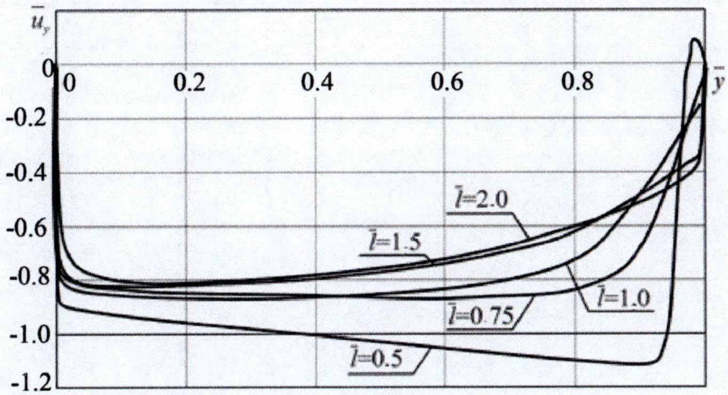

Fig. 3. Transverse velocity component

change the signs, which indicates the presence of a vortex zone at the upper wall of the duct. Further, the velocity profile is practically flattened, there is no vortex zone. At a distance of $\bar{l} > 1.5$, the profiles of the longitudinal and transverse components of velocity become equal, the disturbances generated during the turn do not have an influence. A greater distance between the inlet and the turn does not affect the change in the jet characteristics.

The maximum velocity in the jet sections was used to determine the angle at which the outflow from the inlet opening occurs. Table 1 shows the deflection of the jet outflow angle from 45°:

$$|\Delta\alpha| = \alpha - 45, {}^{\circ} \quad (1)$$

Table 1. The deflection angle

$\bar{l}$, m	0.5	0.75	1.0	1.5	2.0
Δα, °	14.3	11.5	3.8	5.5	5.3

Figure 4 shows a graph of changes in the deflection angle from the jet axis. It also shows the relationship between the angle of jet deflection from the horizon when the inlet is located after the 90-degree turn. Due to the strong influence of the vortex zone after the 90-degree turn, the jet deflects more from the axis and this influence remains significant up to $\bar{l} = 0.4$. In case of a 45-degree turn, the influence of the vortex zone ends already at $\bar{l} = 0.2$. Further, the deflection angle approaches 0 and remains constant. In the area $\bar{l} < 0.2$, the angle change can be described by the following formula:

$$\alpha = 23.834 l^{(-3.106)} \quad (2)$$

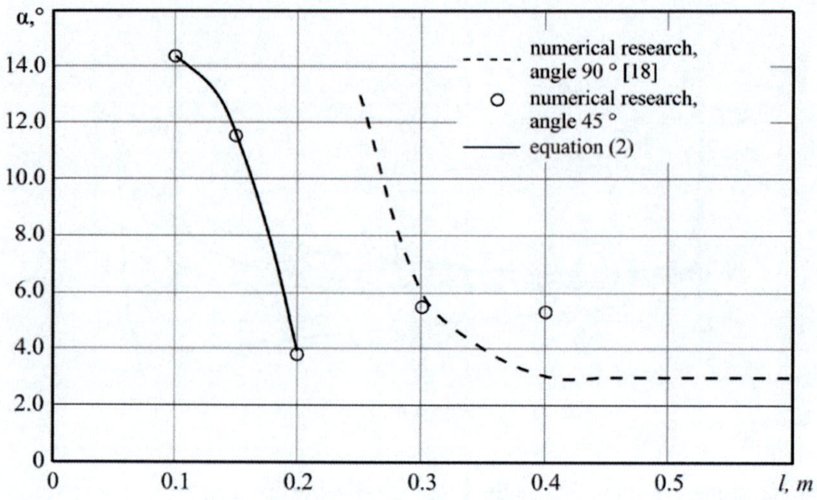

Fig. 4. Jet axis deflection angle

Figure 5 shows the change in average velocity at the jet axis for various distances $\bar{l}$ from the turn. The solid line in Fig. 5b shows the change in the average velocity calculated using the following equation:

$$u_m = \frac{mu_o\sqrt{b_0}}{\sqrt{\eta}}, \tag{3}$$

where $\eta = x \cdot \cos\alpha$ is the dimensionless distance from the center of the outflow along the jet axis.

The change in average velocity at various distances from the turn can be described by Eq. (3). For $\eta < 2$ the results obtained numerically and using Eq. (3) have a significant discrepancy, since this formula is suitable only for describing the change in velocity in the main section of the jet. For $\eta < 2$ the flow is aligned and the main section is formed. The coefficient m for the inlet located after the 45-degree turn of the air duct depends on the distance $\bar{l}$ and it is in the range of 1.45–1.65 (Table 2).

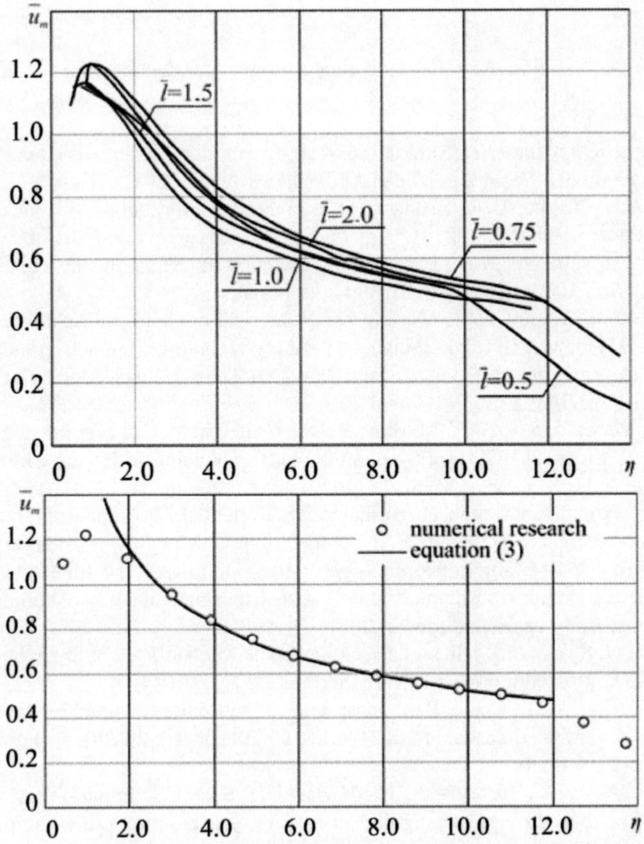

Fig. 5. Average velocity on the jet axis

Table 2. Kinematic coefficient of the inlet opening

$\bar{l}$, m	0.5	0.75	1.0	1.5	2.0
m	1.55	1.65	1.45	1.5	1.6

4 Conclusions

The main geometric and kinematic characteristics of the jet flowing from the inlet located after the 45-degree turn of the air duct were obtained numerically. As a result of the turn, a vortex zone is formed on the upper wall of the air duct, which affects the outflow jet parameters and along its length only at $\bar{l} < 1.0$. It was shown that the jet outflow angle had a deflection from 45°. The change in the average velocity in the main section of the jet is well described by the well-known formula for a free jet. In this case, the kinematic coefficient of the inlet opening is not constant and depends on the distance from the turn.

References

1. Shopayeva, A., Safiullin, R.: CFD-modeling of flow in confluence nodes of ventilation units of multi-storey buildings. IOP Conf. Ser.: Mater. Sci. Eng. (2020). https://doi.org/10.1088/1757-899X/890/1/012157
2. Kocharyants, K.V.: Numerical simulation of air distribution by fair overlaying jets. Selection of turbulence model. Vestnik grazhdanskih inzhenerov **57**, 128 (2016)
3. Naumov, A.L., Kapko, D.V., Brodach, M.M.: Ventilation systems with local recirculation diffusers. Energy Build. 85 (2014). https://doi.org/10.1016/j.enbuild.2014.09.050
4. Ivanov, N.G., Zasimova, M.A.: Large eddy simulation of airflow in a test ventilated room. J. Phys. Conf. Ser. **1038**, 12136 (2018). https://doi.org/c
5. Zasimova, M.A., Ivanov, N.G., Markov, D.: Numerical modeling of air distribution in a test room with 2D sidewall JET. I. Foundations for eddy-resolving approach application based on periodical formulation. St. Petersbg. State Polytech. Univ. J. Phys. Math. **13** (2020). https://doi.org/10.18721/JPM.13305
6. Zasimova, M.A., Ivanov, N.G., Markov, D.: Numerical modeling of air distribution in a test room with 2D Sidewall JET. II. LES-computations for the room with finite width. St. Petersbg. State Polytech. Univ. J. Phys. Math. **13** (2020). https://doi.org/10.18721/JPM.13306
7. Ricci, A., Kalkman, I., Blocken, B., Burlando, M., Repetto, M.P.: Impact of turbulence models and roughness height in 3D steady RANS simulations of wind flow in an urban environment. Build. Environ. **171** (2020). https://doi.org/10.1016/j.buildenv.2019.106617
8. Nguyen, Ch.K., David, D., Kuznik, Fr.: Full-scale experimental study of ceiling turbulent air jets in mechanically ventilated rooms (2017)
9. Xie, H., Zhang, S., Guan, X.: Shanghai Ligong Daxue Xuebao/J. Univ. Shanghai Sci. Technol. **39**, 81 (2017). https://doi.org/10.13255/j.cnki.jusst.2017.01.014
10. Teodosiu, C., Ilie, V., Teodosiu, R.: Appropriate CFD turbulence model for improving indoor air quality of ventilated spaces. Math. Model. Civil Eng. **10**, 28 (2015). https://doi.org/10.2478/mmce-2014-0020
11. Mularski, J., Arora, A., Saeed, M.A., Niedźwiecki, Ł., Saeidi, S.: Impact of turbulence models on the air flow in a confined rectangular space. Eng. Sci. Technol. (2020). https://doi.org/10.37256/est.212021627
12. Teodosiu, R., Niculiţă, L., Teodosiu, C.: Computational fluid dynamics based modeling of a linear heat source. Environ. Eng. Manag. J. **13** (2014). https://doi.org/10.30638/eemj.2014.216
13. Tabunshchikov, Y.A., Shkarpet, V.E., Kapko, D.V., Brodach, M.M., Kochariantc, K.V.: Installing local recirculation air diffusers during building deep renovation reduces energy consumption of ventilation systems. Energy Procedia (2016). https://doi.org/10.1016/j.egypro.2016.09.149
14. Ziganshin, A.M., Batrova, K.E., Logachev, K.I., Mingazeeva, D.N.: Numerical modelling of energy efficient exhaust orifices and hoods in ventilation and air conditioning systems in buildings and facilities of thermal power plants. IOP Conf. Ser.: Earth Environ. Sci. (2019). https://doi.org/10.1088/1755-1315/288/1/012134
15. Logachev, K.I., Ziganshin, A.M., Averkova, O.A., Gol'tsov, A.B.: Reducing power consumption of local exhaust ventilation systems. J. Phys.: Conf. Ser. (2020). https://doi.org/10.1088/1742-6596/1683/4/042015
16. Logachev, K.I., Ziganshin, A.M., Averkova, O.A., Logachev, A.K.: A survey of separated airflow patterns at inlet of circular exhaust hoods. Energy Build. **173** (2018). https://doi.org/10.1016/j.enbuild.2018.05.036
17. Kareeva, J., Bliznyakova, K., Ashadullina, D., Zakieva, R.: Influence of geometrical parameters of air inlet hole on the kinematic characteristics of jet. IOP Conf. Ser.: Mater. Sci. Eng. (2020). https://doi.org/10.1088/1757-899X/890/1/012164

18. Kareeva, J.R., Varsegova, E.V., Bliznyakova, K.A., Zakieva, R.R.: Study of the influence of geometric parameters of the inlet opening on jet characteristics. Izv. KazGASU [News KSUAE] **54**, 104 (2020)
19. Kocharyants, K.V., Denisikhina, D.M.: Constancy of the kinematic coefficient during outlet from modern air distribution devices. Naychnoe obozrenie. **10**, 40 (2017)
20. Posokhin, V.N., Varsegova, E.V., Ziganshin, A.M.: About the numerical determination of the characteristics of the supply nozzle. Izv. KazGASU [News KSUAE]. **39**, 173 (2017)

Gypsum-Fiber Radioprotective Facing Materials

Albert Galautdinov, Rustem Mukhametrakhimov(✉), and Valery Kupriyanov

Kazan State University of Architecture and Engineering, 420043 Kazan, Russia

Abstract. The most basic requirements to structures of buildings are enabling sanitary-hygienic conditions and conditions of human vital activity in indoor applications, and also indoor protection against negative exogenous factor, one of which is electromagnetic waves. It gives rise to topicality of research objective of the present study that consists in gypsum-fiber radioprotective materials production, that can be used inpartitions installation indoors with transmitters of increased electromagnetic interference. Dependences of attenuation ratio of electromagnetic waves on shungite content (0–30%) in tested compositions are determined in the article. It made clear that with shungite content increase we can observe increase of radioprotective properties of facing materials in studied frequency intervals of electromagnetic waves (2–6 GHz). It is found that shungite introduction in the quantity of 5–10% by weight of binder helps to increase flexural strength of gypsum-fiber facing materials by value up to 11.2%. The results of undertaken experimental studies can be used for production of gypsum-fiber radioprotective facing materials meant for protection of different facilities against electromagnetic inference effects.

Keywords: Electromagnetic interference · Composition materials · Gypsum-fiber facing materials · Radioprotective constructions · Radar-absorbent properties · Shungite

1 Introduction

Excess effect of electromagnetic interference and waves (EMI and EMW) has an adverse impact on human. Increasing number of EMI that appears in human environment [1] especially indoors [2] defines topicality of the research focused on human protection against factors dangerous to health by the way of development of radioprotective materials [3]. Growth of electromagnetic field is connected with increase of quantity of radio and television stations, with net extension of high-voltage power transmission lines [4] and nuclear power plants, with rapid evolution of mobile communication, with rise in radar installations and large-scale implementation of radioelectronic devices and devices radiating hyperfrequency waves [5]. Frequently, the emitter of EMI is devices and equipment located in neighboring accommodations where partition-wall is the only barrier between these rooms. With prolonged exposure of the human body to a radiation level exceeding the threshold criteria, persistent changes occur in the nervous, cardiovascular and endocrine systems [6, 7]. In this vein, problems focused on increase of radioprotective properties of enclosing structures and partition-walls under the influence of

N. Vatin (Ed.): STCCE 2021, LNCE 169, pp. 372–381, 2021.
https://doi.org/10.1007/978-3-030-80103-8_40

electromagnetic waves in accommodations with sources of high EMIgrow more urgent [8].

In the capacity of powder extenders of radioprotective composition materials, the materials with magnetic properties and high electrical conductivity (shungite) [9], dielectrics with high values of magnetic permittivity and semiconductors have widespread use. The most commonly used powders are metal and ferrite powders, technical carbon and fullerenes [10]. Introduction of stated materials in the form of powders to binders helps to produce radioprotective products with required values of EMI's energy reducing in working frequencies and temperatures.

There are different materials and methods for protection against EMI [12–14]. Application of radioprotective structure concrete what includes carbon-containing radar-absorbent filler in the form of structured gel is reported [15]. However, application of such material is more preferable for construction of radioprotective carrying walls. Special aspects of cement stone structure formation with carbon-siliciferous nanodispersed admixture based on shungite, reviewed in the study [16], shows their positive influence on early and long-time strength properties. Application of known porous radioprotective materials such as glass foam, foaming gypsum, and cellular concrete, foam composites, fillers based on porous or discrete aggregates is possible for construction of internal partitions.Radioprotective properties of these materials are achieved by means of introduction of grinded (discrete) carbon-contained components or radar-absorbent water-based compositions [17, 18]. For instance, coal soot is added into melt of glass during foam glass production. Moreover, there is well-known radioprotective multilayer material that is used in the capacity of polyfunctional glazing [19], containing the layer of organic glass and two layers of polymeric film with functional conductive thermal protection coating. There are also known radioprotective facing plasters with sand-cement and gypsum base intended for EMI level reduction by the way of screening. Their properties of EMI attenuation depend on radar-absorbent compound concentration and layer thickness.

There are already a number of EMI shield materials based on shungite rocks and also plates for radiation protection. It should, however, be noted that nowadays the question of production of facing materials for interior finish of buildings that consist of a combination of strength, ecological, special (protection against EMI) propertiesand high degree of industrialization with application of rapid-hardening binder based on gypsum is poorly explored [10, 11]. With the use of such materials construction of light radioprotective partition walls is possible.

The purpose of the present study is production of gypsum radioprotective facing materials that can be used during construction of partition walls in accommodations with emitter of increased EMI.

2 Materials and Methods

In the course of investigations in the capacity of initial raw components for radioprotective materials production we used the following products:

- gypsum G6BII produced by «Arakchinskijgips» LLC, GOST (State Standard) 125–79;

- portland cement (PC) produced by Belgorod cement plant graded PC500-D0-N;
- active mineral admixture (AMA) naturally occurring – metakaolin with specific surface no less than 1300 m^2/kg, produced by means of burning of kaolin at the temperature of 700 °C within an hour. This AMA is chosen as best-performing in the light of literature data analysis and previous undertaken studies [20–23];
- fractionated shungite from Zazhoginskoe field with specific surface area of 200 m^2/kg and produced by «Nauchno-proizvodstvennyj kompleks «KARBON-SHUNGIT» LLC, Petrozavodsk. Chemical composition of shungite: SiO_2 – 57%; TiO_2 – 0,2%; Al_2O_3 – 4,3%; MgO – 1,2%; CaO – 0,3%; Na_2O – 0,2%; K_2O – 1,5%; S – 1,5%; C – 28%; H_2O_{cryst} – 3%.

In the capacity of chemical admixtures we used polyfunctional modifying additive [24] that includes:

- plasticizer of the first group based on carboxylates «Odolit-K» produced as per technical conditions (TC) 5745-01-96326574-08 by «Servis-Grupp» LLC. It has the form of semipellucid viscous liquid in hydrous solution and chloride-free, its density is 1.06 g/cm^3, pH at a temperature of 20 °C – 6.9;
- superplasticizing agent of the first group «BEST-SPl» produced by «Innovacionnye Tekhnologii» LLC. It has the form of aqueous solution of organic acids salts in the liquid chloride-free state, brown liquid, its density at a temperature of 20 °C is 1.1–1.3 g/cm^3, dry solid weight ratio is 30–40%.

Choice of active mineral and chemical additives is based on the results of work on modifying the structure of composite building materials based on gypsum [25, 26], cement [27–31] and mixed [32, 33] binders. The effectiveness of above-mentioned modifying agents is showed in studies [34–36].

For dispersed reinforcement of composition material under investigation we used polypropylene fibers graded VSM-II, the effectiveness of which is showed in the study [37], produced as per technical conditions (TC) 5458-001-82255741–2008 by «Si Ajrlajd» LLC, Chelyabinsk. Fibre characteristics: length – 6 mm, diameter – 20–22 μm, rupture resistance – 550 MPa, fiber elongation – 20%.

At the first stage of the studies of gypsum-fiber radioprotective facing materialsproduction minimum requirement quantity of the AMA, that excludes formation of high-sulphatic form of calcium aluminate sulfate hydrate, was measured. According to received experimental data it has been found that required metakaolin content in gypsum-cement composition amounts to 20% by weight of PC or 4% by weight of gypsum-cement-puzzolan binder (GCPB).

Quantity of gauged water was chosen on the assumption of having normal consistency of modified gypsum-cement-puzzolan mix as per GOST 23789-79.We added chemical admixtures (Odolit-K, Best-SPl) and reinforcing fibres for their preliminary opening to estimated quantity of water. After that, we made weight batching and intimate mixing of the binder components in the dry state (gypsum, portland cement, metakaolin) and shungite. We made samples of radioprotective facing materials with thickness of 10 mm from gypsum-cement fibrous mixes by the way of molding process.

Rheological test of GCPB was made according to GOST 23789-79 «Gypsum binders. Test methods». Bending strength test of gypsum-fiber radioprotective facing materials samples and their water absorption were made according to GOST R 51829-2001 «Gypsum fiber sheets. Specifications».

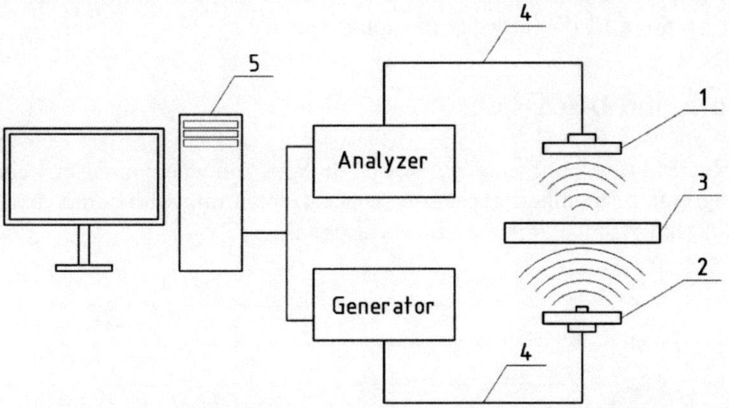

Fig. 1. Structural scheme of measuring equipment: 1 – receive antenna; 2 – transmission antenna; 3 – sample under investigation; 4 – interconnecting cable; 5 – computer with required computer software.

Fig. 2. Measuring equipment for analysis of EMI passing through the gypsum-fiber radioprotective facing materials under investigation.

Analysis of EMW that is passing through the gypsum-fiber radioprotective facing materials was made by means of FSH8 produced by ROHDE & SCHWARZ company. For these investigations we used the samples with thickness of 10 mm. The research was done at frequency interval of EMWs amount to 2–6 GHz. Structural scheme of used measuring equipment is shown in the Fig. 1. Measuring equipment for analysis of EMI passing through the gypsum-fiber radioprotective facing materials is shown in the Fig. 2.

In the capacity of antennas of the measuring equipment we used emitter in form of open end of rectangular waveguide that provide standing-wave factor in the range of 2 to 6 GHz. Experimental results are presented graphically, and show the dependence of attenuation ratio of EMW (dB) on radiation frequency.

3 Results and Discussion

Results obtained in experimental studies of shuhgite influence on normal consistency and setting time of modified gypsum-cement-puzzolan mix with components ratio of 76:20:4 are shown in the Figs. 3 and 4 correspondingly.

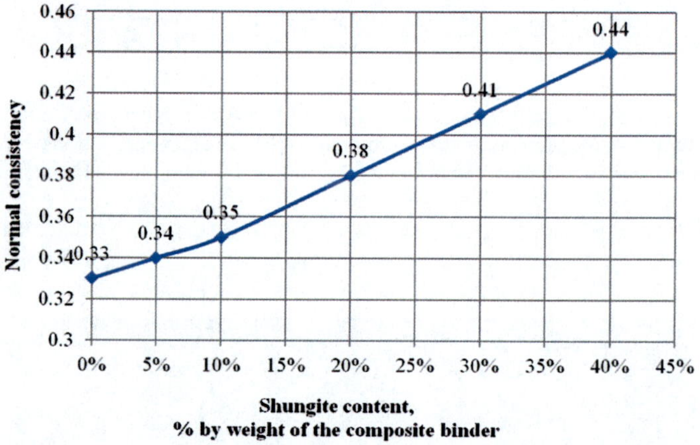

Fig. 3. Dependence of normal consistency of gypsum-cement-puzzolan mix on shungite content in composition of GCPB.

Experimental results of the dependence of flexural strength of gypsum-fiber facing materials on shungite content (5, 10, 20, 30, 40% by weight of GCPB) are shown in the Fig. 5.

As you can see in the Fig. 5, introduction of shungite to gypsum-fiber mix composition leads to slight increase (to 11.2% at the content of 5–10%), and then – to decrease of flexural strength of gypsum-fiber facing materials (at the content of more than 10%). Strength reduction is conditional upon effect of binder dilution at the high content of shungite in mix composition, and also upon its increased water requirement. According to the GOST R 51829-2001 «Gypsum fiber sheets. Specifications» the lowest flexural strength of gypsum-fiber facing materials up to and including 10 mm thick should amount at least 5.4 MPa. In this regard the study samples with shungite content in the quantity of 40% by the weight of binder that have flexural strength with the value of 4.77 MPa, were excluded from further research.

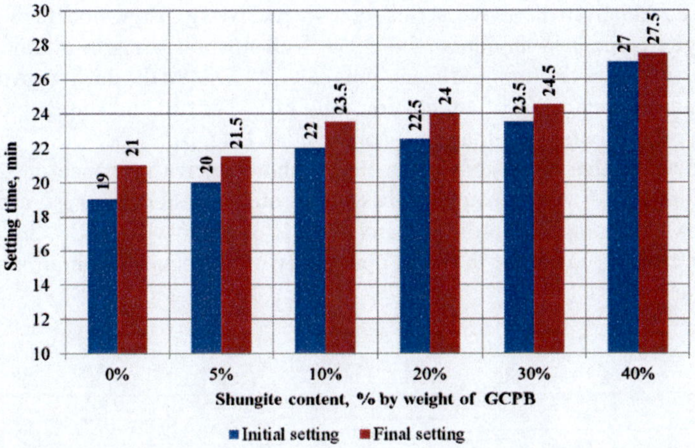

Fig. 4. Dependence of setting time of gypsum-cement-puzzolan mix on shungite content in composition of GCPB.

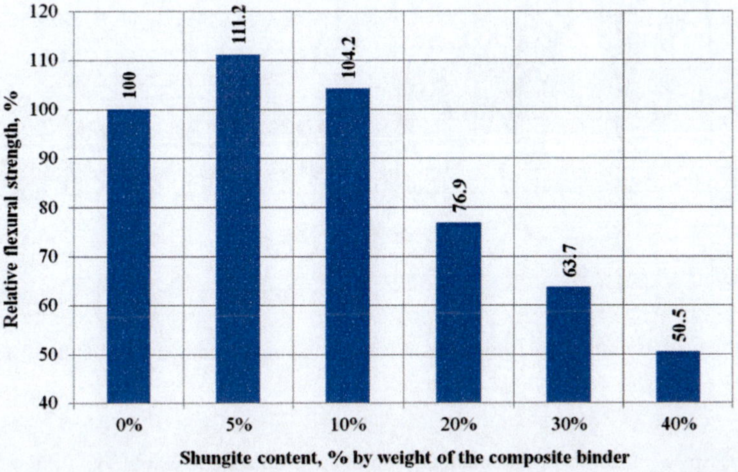

Fig. 5. Dependence of flexural strength of gypsum-fiber facing materials on shungite content.

Experimental results of dependence of water requirement of gypsum-fiber facing materials on shungite content (5, 10, 20, 30% by weight of GCPB) are described by linear function with probability belief of 98.6% on the following form:

$$B = -0.0421x^3 + 0.6967x^2 - 2.9431x + 14.836.$$

Introduction of shungite in the quantity of 5–10% leads to decrease of water requirement of gypsum-fiber facing materials by 2.9%, with further increase of its content in GCPB composition leads to water requirement increase of products.

As can be seen from the above, at this stage we received gypsum-fiber facing materials with shungite content in amounts of 0–30%, with flexural strength in amounts of 6–10.5 MPa, with water requirement in amounts of 11.2–13%. At the third stage of research, we determined dependence of attenuation ratio of EMW (dB) on shungite content (0–30%) in gypsum-fiber facing materials composition (Fig. 6).

Data analysis that are listed in the Fig. 6 shows active influence of shungite on attenuation ratio of EMW during their passing through the studied gypsum-fiber facing materials, which compares to the research results of other writers [38, 39]. Values of attenuation ratio of EMW for the most commonly used frequencies in urban realm are shown in the Table 1.

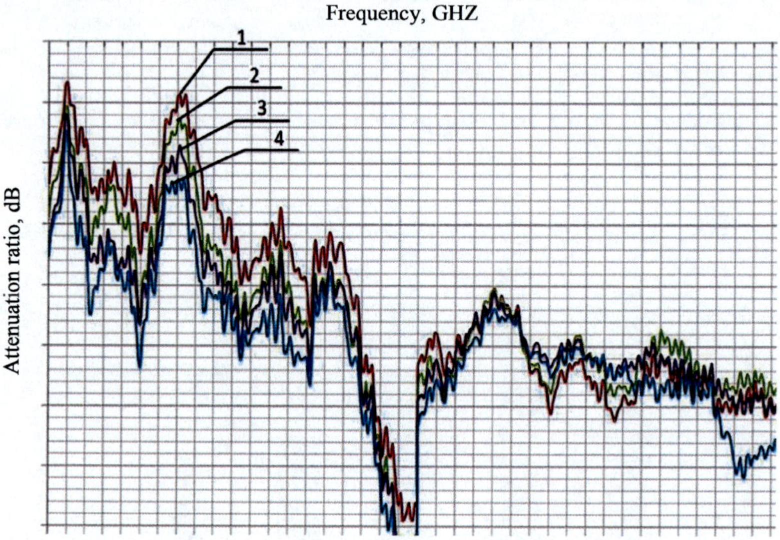

Fig. 6. Attenuation ratio of electromagnetic waves at shungite content: 1–0%; 2–10%; 3–20%; 4–30%.

Table 1. Attenuation ratio of electromagnetic waves.

Frequency, GHz	Shungite content in gypsum fiber facing materials			
	0%	10%	20%	30%
2.1	−24.35	−25.17	−25.59	−26.61
2.45	−28.55	−30.1	−30.94	−31.92
5.2	−33.31	−33.73	−33.73	−34.59

Received experimental data has allowed to establish that best performance of attenuation ratio of EMW are achieved at shungite content in quantity of 30% by weight of GCPB.

Gypsum-fiber radioprotective facing materials that can be used during construction of partition walls in accommodations with emitter of increased EMI were produced according to the research results. It is found that with increase of shungite content protection properties of the samples of gypsum-fiber radioprotective facing materials also increase in all frequency intervals of EMW. With increase of EMW frequency rotecttive properties of the samples are also improved. During increase of shungite content in mix composition we can observe protective electromagnetic properties improvement by the value of 2 dB.

4 Conclusions

1. Gypsum fiber radioprotective facing materials are produced andcan be used during partitions installation indoors with transmitters of increased EMI.
2. It is found that introduction of shungite together with used chemical admixtures to gypsum cement fiber mixes leads to decrease of initial and final setting time, which helps to mould products without the use of retarders.
3. It was demonstrated that introduction of shungite in the quantity of 5–10% by weight of GCPB leads to increase of flexural strength of gypsum-fiber facing materials by value up to 11.2%. Shungite content over 30% leads to substantial increase of water requirement of the mix, binder dilution and strength reduction of products over allowed values according to GOST R 51829-2001, which defines desired content of shungite (0–30%) and allows producing gypsum-fiber radioprotective facing materials with required attenuation ratio of EMI.
4. It is found that with increase of shungite content in composition of gypsum-fiber facing materials increase of radioprotective properties in the frequency intervals of EMW under investigation (2–6 GHz) can be observed. Best performances upon criterion of EMW attenuation are achieved at shungite content in the quantity 30% by weight of GCPB.

References

1. Shafigullin, R.I., Kupriyanov, V.N.: Environmental safety of the urban environment when exposed to electromagnetic fields. Izvestiya KGASU **1**, 171–178 (2015)
2. Schooneveld, H., Van Bijnen, J., Van, P.: Electromagnetic field reduction restores health of electro-sensitive people. Electrical Hypersensitivity Foundation 3–6 (2016)
3. Zbojovsky, J., Pavlík, M.: Calculating the shielding effectiveness of materials based on the simulation of electromagnetic field. In: Proceedings of the 9th International Scientific Symposium on Electric Power Engineering, Stara Lesna, Slovakia, pp. 369–372 (2017)
4. Liptai, P., Moravec, M., Lumnitzer, E., Lukáčová, K.: Impact analysis of the electromagnetic fields of transformer stations close to residential buildings. In: Proceedings of International Multidisciplinary Scientific Geoconference, SGEM, pp. 355–360 (2014)
5. Panagopoulos, D.J., Chrousos, G.P.: Shielding methods and products against man-made electromagnetic fields: protection versus risk. Sci. Total Environ. **667**, 255–262 (2019)
6. Hutter, H.-P., Moshammer, H., Wallner, P., Kundi, M.: Subjective symptoms, sleeping problems, and cognitive performance in subjects living near mobile phone base stations. Occup. Environ. Med. **63**, 307–313 (2006)

7. Havas, M.: Radiation from wireless technology affects the blood, the heart, and the autonomic nervous system. Rev. Environ. Health **28**, 75–84 (2013)
8. Kupriyanov, V.N., Shafigullin, R.I.: On the issue of attenuation of electromagnetic waves by the enclosing structures of buildings. Izvestiya Vysshih Uchebnyh Zavedenij **4**, 229–234 (2016)
9. Belousova, E.S., Mahmud, M., Lynkov, L.M., Nasonova, N.V.: Radio-shielding properties of concretes based on shungite-containing nanomaterials. Nanotekhnologii v Stroitelstve **5**, 56–67 (2016)
10. Lynkov, L.M., Mahmud, M.Sh., Krishtopova, E.A.: Screens of electromagnetic radiation based on powdered shungite. Vestnik Polockogo Gosudarstvennogo Universiteta, pp. 103–108 (2012)
11. Khaliullin, M., Dimieva, A.: Composite gypsum binder under introducing thermallyactivated clay as a pozzolanic component and adding groundlimestone. IOP Conf. Ser.: Mater. Sci. Eng. 890 (2020)
12. Bibikov, S.B., Prokof'ev, M.V.: Composite Materials for Some Radiophysics Applications, Metal, Ceramic and Polymeric Composites for Various Uses, pp. 525–544 (2011)
13. Chung, D.D.L.: Electromagnetic interference shielding effectiveness of carbon materials. Carbon **39**, 279–285 (2001)
14. Geetha, S., Kumar, K.K.S., Rao, C.R.K., Vijayan, M., Trivedi, D.C.: EMI shielding: methods and materials. J. Appl. Polym. Sci. **112**, 2073–2086 (2009)
15. Polivkin, V.V., Gurnenko, I.N., Gulbin, V.N., Kolpakov, N.S.: Radioprotective building concrete and method of its manufacture, RU pat. No. 2545585, 10 April 2015
16. Lukutcova, N.P., Pykin, A.A., Karpikov, E.G.: Features of structure formation of cement stone with carbon-silica nanodispersed additive. Stroitelnye Materialy 66–67 (2011)
17. Polivkin, V.V., Gulbin, V.N., Miheev, V.A., Kolpakov, N.S.: Carbon-containing composition for radio protective materials, RU pat. No. 2519244, 10 June (2014)
18. Tymeneva, A.S., Polivkin, V.V., Gulbin, V.N., Alexandrov, Yu.K., Miheev, V.A., Kolpakov, N.S.: Press composition for radio protective plate materials and method of its production, RU pat. No. 2531817, 27 October (2014)
19. Berezin, N.M., Bogatov, V.A., Grachev, V.P., Kondrashov, S.V., Romanov, A.M., Hohlov, Yu.A.: Radio protective layered material, RU pat. No. 2433916, 20 November (2011)
20. Mukhametrakhimov, R., Galautdinov, A.R.: The role of active mineral additives of natural origin in the formation of the structure and properties of gypsum-cement-pozzolanic binder. Vestnik Tekhnologicheskogo Universiteta **6**, 60–63 (2017)
21. Kocak, Y.: Effects of metakaolin on the hydration development of Portland-composite cement. J. Build. Eng. 62–68 (2020)
22. Mukhametrakhimov, R., Galautdinov, A.R.: Gypsum-cement-pozzolanic binder based on low-grade raw materials and industrial waste. Vestnik Tekhnologicheskogo Universiteta **19**, 56–59 (2016)
23. Li, C., Fan, Y., Yu, J.: Compressive strength and pore structure of cement paste with metakaolin. IOP Conf. Ser.: Earth Environ. Sci. 531 (2020)
24. Izotov, V.S., Mukhametrakhimov, R.Kh., Galautdinov, A.R.: Complex additive, RU pat. No. 2519313, 10 July 2014
25. Khaliullin, M., Dimieva, A.: Composite gypsum binder under introducing thermally activated clay as a pozzolanic component and adding ground limestone. IOP Conf. Ser.: Mater. Sci. Eng. (2020). https://doi.org/10.1088/1757-899X/890/1/012093
26. Khaliullin, M., Rakhimov, R., Faizrakhmanov, I.: The influence of thermally activated clay additives on the properties of composite gypsum binder. ZKG ZEMENT – KALK – GIPS Int. Bauverlag BV GmbH. Gutersloh, pp. 58–63 (2017)

27. Bulanov, P.E., Ermilova, E.U., Mavliev, L.F.: Structure and mineral composition of soil-cement with complex additive. Mag. Civ. Eng. **83**, 38–48 (2018). https://doi.org/10.18720/MCE.83.4
28. Ermilova, E., Kamalova, Z., Ravil, R.: Influence of clay mineral composition on properties of blended portland cement with complex additives of clays and carbonates. IOP Conf. Ser.: Mater. Sci. Eng. Institute of Physics Publishing (2020). https://doi.org/10.1088/1757-899X/890/1/012087
29. Khozin, V., Khokhryakov, O., Nizamov, R.: A «carbon footprint» of low water demand cements and cement-based concrete. IOP Conf. Ser.: Mater. Sci. Eng. 890 (2020). https://doi.org/10.1088/1757-899X/890/1/012105
30. Krasinikova, N., Stepanov, S., Makarov, D.: Cement stone, modified by chemical water treatment sludge. IOP Conf. Ser.: Mater. Sci. Eng. 890 (2020). https://doi.org/10.1088/1757-899X/890/1/012099
31. Pimenov, S.I.: Heavyweight concrete based on hydromechanochemically activated binder. IOP Conf. Ser.: Mater. Sci. Eng. 890 (2020). https://doi.org/10.1088/1757-899X/890/1/012098
32. Vdovin, E.A., Stroganov, V.F.: Modification of cement-bound mixtures with sodium formate additives for the construction of pavement bases at low air temperatures. IOP Conf. Ser.: Mater. Sci. Eng. 890 (2020). https://doi.org/10.1088/1757-899X/786/1/012065
33. Mavliev, L., Vdovin, E.: Structure of road soil cement compositions modified by complex additive based on organosilicon compounds and electrolytes. E3S Web of Conferences (2019). https://doi.org/10.1051/e3sconf/201914002016
34. Mukhametrakhimov, R.Kh., Galautdinov, A.R.: Organomineral modifier for gypsum mixtures and radioprotective products based on them, RU pat. No. 2710245, 25 December (2019)
35. Mukhametrakhimov, R.Kh., Galautdinov, A.R.: Organomineral modifier for gypsum compositions and radioprotective products based on them, RU pat. No. 2710245, 30 December (2019)
36. Mukhametrakhimov, R.Kh., Galautdinov, A.R., Kupriyanov, V.N., Shafigullin, R.I.: Organomineral modifier for gypsum binders and radioprotective products based on them, RU pat. No. 2710647, 30 December (2019)
37. Izotov, V.S., Mukhametrakhimov, R., Galautdinov, A.R.: Influence of polypropylene fibers on the main properties of gypsum-cement-pozzolanic binder. Vestnik Tekhnologicheskogo Universiteta **18**, 135–137 (2015)
38. Kupriyanov, V.N., Shafigullin, R.I.: To the calculation of the attenuation of the energy of an electromagnetic wave when it passes through the enclosing structures. Fundamental, Prospecting and Applied Research of RAACN on Scientific Support for the Development of Architecture, Urban Planning and Construction Industry of the Russian Federation in 2017, Moscow, pp. 283–290 (2018)
39. Molchanov, S., Ushakov, N.M.: Microwave dielectric constant and attenuation of electromagnetic waves in polymer nanocomposite media based on iron nanoparticles and cadmium sulfide. Fiziko-Himicheskie Aspekty Izucheniya Klasterov Nanostruktur i Nanomaterialov **4**, 195–199 (2012)

Author Index

A
Abdikarimov, Rustamkhan, 72
Abdrakhmanova, Lyaylya, 14, 318
Amel'chenko, Maxim, 269, 289
Antipin, Igor, 22
Ayupov, Damir, 325

B
Bliznjakova, Kseniya, 364
Borovskikh, Olga, 47

D
Diakov, Grigory, 343

E
Eremeev, Daniil, 142
Eremeev, Pavel, 142
Eremeev, Valery, 142
Eremina, Svetlana, 213
Eskin, Aleksey, 38
Evstigneeva, Yuliya, 90

F
Fakhrutdinova, Venera, 261
Fazleev, Ilshat, 289
Furer, Victor, 22

G
Galautdinov, Albert, 372
Garipov, Aidar, 351
Gerkina, Zhanna, 223
Gilmanshina, Alsu, 334
Gimadieva, Liliya, 122
Gimranov, Linur, 58
Goryachev, Mikhail, 343
Grishina, Maria, 101

I
Ibragimov, Ruslan, 90
Islamov, Anvar, 261
Ivantsov, Aleksey, 80

K
Kareeva, Juliya, 364
Kayumov, Rashit, 239
Kazakulov, Rauf, 325
Kazymova, Tatiana, 47
Khaliullin, Marat, 334
Khantimirov, Ayaz, 318
Khodzhaev, Dadakhan, 72
Khozin, Vadim, 351
Khozinv, Vadim, 318
Khuziakhmetova, Karina, 14
Koklugina, Lyudmila, 90
Kolchin, Maksim, 250
Konovalov, Nikita, 308
Kraus, Eduard, 269, 289
Krutova, Iraida, 194
Kupriyanov, Valery, 372
Kuznetsov, Dmitry, 297
Kuznetsov, Ivan, 58

L
Lashin, Maxim, 279
Logachev, Konstantin, 213
Lukmanova, Liliya, 1

N. Vatin (Ed.): STCCE 2021, LNCE 169, pp. 383–384, 2021.
https://doi.org/10.1007/978-3-030-80103-8

M
Makarov, Dmitry, 325, 351
Mavliev, Lenar, 308
Medyanik, Yulia, 122
Mirsaidov, Mirziyod, 72
Mirsayapov, Ilizar, 150, 160, 167, 175, 183, 203
Morozov, Vladimir, 38
Mukhametrakhimov, Rustem, 1, 372

N
Nizamov, Rashit, 14
Nizamova, Adilya, 47
Normuminov, Bakhodir, 72
Nurmuhametov, Karim, 31

P
Petrov, Artem, 250
Popova, Elena, 22
Posokhin, Vladimir, 230

R
Rakhimova, Nailia, 38

S
Sabirzyanov, Danil, 175
Sadykov, Denis, 31
Safiullin, Rinat, 230
Safiullina, Guzel, 213
Sagdiev, Ruslan, 135
Serova, Valentina, 223
Shaehov, Almaz, 90
Shagiakhmetova, Elvira, 47, 122
Shakirov, Almaz, 113
Shakirov, Marat, 175, 203
Shelikhov, Nikolay, 135
Shiryaev, Artem, 297
Shmelev, Gennady, 142
Sirazitdinov, Vadil, 80
Solovieva, Svetlana, 22
Stepanov, Sergey, 351
Strakhov, Dmitry, 239
Stroganov, Victor, 269, 289
Stroganov, Viktor, 223
Sulejmanov, Alfred, 113, 239

T
Tabaeva, Radmilla, 269

U
Ushakov, Viktor, 343

V
Vachagina, Ekaterina, 194
Vakhitova, Dilyara, 122
Vandyukov, Alexandr, 22
Vatin, Nikolay, 72
Vdovin, Evgenii, 269, 289, 297, 308
Vdovin, Evgeniy, 279
Vysotskaya, Marina, 279, 297

Y
Yarkin, Sergey, 343

Z
Zakieva, Raushan, 364
Ziganshin, Arslan, 213
Zinnurov, Tagir, 31
Zolotonosov, Yakov, 194